VECTORS AND TENSORS

for engineers and scientists

VECTORS and TENSORS for engineers and scientists

FRED A HINCHEY

A HALSTED PRESS BOOK

JOHN WILEY & SONS
New York London Sydney Toronto

Published in the Western Hemisphere
by Halsted Press, a division of
John Wiley & Sons, Inc., New York

Library of Congress Cataloging in Publication Data

Hinchey, Fred A
Vectors and tensors for engineers and scientists.

1. Vector analysis. 2. Calculus of tensors.
I. Title.
TA 347.V4H56 515′.63′02462 76-21725
ISBN 0-470-15194-3

Printed at Prabhat Press, Meerut 250002, India

BST
R

To
Anil, Dilip and
Manik Tara

9.95–

Preface

The great utility of vectors and tensors in the analysis of physical problems depends to a large extent on two properties of these entities. First of all, the use of vectors and tensors provides a very compact notation with concomitant ease of mathematical manipulation. Secondly, and this property is probably more important in physics than in engineering, vector and tensor quantities are ideally suited for exploiting symmetry and invariance properties which exist in nature.

The first five chapters of the present book provide the necessary formalism for dealing with vector and tensor quantities. The notation and rules for the manipulation of vectors are given in the first two chapters in terms of orthogonal cartesian coordinate. These concepts and rules are then extended to curvilinear coordinate systems. This extension is first done in very general terms and then specialized to orthogonal curvilinear systems, since the latter are the only systems which find any particular use in elementary applications.

The transformation properties of vectors under both the orthogonal and affine groups are studied with two objectives in mind. First, these transformation properties are necessary in many of the applications, and secondly, such a study provides an introductory background for the tensor formalism.

The sudy of the tensor formalism and its application to vector analysis is of necessity brief and incomplete in many aspects. However, it provides an introduction to the subject and will prepare the interested reader for further reading in tensor analysis, differential geometry, and the general theory of relativity.

Like most texts of this type, the present work is based on lecture notes. These notes were originally prepared for a one-term course, at the advanced undergraduate level, in vector and tensor analysis at New Mexico State University. Portions of the material have also been used with undergraduate students at Oberlin College, Oberlin, Ohio, and The American University in Cairo, Cairo, Egypt.

I should like to express my deep appreciation to the many students, too numerous to mention by name; several of my colleagues, in particular Dr. C.S. Shastry; and Mr. Vinod Kumar whose suggestions have made this a better book than it would have otherwise been. However, all errors, both of omission and commission remain my sole responsibility. Finally, I should like to thank my wife for her almost infinite patience and understanding.

F.A.H.

Bangalore
October, 1975

Contents

CHAPTER 1

Vector Algebra

1-1. Elementary Definitions. Although a vector is a precisely defined mathematical entity, it is difficult to obtain an intuitive feeling for a vector from the exact mathematical definition. Since our primary aim is to make use of the vector concept for the formulation and solution of physical problems, we can initially make use of a more intuitive concept than that required for complete mathematical rigour. We shall, however, soon discover that this intuitive concept does not permit us to make full use of the vector calculus. At this time, we shall introduce the concept of a vector in a more rigorous fashion.

The most elementary concept of a vector is in terms of directed line segments. That is to say, we shall represent vectors by directed line segments in a three-dimensional Euclidean space, the length of the line representing the magnitude of the vector, and the direction of the line being the same as the direction of the vector in space. It is important to emphasize that these directed line segments are not vectors, but are merely a convenient description of them. Although we can formulate all the operations of the calculus of vectors in terms of this simple description, we soon find that other representations are more suitable for the description of physical phenomena. The description of vectors in terms of directed line segments, however, provides a key to their utility in the formulation and solution of many physical problems. Many quantities of physical interest such as force, momentum, electric and magnetic intensity, inherently have both a magnitude and direction associated with them. Since the description of vectors by directed line segments also involves both a magnitude and direction, these physical quantities are conveniently described by vectors. This does not imply that the physical quantities are vectors, but merely indicates that the mathematical description of such quantities is best made in terms of vectors.

In contrast to those physical quantities which we shall describe by vectors, there are many physical quantities such as mass, temperature, and electric charge which require only a magnitude for their description. Such physical quantities are described by what we call scalars. Although scalars also have a precise mathematical definition, it is adequate at this time to regard them as simply real numbers. In order to distinguish between vector and scalar quantities, the former will be denoted by bold face type, whereas the latter will be denoted by light face type.

Before constructing an algebra for vector quantities, we shall introduce the concept of the norm of a vector. With each vector **a**, we associate a positive real number a, defined to be the norm of **a**, and usually written as follows: $a = |\mathbf{a}|$. This positive real number is numerically equal to the magnitude of the vector, i.e. equal to the length of the directed line segment which represents the vector. Any vector having a norm equal to unity is called a unit vector, and any vector with zero norm is known as a zero or null vector.

1-2. The Addition of Vectors. The first concept required in order to form an algebra of vectors is the idea of equality. This is the only one of the order relations which exist among the real numbers, which is meaningful in the case of vector quantities. The concepts of "less than" and "greater than" cannot be applied to vectors, since there is no one property of the vector with which we can associate the two concepts. Two vectors are defined to be equal if and only if they have the same norm, the directed line segments representing them are parallel, and if they have the same sense of positive direction. The initial and terminal points of the directed line segments representing the two vectors are immaterial, since it is only the length and direction of these two segments which determines the quantity. This property is illustrated in Fig. 1-1. An important trans-

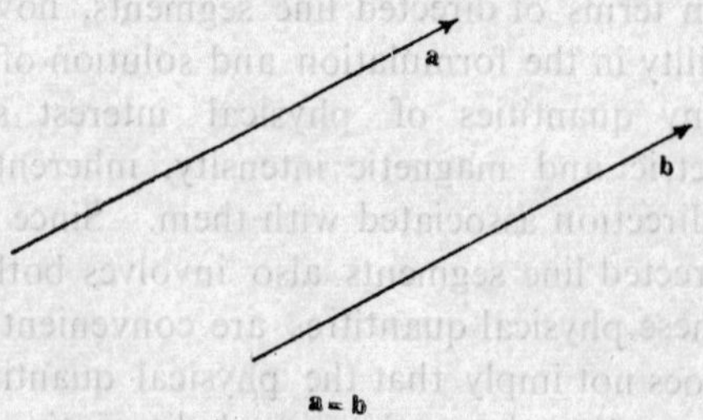

Figure 1-1.

formation property of vectors follows immediately from the definition of equality. A vector is invariant, i.e. remains unchanged, whenever it is displaced parallel to itself. In the subsequent development, we shall make frequent use of this invariance. In later chapters, we shall consider the invariance of vectors under more general transformations than parallel displacement.

The negative of a given vector may be defined in terms of the equality of two vectors. If **a** is an arbitrary vector, the negative of **a**, written —**a**, is defined to be the vector which is parallel to **a**, with norm *a*, and the opposite sense of direction. This concept is illustrated in Fig. 1-2.

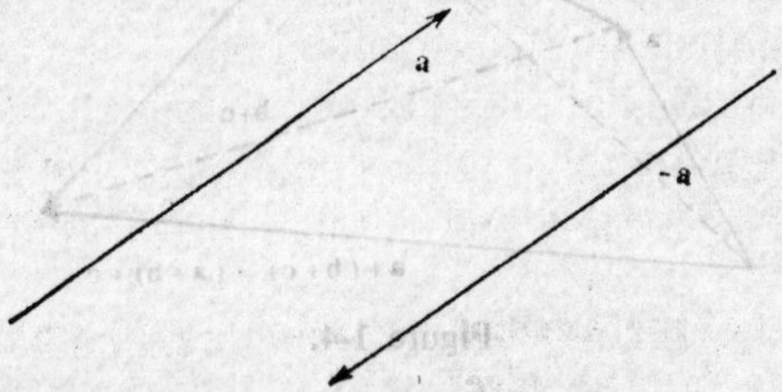

Figure 1-2.

The fundamental algebraic operation on vectors is addition. We shall first define this operation for two vectors, and then extend the definition to include an arbitrary number of vectors. Let **a** and **b** be two arbitrary vectors. We form a third vector **c** by the following geometrical construction: displace *b* parallel to itself so that the initial point of *b* is coincident with the terminal point of **a**. The vector **c** is represented by the directed line segment from the initial point of **a** to the terminal point of **b** (see Fig. 1-3). The vector **c** defined in this way

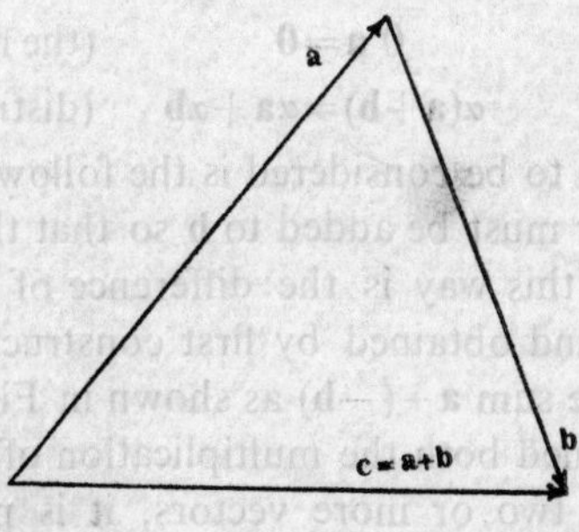

Figure 1-3.

is called the vector sum of the two vectors **a** and **b**, usually written

$$\mathbf{c}=\mathbf{a}+\mathbf{b}$$

The extension of this construction to an arbitrary number of vectors is trivial. If **a**, **b**, and **c** are arbitrary vectors, it is easily shown that

$$\mathbf{a}+\mathbf{b}=\mathbf{b}+\mathbf{a} \qquad \text{(commutative law)} \qquad (1\text{-}1)$$

$$(\mathbf{a}+\mathbf{b})+\mathbf{c}=(\mathbf{a}+\mathbf{b})+\mathbf{c} \qquad \text{(associative law)} \qquad (1\text{-}2)$$

The proof of the commutative law is obvious, and the associative law is demonstrated geometrically in Fig. 1-4.

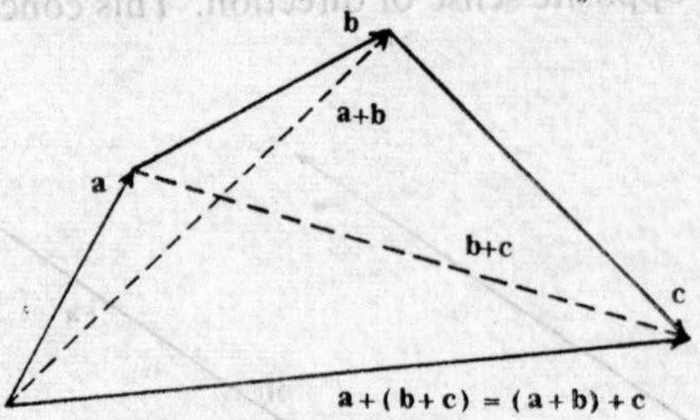

Figure 1-4.

A second fundamental algebraic operation is the multiplication of a vector by a scalar. If **a** is an arbitrary vector and α is an arbitrary scalar, the product α **a** is the vector parallel to **a** with norm

$$|\alpha\mathbf{a}|=|\alpha|\,|\mathbf{a}| \qquad (1\text{-}3)$$

If α is positive, the sense of the product is the same as that of the vector **a**. On the other hand, if α is negative, the sense of the product is opposite to that of **a**. If **a** and **b** are both arbitrary vectors, and if α and β are arbitrary scalars, it follows that

$$\alpha(\beta\mathbf{a})=(\alpha\beta)\,\mathbf{a}=(\beta\alpha)\,\mathbf{a}=\beta(\alpha\mathbf{a}) \qquad (1\text{-}4)$$

$$(\alpha+\beta)\,\mathbf{a}=\alpha\mathbf{a}+\beta\mathbf{a} \qquad (1\text{-}5)$$

$$0\,\mathbf{a}=\mathbf{0} \qquad \text{(the null vector)} \qquad (1\text{-}6)$$

$$\alpha(\mathbf{a}+\mathbf{b})=\alpha\mathbf{a}+\alpha\mathbf{b} \qquad \text{(distributive law)} \qquad (1\text{-}7)$$

The next question to be considered is the following: Given vectors **a** and **b**, what vector must be added to **b** so that the result is **a**? The vector **c** defined in this way is the difference of the vectors **a** and **b**, written $\mathbf{c}=\mathbf{a}-\mathbf{b}$, and obtained by first constructing the vector $-\mathbf{b}$, and then forming the sum $\mathbf{a}+(-\mathbf{b})$ as shown in Fig. 1-5.

Since we have defined both the multiplication of a vector by a scalar and the addition of two or more vectors, it is possible to consider linear combinations of vectors. Consider any two non-zero, non-

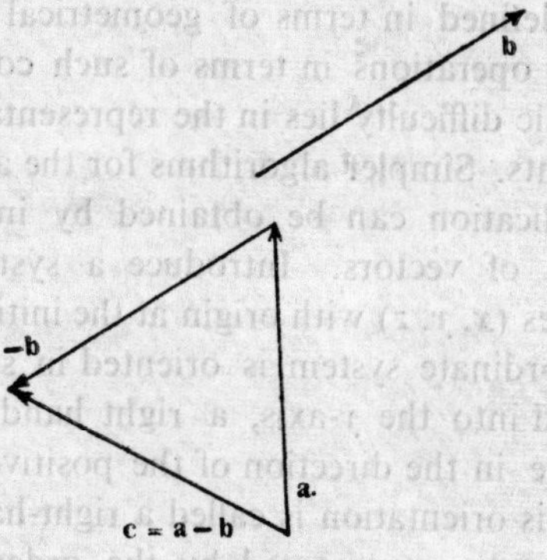

Figure 1-5.

parallel vectors **a** and **b**. These two vectors form a basis for a plane, in the sense that any vector which lies in the same plane as the two given vectors can be expressed as a linear combination

$$\mathbf{c}=\alpha\mathbf{a}+\beta\mathbf{b} \tag{1-8}$$

for suitably chosen scalar constants α and β. The validity of Eq. (1-8) is easily seen by the construction shown in Fig. 1-6. In a similar way, any three non-zero, non-coplanar vectors **a, b,** and **c** form *a* basis for a three dimensional space, since any vector in the three-dimensional space can be represented by the linear combination

$$\mathbf{d}=\alpha a+\beta b+\gamma c \tag{1-9}$$

for suitably chosen scalar constants α, β, and γ.

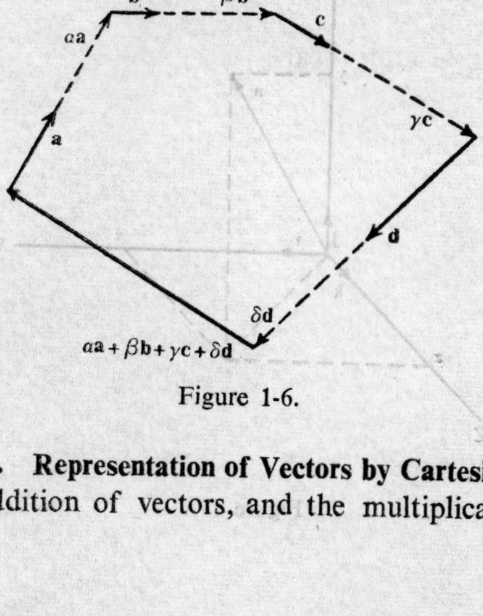

Figure 1-6.

1-3. Representation of Vectors by Cartesian Components. Although the addition of vectors, and the multiplication of a vector by a scalar can be defined in terms of geometrical constructions, the application of these operations in terms of such constructions is very awkward. The basic difficulty lies in the representation of vectors by directed line segments. Simpler algorithms for the addition of vectors and scalar multiplication can be obtained by introducing a more elegant description of vectors. Introduce a system of orthogonal cartesian coordinates (x, y, z) with origin at the initial point of a given vector **a**. The coordinate system is oriented in such a way that if the x-axis is rotated into the y-axis, a right hand screw fixed to the system will advance in the direction of the positive z-axis. A set of coordinates with this orientation is called a right-handed system. The given vector **a** may be represented by the ordered triplet of real numbers,

$$\mathbf{a}=(a_1, a_2, a_3) \tag{1-10}$$

where a_1, a_2 and a_3 are the projections of the directed line segment which represents **a** on the x-, y-, and z-axis respectively as shown in Fig. 1-7. An alternate description of the vector **a** may be obtained by introducing unit vectors **i**, **j**, and **k** along the x-, y-, and z-axis respectively as shown. These unit vectors are known as the cartesian unit vectors. In terms of the cartesian unit vectors, the given vector **a** may be represented by the linear combination

$$\mathbf{a}=a_1\mathbf{i}+a_2\mathbf{j}+a_3\mathbf{k} \tag{1-11}$$

The real numbers a_1, a_2, and a_3 are known as the cartesian components of the given vector **a**. It is clear that the description of a given

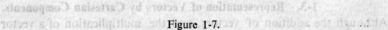

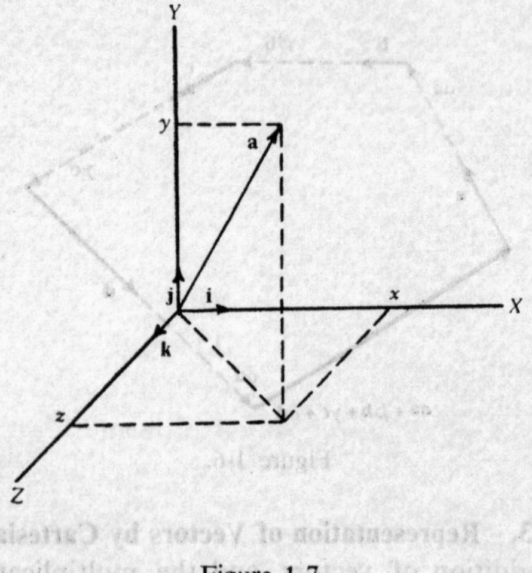

Figure 1-7.

vector in terms of its cartesian components is unique, that is, if **a** and **b** are two vectors which have the same set of cartesian components, then **a**=**b**. It follows from the analytic geometry of cartesian coordinates that the length or norm of the vector **a** is given by

$$|\mathbf{a}|=(a_1^2+a_2^2+a_3^2)^{\frac{1}{2}} \tag{1-12}$$

The null vector has all of its cartesian components identically equal to zero.

We shall now examine the rules for forming the scalar product and the sum of two vectors, when the vectors are expressed in terms of their cartesian components. If **a** is an arbitrary vector with cartesian components (a_1, a_2, a_3), and if α is an arbitrary scalar, the product α **a** is defined to be the vector with cartesian components $(\alpha a_1, \alpha a_2, \alpha a_3)$. It is not difficult to see that this is equivalent to the definition in terms of directed line segments. If **a** and **b** are two vectors with cartesian components (a_1, a_2, a_3) and (b_1, b_2, b_3) respectively, then the vector sum **a**+**b** has the cartesian components $(a_1+b_1, a_2+b_2, a_3+b_3)$. Again, it is simple to show that this definition of the sum is equivalent to the definition in terms of directed line segments. The commutative, associative, and distributive laws for addition and multiplication by a scalar follow immediately from the definitions,

$$\begin{aligned}\mathbf{a}+\mathbf{b}&=(a_1+b_1,\ a_2+b_2,\ a_3+b_3)\\&=(b_1+a_1,\ b_2+a_2,\ b_3+a_3)\\&=\mathbf{b}+\mathbf{a}\end{aligned} \tag{1-13}$$

$$\begin{aligned}(\mathbf{a}+\mathbf{b})+\mathbf{c}&=[(a_1+b_1)+c_1,\ (a_2+b_2)+c_2,(a_3+b_3)+c_3]\\&=[a_1+(b_1+c_1),\ a_2+(b_2+c_2),\ a_3+(b_3+c_3)]\\&=\mathbf{a}+(\mathbf{b}+\mathbf{c})\end{aligned} \tag{1-14}$$

$$\begin{aligned}\alpha(\mathbf{a}+\mathbf{b})&=\alpha\ (a_1+b_1,\ a_2+b_2,\ a_3+b_3)\\&=(\alpha a_1+\alpha b_1,\ \alpha a_2+\alpha b_2,\ \alpha a_3+\alpha b_3)\\&=\alpha\mathbf{a}+\alpha\mathbf{b}\end{aligned} \tag{1-15}$$

In terms of the cartesian unit vectors, the rules for the multiplication by a scalar and the addition of two vectors are expressed as

$$\alpha\mathbf{a}=\alpha a_1\mathbf{i}+\alpha a_2\mathbf{j}+\alpha \mathbf{a}_3 k \tag{1-16}$$

$$\mathbf{a}+\mathbf{b}=(a_1+b_1)\,\mathbf{i}+(a_2+b_2)\,\mathbf{j}+(a_3+b_2)\,\mathbf{k}$$

Although there are applications of elementary vector algebra in almost all areas of analysis, the most direct and immediate applications are in geometry. In order to appreciate this, let us consider the following examples.

EXAMPLE 1. Consider the quadrilateral with vertices $ABCD$ as shown in Fig. 1-8. We can use simple vector addition to show that if we connect the midpoints of the consecutive sides of the quadrilateral by straight lines, the resulting quadrilateral is a parallelogram. Denoting the midpoints of the sides by P, Q, R, S, we have

$$\mathbf{PQ}=\tfrac{1}{2}(\mathbf{a}+\mathbf{b}),\ \mathbf{QR}=\tfrac{1}{2}(\mathbf{b}+\mathbf{c}),\ \mathbf{RS}=\tfrac{1}{2}(\mathbf{c}+\mathbf{d}),\ \mathbf{SP}=\tfrac{1}{2}(\mathbf{d}+\mathbf{a})$$

Now,

$$\mathbf{a}+\mathbf{b}+\mathbf{c}+\mathbf{d}=\mathbf{PQ}+\mathbf{QR}+\mathbf{RS}+\mathbf{SP}=0$$

and hence,

$$\mathbf{PQ}=-\mathbf{RS},\ \mathbf{QR}=-\mathbf{SP}$$

Therefore the opposite sides are equal and anti-parallel, which obtains the result.

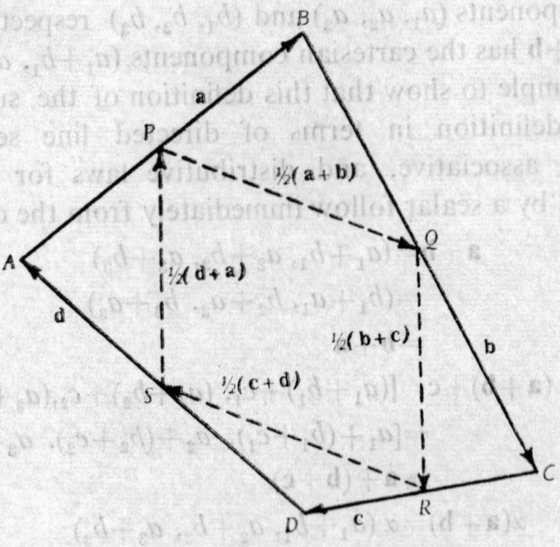

Figure 1-8.

EXAMPLE 2. The corner reflector, which is frequently used in the alignment and calibration of radar systems, consists of three highly reflecting, mutually perpendicular planes. We can use a rather simple invariance property of vectors to show that the reflected radar beam is anti-parallel to the incident radar beam. Under the geometrical transformation of reflecting a vector in a plane, the component of the vector parallel to the plane is invariant, and the component perpendicular to the plane is multiplied by (-1).

Choose a coordinate system so that the three surfaces of the corner reflector coincide with the XY-, YZ-, and ZX-planes respectively.

Let **A** be a unit vector in the direction of the incident radar beam. Then,

$$\mathbf{A}=A_x\mathbf{i}+A_y\mathbf{j}+A_z\mathbf{k}$$

Let us assume that the incident beam is first reflected in the XY-plane, then in the YZ-plane, and finally in the ZX-plane. Let $\mathbf{A}_1$ be the vector resulting from reflecting A in the XY-plane, so that

$$\mathbf{A}_1=(A_x\mathbf{i}+A_y\mathbf{j})-A_z\mathbf{k}$$

Then, if $\mathbf{A}_2$ and $\mathbf{A}_3$ are the vectors which result from reflecting A_1 in the YZ-plane and $\mathbf{A}_2$ in the ZX-plane, we have

$$\mathbf{A}_2=A_y\mathbf{j}-(A_x\mathbf{i}+A_z\mathbf{k})$$

and

$$A_3=-(A_xi+A_yj+A_zk)=-A$$

1-4. The Product of Two Vectors. If we wish to consider the product of two given vectors **a** and **b**, there are two ways in which this product can be formed. The first of these, known as the scalar or inner product, is defined to be

$$\mathbf{a}\cdot\mathbf{b}=|\mathbf{a}|\,|\mathbf{b}|\cos\theta \tag{1-17}$$

where θ is the angle between the directed line segments representing the two given vectors when they are drawn from a common initial point. It is clear from Eq. (1-17) that the inner product of two vectors is commutative, and that the inner product of a vector with itself is equal to the square of its norm. Whenever neither **a** nor **b** is a null vector, the equality

$$\mathbf{a}\cdot\mathbf{b}=0 \tag{1-18}$$

is used to define orthogonality between the two vectors a and b. This definition results from the fact that the inner product of two non-null vectors vanishes if and only if the angle between their directed line segments is an odd multiple of $\pi/2$.

The inner product of **a** and **b** may be interpreted geometrically as the projection of **a** on **b** multiplied by the norm of **b**.

$$\mathbf{a}\cdot\mathbf{b}=(\text{proj }\mathbf{a})_b\,|\mathbf{b}|=(\text{proj }\mathbf{b})_a\,|\mathbf{a}| \tag{1-19}$$

Equation (1-19) may be used to show that the inner product is a distributive operation. Consider

$$\begin{aligned}\mathbf{a}\cdot(\mathbf{b}+\mathbf{c})&=\text{proj }(\mathbf{b}+\mathbf{c})_a\,|\mathbf{a}|\\&=(\text{proj }\mathbf{b})_a\,|\mathbf{a}|+(\text{proj }\mathbf{c})_a\,|\mathbf{a}|\\&=\mathbf{a}\cdot\mathbf{b}+\mathbf{a}\cdot\mathbf{c}\end{aligned}$$

since the projection of the sum is the sum of the projections (see Fig. 1-9). Similarly,

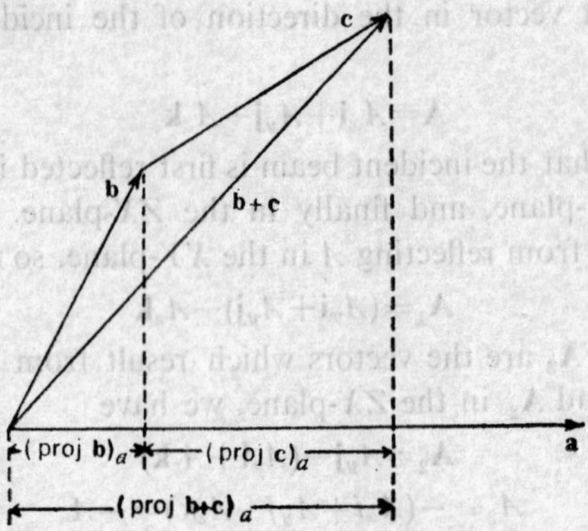

Figure 1-9.

$$\begin{aligned}(\mathbf{a}+\mathbf{b})\cdot(\mathbf{c}+\mathbf{d})&=(\mathbf{c}+\mathbf{d})\cdot(\mathbf{a}+\mathbf{b})\\&=(\mathbf{c}+\mathbf{d})\cdot\mathbf{a}+(\mathbf{c}+\mathbf{d})\cdot\mathbf{b}\\&=\mathbf{a}\cdot(\mathbf{c}+\mathbf{d})+\mathbf{b}\cdot(\mathbf{c}+\mathbf{d})\\&=\mathbf{a}\cdot\mathbf{c}+\mathbf{a}\cdot\mathbf{d}+\mathbf{b}\cdot\mathbf{c}+\mathbf{b}\cdot\mathbf{d}\end{aligned}$$

Now, consider three non-zero vectors **a**, **b**, and **c** such that $\mathbf{c}=\mathbf{b}-\mathbf{a}$. Then,

$$\begin{aligned}\mathbf{c}\cdot\mathbf{c}&=(\mathbf{b}-\mathbf{a})\cdot(\mathbf{b}-\mathbf{a})\\&=\mathbf{b}\cdot\mathbf{b}+\mathbf{a}\cdot\mathbf{a}-2\mathbf{a}\cdot\mathbf{b}\end{aligned}$$

or

$$|\mathbf{c}|^2=|\mathbf{b}|^2+|\mathbf{a}|^2-2|\mathbf{a}||\mathbf{b}|\cos\theta$$

which is the familiar cosine law for plane triangles (see Fig. 1-10).

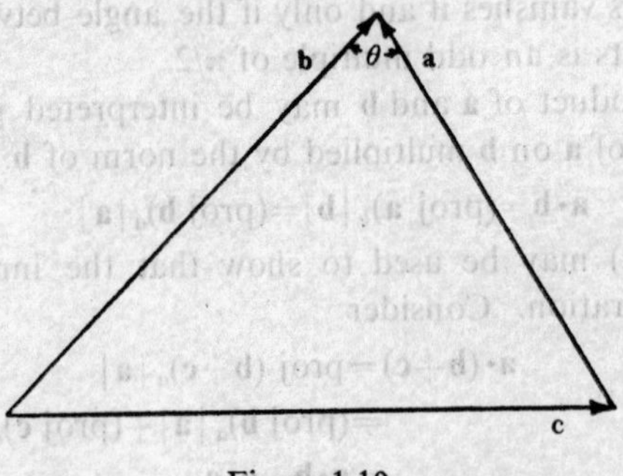

Figure 1-10.

The set of cartesian unit vectors (**i, j, k**) which we defined in Sec.

1-3 form what is known as an orthonormal set, i.e. they satisfy the relations

$$\mathbf{i}\cdot\mathbf{i}=\mathbf{j}\cdot\mathbf{j}=\mathbf{k}\cdot\mathbf{k}=1$$
$$\mathbf{i}\cdot\mathbf{j}=\mathbf{j}\cdot\mathbf{k}=\mathbf{k}\cdot\mathbf{i}=0 \tag{1-20}$$

Now, let **a** and **b** be two vectors with the cartesian representations

$$\mathbf{a}=a_1\mathbf{i}+a_2\mathbf{j}+a_3\mathbf{k}$$
$$\mathbf{b}=b_1\mathbf{i}+b_2\mathbf{j}+b_3\mathbf{k}$$

We then have, making use of the orthogonality relations, Eq. (1-20),

$$\mathbf{a}\cdot\mathbf{b}=(a_1\mathbf{i}+a_2\mathbf{j}+a_3\mathbf{k})\cdot(b_1\mathbf{i}+b_2\mathbf{j}+b_3\mathbf{k})$$
$$=a_1b_1+a_2b_2+a_3b_3 \tag{1-21}$$

EXAMPLE 3. An electric dipole consists of equal and opposite electric charges $(+q, -q)$, separated by a distance a. We define the electric dipole moment $\mathbf{p}=q\mathbf{a}$, where a is the vector of magnitude a directed from $-q$ to $+q$. It follows from Coulomb's law that the electric potential at the point P is

$$\phi=\frac{q}{4\pi\epsilon_0}\left(\frac{1}{r}-\frac{1}{r'}\right)$$

where r and r' are the distances from $-q$ and $+q$ to the point P (see Fig. 1-11). If θ is the angle between **r** and **a**,

$$r'=r-a\cos\theta$$

and

$$\phi=\frac{q}{4\pi\epsilon_0}\left(\frac{1}{r}-\frac{1}{r-a\cos\theta}\right)=\frac{1}{4\pi\epsilon_0}\frac{qa\cos\theta}{r(r-a\cos\theta)}$$

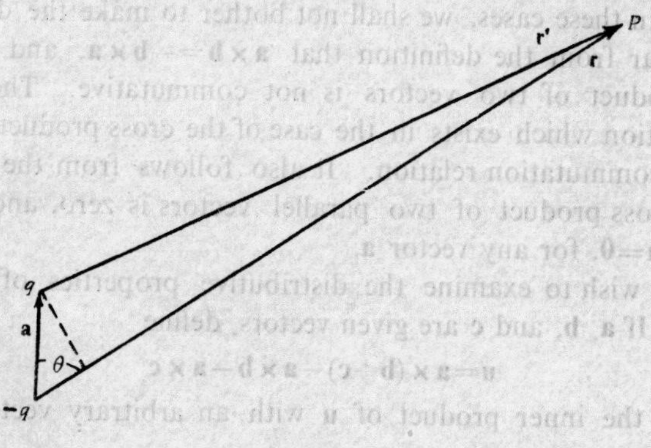

Figure 1-11.

For $r \gg a$, the term $a \cos\theta$ in the denominator can be ignored in comparison with r, and

$$\phi = \frac{1}{4\pi\varepsilon_0}\left[\frac{(qa)r\cos\theta}{r^3}\right] = \frac{1}{4\pi\varepsilon_0}\frac{\mathbf{p}\cdot\mathbf{r}}{r^3}$$

The second product we may form between two vectors results in a vector, and is obtained in the following way: Let **a** and **b** be two arbitrary non-zero, non-parallel vectors. Whenever **a** and **b** are translated so that they have a common initial point, they form two sides of a parallelogram. Define a vector **c**, such that (i) $|\mathbf{c}|$ is the area of the parallelogram defined by the given vectors **a** and **b**, (ii) **c** is in a direction normal to the plane of this parallelogram, and (iii) the positive sense of **c** is in the direction of advance of a right-hand screw as **a** is rotated into **b**. The vector **c** defined in this way is known as the vector, cross, or outer product of the two given vectors, and is written

$$\mathbf{c} = \mathbf{a} \times \mathbf{b} \tag{1-22}$$

Since the area of the parallelogram with sides **a** and **b** is $|\mathbf{a}||\mathbf{b}| \sin\theta$, where θ is the angle from **a** to **b**, we have

$$|\mathbf{c}| = |\mathbf{a}||\mathbf{b}| \sin\theta$$

Although the vector **c** defined above obeys the algebra we have developed for vector quantities, the method we have used to define the sense of **c** is such that **c** does not possess all the properties which we shall later require vectors to have. For this reason, the "vector" **c** is frequently known as an axial vector, pseudovector, screw vector, or vector density. In many of the applications, the distinction between real vectors, also known as polar vectors, and axial vectors is unimportant. In these cases, we shall not bother to make the distinction.

It is clear from the definition that $\mathbf{a} \times \mathbf{b} = -\mathbf{b} \times \mathbf{a}$, and hence the vector product of two vectors is not commutative. The commutation relation which exists in the case of the cross product is known as an anticommutation relation. It also follows from the definition that the cross product of two parallel vectors is zero, and in particular, $\mathbf{a} \times \mathbf{a} = \mathbf{0}$, for any vector **a**.

We now wish to examine the distributive properties of the vector product. If **a**, **b**, and **c** are given vectors, define

$$\mathbf{u} = \mathbf{a} \times (\mathbf{b} + \mathbf{c}) - \mathbf{a} \times \mathbf{b} - \mathbf{a} \times \mathbf{c}$$

and form the inner product of **u** with an arbitrary vector **v**. We obtain

$$\mathbf{v}\cdot\mathbf{u} = \mathbf{v}\cdot\mathbf{a} \times (\mathbf{b} + \mathbf{c}) - \mathbf{v}\cdot(\mathbf{a} \times \mathbf{b}) - \mathbf{v}\cdot(\mathbf{a} \times \mathbf{c})$$

We shall later show that for any three vectors **a**, **b**, and **c**,

$$\mathbf{a}\cdot(\mathbf{b}\times\mathbf{c})=(\mathbf{a}\times\mathbf{b})\cdot\mathbf{c}$$

i.e. we can interchange the scalar and vector product between three factors in a definite manner. Hence,

$$\begin{aligned}\mathbf{v}\cdot\mathbf{u}&=(\mathbf{v}\times\mathbf{a})\cdot(\mathbf{b}+\mathbf{c})-(\mathbf{v}\times\mathbf{a})\cdot\mathbf{b}-(\mathbf{v}\times\mathbf{a})\cdot\mathbf{c}\\&=(\mathbf{v}\times\mathbf{a})\cdot\mathbf{b}+(\mathbf{v}\times\mathbf{a})\cdot\mathbf{c}-(\mathbf{v}\times\mathbf{a})\cdot\mathbf{b}-(\mathbf{v}\times\mathbf{a})\cdot\mathbf{c}\\&=0\end{aligned}\tag{1-23}$$

since the inner product is distributive. Equation (1-23) implies that either $\mathbf{u}=\mathbf{0}$ or else **v** is orthogonal to **u**. However, **v** is arbitrary, so that it can be chosen in such a way that it is not orthogonal to **u**. Hence **u** must be a null vector, and we see that the cross product is distributive,

$$\mathbf{a}\times(\mathbf{b}+\mathbf{c})=\mathbf{a}\times\mathbf{b}+\mathbf{a}\times\mathbf{c}\tag{1-24}$$

The vector products of the cartesian unit vectors are

$$\begin{gathered}\mathbf{i}\times\mathbf{i}=\mathbf{j}\times\mathbf{j}=\mathbf{k}\times\mathbf{k}=\mathbf{0}\\\mathbf{i}\times\mathbf{j}=\mathbf{k},\ \mathbf{j}\times\mathbf{k}=\mathbf{i},\ \mathbf{k}\times\mathbf{i}=\mathbf{j}\end{gathered}\tag{1-25}$$

Now, if **a** and **b** are two vectors which have the cartesian representations

$$\mathbf{a}=a_1\mathbf{i}+a_2\mathbf{j}+a_3\mathbf{k}$$
$$\mathbf{b}=b_1\mathbf{i}+b_2\mathbf{j}+b_3\mathbf{k}$$

respectively, their vector product has the cartesian representation

$$\begin{aligned}\mathbf{a}\times\mathbf{b}=(a_2b_3-a_3b_2)\mathbf{i}+(a_3b_1-a_1b_3)\mathbf{j}+\\(a_1b_2-a_2b_1)\mathbf{k}\end{aligned}\tag{1-26}$$

A careful examination of Eq. (1-26) shows that the cartesian representation of the cross product is equivalent to the formal expansion of the determinant

$$\mathbf{a}\times\mathbf{b}=\begin{vmatrix}\mathbf{i}&\mathbf{j}&\mathbf{k}\\a_1&a_2&a_3\\b_1&b_2&b_3\end{vmatrix}\tag{1-27}$$

The expressions (1-26) and (1-27) are equivalent to the definition of the vector product which we gave in terms of directed line segments, and are generally preferred for computational purposes.

EXAMPLE 4. Consider a light rigid rod pivoted at one end as shown in Fig. 1-12. If we apply a force **F** to the free end of the rod, we find that the effect of this force depends, in addition to the magnitude of the force, on two factors: (i) the distance r from the pivot to the point of application of the force, and (ii) the angle

θ between the rod and the direction of the applied force **F**. Repeated measurements show that the effect of the force is determined by a quantity of magnitude $Fr \sin \theta$. This suggests that we define a vector quantity $\boldsymbol{\tau}$, known as the torque or moment of force by

$$\boldsymbol{\tau} = \mathbf{r} \times \mathbf{F}$$

We note that $\boldsymbol{\tau}$ is perpendicular to the plane containing the rod and the applied force **F**.

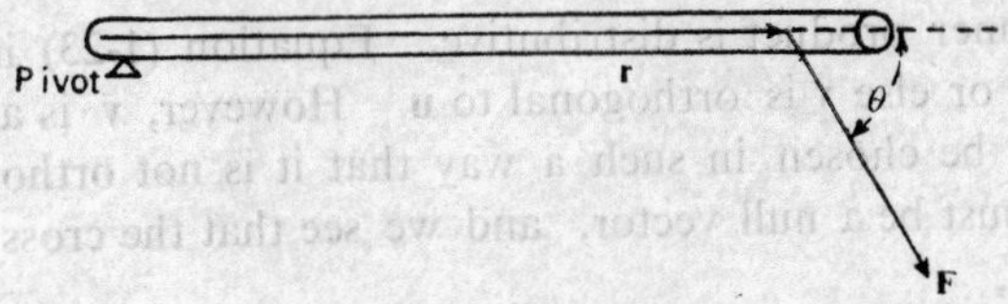

Figure 1-12.

1-5. The Product of Three Vectors. As we have just seen, the product of two vectors may be formed in two distinct ways, one product resulting in a scalar and the other in a vector. We might, therefore, expect that there are several ways in which the product of three vectors may be formed. Although this is the case, there are only two products of three vectors which lead to results that are essentially new. For example, it is quite reasonable to consider the product $(\mathbf{a} \cdot \mathbf{b})\,\mathbf{c}$. However, this is simply the scalar multiple of the vector **c**, and has already been considered. Operations such as $(\mathbf{a} \cdot \mathbf{b}) \times \mathbf{c}$ are meaningless, since the first product is a scalar, and the cross product is defined only when both quantities are vectors.

We now consider the meaningful products of three vectors **a**, **b** and **c**, which we shall assume are non-zero and non-coplanar. Let us also assume that the three given vectors have been translated so that they have a common initial point. Let the angle between the vectors **b** and **c** be denoted by θ, and let α be the angle between the vector **a** and the normal to the plane of **b** and **c**, as shown in Fig. 1-13. The three vectors form three sides of a parallelepiped with value given by

$$\mathbf{a} \cdot (\mathbf{b} \times \mathbf{c}) = |\mathbf{a}|\,|\mathbf{b}|\,|\mathbf{c}| \sin \theta \cos \alpha \tag{1-28}$$

Equation (1-28) may be regarded as the definition of the triple mixed product or triple scalar product of the three given vectors. Since the volume of the parallelepiped is independent of the order of the factors, and the cross product anti-commutes, it is easily seen that

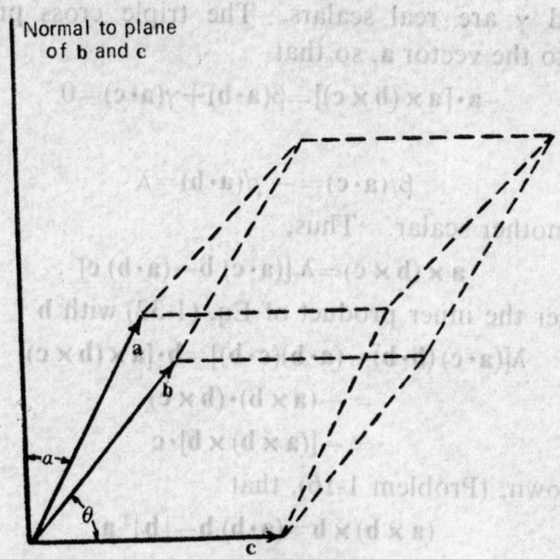

Figure 1-13.

$$\mathbf{a}\cdot(\mathbf{b}\times\mathbf{c})=\mathbf{c}\cdot(\mathbf{a}\times\mathbf{b})=\mathbf{b}\cdot(\mathbf{c}\times\mathbf{a})$$
$$=-\mathbf{b}\cdot(\mathbf{a}\times\mathbf{c})=-\mathbf{c}\cdot(\mathbf{b}\times\mathbf{a})=-\mathbf{a}\cdot(\mathbf{c}\times\mathbf{b}) \qquad (1\text{-}29)$$

We have already made use of Eq. (1-29) in establishing the distributive law for the cross product.

In terms of the cartesian representations of the vectors **a**, **b**, and **c**, we obtain by way of Eqs. (1-21) and (1-26),

$$\mathbf{a}\cdot(\mathbf{b}\times\mathbf{c})=(a_1b_2c_3-a_1b_3c_2)+(a_2b_3c_1-a_2b_1c_3)+(a_3b_1c_2-a_3b_2c_1) \qquad (1\text{-}30)$$

which is the expansion of the determinant

$$\mathbf{a}\cdot(\mathbf{b}\times\mathbf{c})=\begin{vmatrix} a_1 & a_2 & a_3 \\ b_1 & b_2 & b_3 \\ c_1 & c_2 & c_3 \end{vmatrix} \qquad (1\text{-}31)$$

If we choose Eq. (1-31) as the definition of the triple mixed product, the commutation rules, Eqs. (1-29) follow immediately from the properties of the determinant.

The triple cross product $\mathbf{a}\times(\mathbf{b}\times\mathbf{c})$ plays a very important role in the further development of vector analysis. This product is clearly a vector, since it is the cross product of the vector **a** with the vector $\mathbf{b}\times\mathbf{c}$. The triple cross product is orthogonal to the vector $\mathbf{b}\times\mathbf{c}$, so that it lies in the plane determined by the vectors **b** and **c**. Then, if **b** is not parallel to **c**,

$$\mathbf{a}\times(\mathbf{b}\times\mathbf{c})=\beta\mathbf{b}+\gamma\mathbf{c}$$

where β and γ are real scalars. The triple cross product is also orthogonal to the vector **a**, so that

$$\mathbf{a}\cdot[\mathbf{a}\times(\mathbf{b}\times\mathbf{c})]=\beta(\mathbf{a}\cdot\mathbf{b})+\gamma(\mathbf{a}\cdot\mathbf{c})=0$$

and hence

$$\beta/(\mathbf{a}\cdot\mathbf{c})=-\gamma/(\mathbf{a}\cdot\mathbf{b})=\lambda$$

where λ is another scalar. Thus,

$$\mathbf{a}\times(\mathbf{b}\times\mathbf{c})=\lambda\,[(\mathbf{a}\cdot\mathbf{c})\,\mathbf{b}-(\mathbf{a}\cdot\mathbf{b})\,\mathbf{c}] \qquad (1\text{-}32)$$

Now, consider the inner product of Eq, (1-32) with **b**

$$\begin{aligned}\lambda[(\mathbf{a}\cdot\mathbf{c})(\mathbf{b}\cdot\mathbf{b})-(\mathbf{a}\cdot\mathbf{b})(\mathbf{c}\cdot\mathbf{b})]&=\mathbf{b}\cdot[\mathbf{a}\times(\mathbf{b}\times\mathbf{c})]\\&=-(\mathbf{a}\times\mathbf{b})\cdot(\mathbf{b}\times\mathbf{c})\\&=-[(\mathbf{a}\times\mathbf{b})\times\mathbf{b}]\cdot\mathbf{c}\end{aligned}$$

It can be shown, (Problem 1-16), that

$$(\mathbf{a}\times\mathbf{b})\times\mathbf{b}=(\mathbf{a}\cdot\mathbf{b})\,\mathbf{b}-|\mathbf{b}|^2\,\mathbf{a}$$

and hence

$$[(\mathbf{a}\times\mathbf{b})\times\mathbf{b}]\cdot\mathbf{c}=|\mathbf{b}|^2\,(\mathbf{a}\cdot\mathbf{c})-(\mathbf{a}\cdot\mathbf{b})(\mathbf{b}\cdot\mathbf{c})$$

which implies that λ is unity. Referring to Eq. (1-32), we see that

$$\mathbf{a}\times(\mathbf{b}\times\mathbf{c})=(\mathbf{a}\cdot\mathbf{c})\,\mathbf{b}-(\mathbf{a}\cdot\mathbf{b})\,\mathbf{c} \qquad (1\text{-}33)$$

On the other hand,

$$\begin{aligned}(\mathbf{a}\times\mathbf{b})\times\mathbf{c}&=-\mathbf{c}\times(\mathbf{a}\times\mathbf{b})\\&=[(\mathbf{a}\cdot\mathbf{c})]\,\mathbf{b}-[(\mathbf{b}\cdot\mathbf{c})]\,\mathbf{a}\\&\neq\mathbf{a}\times(\mathbf{b}\times\mathbf{c})\end{aligned}$$

so that the triple cross product is not associative.

EXAMPLE 5. According to the Biot-Savart law, a charge q_1 moving with velocity $\mathbf{v}_1$ generates a magnetic field $\mathbf{B}_1$ given by

$$\mathbf{B}_1=\frac{\mu_0}{4\pi}q_1\frac{\mathbf{v}_1\times\mathbf{r}}{r^2}$$

where r is the vector from q_1 to the point at which $\mathbf{B}_1$ is measured. If there is a second charge q_2 moving with velocity $\mathbf{v}_2$ a distance $\mathbf{r}_0$ from q_1, the magnetic part of the Lorentz force on q_2 due to q_1 is

$$\mathbf{F}_{21}=\frac{\mu_0}{4\pi}\frac{q_1q_2}{r_0^2}\,\mathbf{v}_2\times(\mathbf{v}_1\times\mathbf{r}_0)$$

On the other hand, a similar argument shows that the force on q_1 due to q_2 is

$$\mathbf{F}_{12}=-\frac{\mu_0}{4\pi}\frac{q_1q_2}{r_0^2}\,\mathbf{v}_1\times(\mathbf{v}_2\times\mathbf{r}_0)\neq-\mathbf{F}_{21}$$

which is a contradiction of Newton's law of action and reaction.

Let $\mathbf{e}_1$ and $\mathbf{e}_2$ be unit vectors along $\mathbf{v}_1$ and $\mathbf{v}_2$ respectively. Then, expanding the triple vector products according to Eq. (1-33),

$$\mathbf{F}_{21}=\frac{\mu_0}{4\pi}\frac{q_1q_2}{r_0^2}\,v_1v_2\,[(\mathbf{e}_2\cdot\mathbf{r}_0)\,\mathbf{e}_1-(\mathbf{e}_2\cdot\mathbf{e}_1)\,\mathbf{r}_0]$$

$$\mathbf{F}_{12}=-\frac{\mu_0}{4\pi}\frac{q_1q_2}{r_0^2}\,v_1v_2\,[(\mathbf{e}_1\cdot\mathbf{r}_0)\,e_2-(\mathbf{e}_1\cdot\mathbf{e}_2)\,\mathbf{r}_0]$$

and we see that $\mathbf{F}_{12}=-\mathbf{F}_{21}$ if and only if $\mathbf{e}_1$ and $\mathbf{e}_2$ are parallel.

Many rather complicated products of vectors can be simplified by the use of Eq. (1-33) for the triple cross product. For example,

$$\begin{aligned}(\mathbf{a}\times\mathbf{b})\times(\mathbf{c}\times\mathbf{d})&=[(\mathbf{a}\times\mathbf{b})\cdot\mathbf{d}]\,\mathbf{c}-[(\mathbf{a}\times\mathbf{b})\cdot\mathbf{c}]\,\mathbf{d}\\&=(\mathbf{a}\cdot\mathbf{b}\times\mathbf{d})\,\mathbf{c}-(\mathbf{a}\cdot\mathbf{b}\times\mathbf{c})\,\mathbf{d}\end{aligned}$$

1-6. Summary of Vector Identities

(*i*) *Scalar Multiplication and Addition*:

$$\alpha(\beta\mathbf{a})=(\alpha\beta)\,\mathbf{a}=(\beta\alpha)\,\mathbf{a}=\beta(\alpha\mathbf{a})$$

$$(\alpha+\beta)\,\mathbf{a}=\alpha\mathbf{a}+\beta\mathbf{a}$$

$$0\mathbf{a}=\mathbf{0}$$

$$\mathbf{a}+\mathbf{b}=\mathbf{b}+\mathbf{a}$$

$$(\mathbf{a}+\mathbf{b})+\mathbf{c}=\mathbf{a}+(\mathbf{b}+\mathbf{c})$$

$$\alpha(\mathbf{a}+\mathbf{b})=\alpha\mathbf{a}+\alpha\mathbf{b}$$

(*ii*) *The Inner Product*:

$$\mathbf{a}\cdot\mathbf{b}=\mathbf{b}\cdot\mathbf{a}$$

$$\mathbf{a}\cdot\mathbf{a}=|\mathbf{a}|^2$$

$$\mathbf{a}\cdot(\mathbf{b}+\mathbf{c})=\mathbf{a}\cdot\mathbf{b}+\mathbf{a}\cdot\mathbf{c}$$

$$(\mathbf{a}+\mathbf{b})\cdot(\mathbf{c}+\mathbf{d})=\mathbf{a}\cdot\mathbf{c}+\mathbf{a}\cdot\mathbf{d}+\mathbf{b}\cdot\mathbf{c}+\mathbf{b}\cdot\mathbf{d}$$

(*iii*) *The Cross Product*:

$$\mathbf{a}\times\mathbf{b}=-\mathbf{b}\times\mathbf{a}$$

$$\mathbf{a}\times(\mathbf{b}+\mathbf{c})=\mathbf{a}\times\mathbf{b}+\mathbf{a}\times\mathbf{c}$$

(*iv*) *Products of More than Two Factors*:

$$\mathbf{a}\cdot(\mathbf{b}\times\mathbf{c})=\mathbf{c}\cdot(\mathbf{a}\times\mathbf{b})=\mathbf{b}\cdot(\mathbf{c}\times\mathbf{a})$$

$$\mathbf{a}\times(\mathbf{b}\times\mathbf{c})=(\mathbf{a}\cdot\mathbf{c})\,\mathbf{b}-(\mathbf{a}\cdot\mathbf{b})\,\mathbf{c}$$

$$(\mathbf{a}\times\mathbf{b})\times\mathbf{c}=(\mathbf{a}\cdot\mathbf{c})\,\mathbf{b}-(\mathbf{b}\cdot\mathbf{c})\,\mathbf{a}$$

$$(\mathbf{a}\times\mathbf{b})\times(\mathbf{c}\times\mathbf{d})=(\mathbf{a}\cdot\mathbf{b}\times\mathbf{d})\,\mathbf{c}-(\mathbf{a}\cdot\mathbf{b}\times\mathbf{c})\,\mathbf{d}$$

$$(\mathbf{a}\times\mathbf{b})\cdot(\mathbf{c}\times\mathbf{d})=(\mathbf{a}\cdot\mathbf{c})\,(\mathbf{b}\cdot\mathbf{d})-(\mathbf{a}\cdot\mathbf{d})\,(\mathbf{b}\cdot\mathbf{c})$$

PROBLEMS

1-1. A boat with a maximum speed of 15 miles per hour is crossing a stream which has a current of 7 miles per hour parallel to the bank. If the operator of the boat wishes to go directly across the stream, at what angle relative to the bank should the boat be launched?

1-2. A mass of 10 kg is suspended from two light strings as shown in Fig. 1-14. The forces acting on the mass are its weight mg ($g=9.8$ nt/kg) and the tension in the two strings. According to Newton's law, the mass will remain at rest if the vector sum of the forces acting on it is zero. Determine the tension in the two strings.

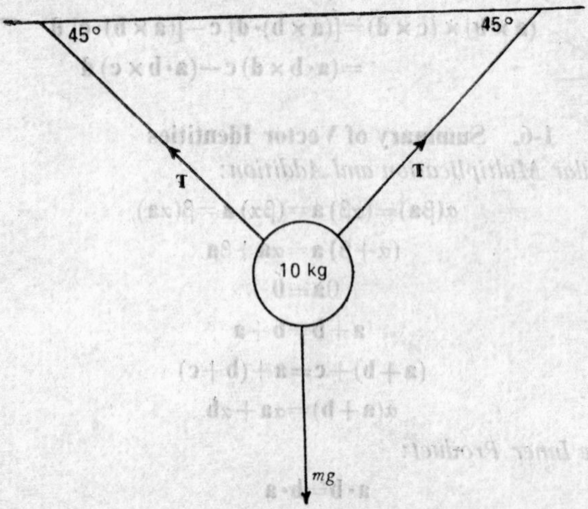

Figure 1-14.

1-3. Let **a** and **b** be two vector which have a common origin 0, and let **c** be any vector whose origin is 0 and whose terminal point lies on the line joining the terminal points of **a** and **b** as shown in Fig. 1-15. If the point C divides the line segment BA in the ratio $\alpha:\beta$, where $\alpha+\beta=1$, show that

$$\mathbf{c}=-\alpha\mathbf{a}+\beta\mathbf{b}$$

1-4. Consider the sum of N unit vectors, each of which makes an angle $2\pi/N$ with the preceeding vector. Show that

$$\sum_{n=0}^{N-1} \sin(2n\pi/N)=\sum_{n=0}^{N-1}\cos(2n\pi/N)=0$$

1-5. If **a**, **b**, and **c** are any three non-coplanar vectors, show that the relation

$$\alpha\mathbf{a}+\beta\mathbf{b}+\gamma\mathbf{c}=0$$

implies that

$$\alpha=\beta=\gamma=0$$

where α, β, and γ are any scalars.

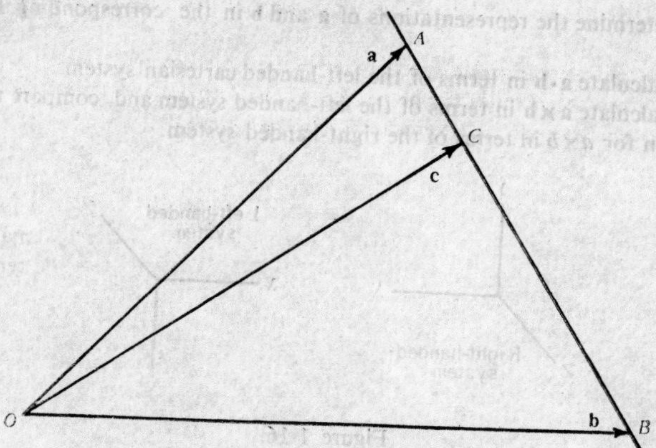

Figure 1-15.

1-6. Let **a** be any vector, and let (**i**, **j**, **k**) be the set of cartesian unit vectors. Show that the cartesian representation of **a** can be written as

$$\mathbf{a}=(\mathbf{a}\cdot\mathbf{i})\,\mathbf{i}+(\mathbf{a}\cdot\mathbf{j})\,\mathbf{j}+(\mathbf{a}\cdot\mathbf{k})\,\mathbf{k}$$

1-7. In the quantum theory of the hydrogen atom, the angular momentum **L** of the electron has a magnitude $|\mathbf{L}|^2=l\,(l+1)\,\hbar$, where l is a non-negative integer and $\hbar$ is Planck's constant divided by 2π. The direction of **L** is such that $L_z=\pm m\hbar$, where $m=0, 1,\ldots, l$. If $l=3$, determine, so far as possible, the allowed orientations of **L**.

1-8. The work done by a force **F** in moving a particle through the displacement **r** is defined to be $W=\mathbf{F}\cdot\mathbf{r}$. Determine the work done by the force $\mathbf{F}=3\mathbf{i}+2\mathbf{j}-6\mathbf{k}$ in moving a particle along the straight line connecting the points (1, 0, 0) and (−3, 3, −1).

1-9. Prove, by vector methods, the law of sines for plane triangles.

1-10. Show that the area of the triangle defined by the vectors **A** and **B** is $\frac{1}{2}|\mathbf{A}\times\mathbf{B}|$.

1-11. Use the three unit vectors

$$\mathbf{A}=\cos\theta\,\mathbf{i}+\sin\theta\,\mathbf{j}$$
$$\mathbf{B}=\cos\varphi\,\mathbf{i}-\sin\varphi\,\mathbf{j}$$
$$\mathbf{C}=\cos\varphi\,\mathbf{i}+\sin\varphi\,\mathbf{j}$$

to show that

(i) $\sin(\theta+\varphi)=\sin\theta\cos\varphi+\cos\theta\sin\varphi$

(ii) $\cos(\theta+\varphi)=\cos\theta\cos\varphi-\sin\theta\sin\varphi$

1-12. Given a set of right-handed cartesian coordinates (x, y, z), we can form a left-handed system by reversing the positive direction of each of the coordinate axes as shown in Fig. 1-16. In the right-handed system, let the vectors **a** and **b** have the representations

$$\mathbf{a}=a_1\,\mathbf{i}+a_2\,\mathbf{j}+a_3\,\mathbf{k}$$
$$\mathbf{b}=b_1\,\mathbf{i}+b_2\,\mathbf{j}+b_3\,\mathbf{k}$$

(i) Determine the representations of **a** and **b** in the corresponding left-hand system.

(ii) Calculate $\mathbf{a \cdot b}$ in terms of the left-handed cartesian system.

(iii) Calculate $\mathbf{a \times b}$ in terms of the left-handed system and compare this with expression for $a \times b$ in terms of the right-handed system.

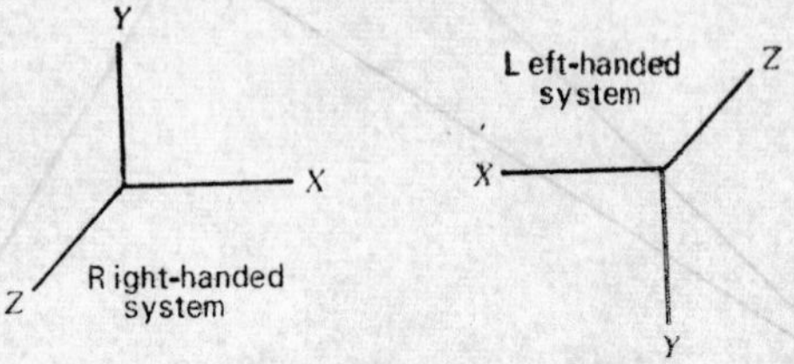

Figure 1-16.

1-13. An electric charge q moving with velocity $\mathbf{v}$ in a magnetic field $\mathbf{B}$ is subject to a Lorentz force

$$F = q\,(\mathbf{v} \times \mathbf{B})$$

In a magnetic field $\mathbf{B} = \mathbf{B}_z\,\mathbf{k}$, determine the force on an electron (charge e) moving with a velocity $\mathbf{v} = -v_0 \sin\theta\ \mathbf{i} + v_0 \cos\theta\ \mathbf{j}$.

1-14. Given a set of three non-coplanar vectors **a**, **b**, **c**, we define a set of reciprocal vectors by the relations

$$\mathbf{a}' \cdot \mathbf{a} = \mathbf{b}' \cdot \mathbf{b} = \mathbf{c}' \cdot \mathbf{c} = 1$$

$$\mathbf{a}' \cdot \mathbf{b} = \mathbf{b}' \cdot \mathbf{c} = \mathbf{c}' \cdot \mathbf{a} = \mathbf{a} \cdot \mathbf{b}' = \mathbf{b} \cdot \mathbf{c}' = \mathbf{c} \cdot \mathbf{a}' = 0$$

Show that

$$\mathbf{a}' = \frac{\mathbf{b \times c}}{\mathbf{a \cdot b \times c}}, \quad \mathbf{b}' = \frac{\mathbf{c \times a}}{\mathbf{b \cdot c \times a}}, \quad \mathbf{c}' = \frac{\mathbf{a \times b}}{\mathbf{c \cdot a \times b}}$$

1-15. Show that the only right-handed self-reciprocal set of vectors are the cartesian unit vectors (**i**, **j**, **k**).

1-16. Show that

$$(\mathbf{a \times b}) \times \mathbf{b} = (\mathbf{a \cdot b})\,\mathbf{b} - |\mathbf{b}|^2\,\mathbf{a}$$

1-17. Show that

$$(\mathbf{a \times b}) \cdot (\mathbf{c \times d}) = (\mathbf{a \cdot c})\,(\mathbf{b \cdot d}) - (\mathbf{a \cdot d})\,(\mathbf{b \cdot c})$$

1-18. If $\mathbf{a \cdot b \times c} = 0$, show that either

(i) **a**, **b**, and **c** are coplanar, but no two of them are collinear, or

(ii) two of the vectors are colinear, or

(iii) all of the vectors are colinear

1-19. Show that a necessary and sufficient condition for

$$\mathbf{a \times (b \times c)} = (\mathbf{a \times b}) \times \mathbf{c}$$

is that

$$(\mathbf{a \times c}) \times \mathbf{b} = 0$$

CHAPTER 2

Differential and Integral Calculus of Vectors

2-1. Vector and Scalar Fields. We have developed an algebra for vectors by considering quantities which are defined at a single point. However, in order to develop a differential and integral calculus for these quantities, we must introduce the concept of vector and scalar fields. Consider a Euclidean space of three dimensions, the points of which are denoted by P_i. At each point of the space associate a scalar number a_i such that

$$a_i = f(P_i)$$

where f is a continuous, single-valued function of the points of the space. If $f(P_i)$ is defined at every point in some region of the space, we say that the set of function values a_i defines a scalar field over that region of the space. If we introduce an orthogonal cartesian coordinate system in order to describe the points of the space, the point P_i is defined by the coordinate triplet (x_i, y_i, z_i), and the scalar field can be represented by

$$a = f(x, y, z)$$

The point scalars which comprise the field are the functional values of f at the points (x_i, y_i, z_i).

Now, suppose that at each point P_i in the space, we assign a definite vector $\mathbf{u}_i$. The totality of such vectors we shall denote by $\mathbf{u}$, and say that $\mathbf{u}$ is a vector field. In terms of the cartesian representation of $\mathbf{u}$ and the points of the space, we may write

$$\begin{aligned}\mathbf{u} &= \mathbf{u}(x, y, z)\\ &= u_1(x, y, z)\,\mathbf{i} + u_2(x, y, z)\,\mathbf{j} + u_3(x, y, z)\,\mathbf{k}\end{aligned}$$

where u_1, u_2, and u_3 are specified functions of the coordinates. In a great many of the physical applications, the functions u_1, u_2, and u_3 are also explicitly functions of the time, and the vector field is expressed as

$$\mathbf{u} = \mathbf{u}(x, y, z, t)$$

2-2. Differentiation of Vector Fields. The time dependent vector field whose cartesian representation is

$$\mathbf{u}(x, y, z, t)=u_1(x, y, z, t)\,\mathbf{i}+u_2(x, y, z, t)\,\mathbf{j}+u_3(x, y, z, t)\,\mathbf{k}$$

defines a vector at any space point $P(x, y, z)$, and at any time t. At the given space point $P(x_0, y_0, z_0)$, the vector changes from time to time due to the explicit time dependence of its components. Similarly, at a given time t_0, the vector at the space point $P(x, y, z)$ is in general different from the vector at the space point $Q(x+dx, y+dy, z+dz)$. We encounter no difficulty in calculating the change in the vector $\mathbf{u}$, since it changes if and only if its components change. We define the cartesian representation of the differential

$$\begin{aligned}\mathbf{du}&=du_1\,\mathbf{i}+du_2\,\mathbf{j}+du_3\,\mathbf{k}\\&=\left(\frac{\partial u_1}{\partial x}\,dx+\frac{\partial u_1}{\partial y}\,dy+\frac{\partial u_1}{\partial z}\,dz+\frac{\partial u_1}{\partial t}\,dt\right)\mathbf{i}+\\&\quad\left(\frac{\partial u_2}{\partial x}\,dx+\frac{\partial u_2}{\partial y}\,dy+\frac{\partial u_2}{\partial z}\,dz+\frac{\partial u_2}{\partial t}\,dt\right)\mathbf{j}+\\&\quad\left(\frac{\partial u_3}{\partial x}\,dx+\frac{\partial u_3}{\partial y}\,dy+\frac{\partial u_3}{\partial z}\,dz+\frac{\partial u_3}{\partial t}\,dt\right)\mathbf{k}\end{aligned}\tag{2-1}$$

In the fortunate, but rather infrequently encountered, event that the vector $\mathbf{u}$ is a function of a single variable, say t, the derivative of the vector field with respect to the independent variable is obtained by the usual limiting process, which is illustrated in Fig. 2-1,

$$\frac{d\mathbf{u}}{dt}=\lim_{\Delta t\to 0}\frac{\mathbf{u}(t+\Delta t)-\mathbf{u}(t)}{\Delta t}\tag{2-2}$$

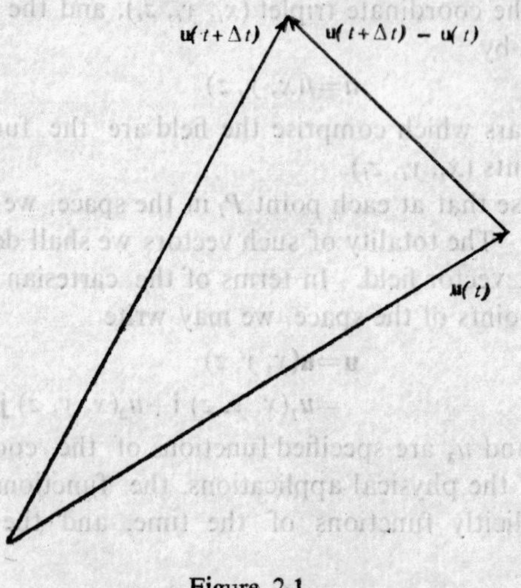

Figure 2-1.

EXAMPLE 1. An elementary example occurs in the description of the motion of a mass particle. At any time t, the instantaneous location of the particle is given by the triplet of numbers (x, y, z), that is, by the description of some point in space. The location of this space point may equally well be specified by the vector **r**, which is represented by the directed line segment from the origin to the point in question. This vector is known as the radius or position vector of the particle, and has the cartesian representation

$$\mathbf{r}(x, y, x)=x\,\mathbf{i}+y\,\mathbf{j}+z\,\mathbf{k}$$

The radius vector, as a function of the coordinate triplet is a vector field. If the particle is moving in space, the cartesian components are explicit functions of the time. It then follows from Eq. (2-1) that the differential of the radius vector is

$$d\mathbf{r}=dx\,\mathbf{i}+dy\,\mathbf{j}+dz\,\mathbf{k}$$

Whenever the cartesian components of **r** are twice differentiable functions of the time, we define two new vectors **v** and **a**, known as the velocity and acceleration vectors of the particle respectively

$$\mathbf{v}=d\mathbf{r}/dt=(dx/dt)\,\mathbf{i}+(dy/dt)\,\mathbf{j}+(dz/dt)\,\mathbf{k}$$
$$\mathbf{a}=d^2\mathbf{r}/dt^2=(d^2x/dt^2)\,\mathbf{i}+(d^2y/dt^2)\,\mathbf{j}+(d^2z/dt^2)\,\mathbf{k}$$

In defining the velocity and acceleration vectors, we have tacitly assumed that the set of cartesian unit vectors (**i**, **j**, **k**) is constant in time. This is not always the case, and when this does not obtain, the analysis becomes much more difficult.

EXAMPLE 2. Consider a particle constrained to move on a circle of radius r_0. Introduce a two-dimensional cartesian coordinate system (x, y) oriented so that at time $t=0$, the position vector of the particle is $\mathbf{r}(0)=r_0\mathbf{i}$. Then, for all time t, the position vector is given by

$$\mathbf{r}(t)=r_0\cos\theta(t)\,\mathbf{i}+r_0\sin\theta(t)\,\mathbf{j}$$

where $\theta(t)$ is the instantaneous angle between $\mathbf{r}(t)$ and the positive x-axis. The velocity of the particle is

$$\begin{aligned}\mathbf{v}(t)&=-r_0\sin\theta\,(d\theta/dt)\,\mathbf{i}+r_0\cos\theta\,(d\theta/dt)\,\mathbf{j}\\&=-r_0\,\omega\,(-\sin\theta\,\mathbf{i}+\cos\theta\,\mathbf{j})\end{aligned}$$

where $\omega=d\theta/dt$ is known as the angular velocity. Differentiating a second time, we find that the particle has an acceleration

$$\begin{aligned}\mathbf{a}(t)=&-r_0\,\omega^2\,(\cos\theta\mathbf{i}+\sin\theta\mathbf{j})+\\&r_0\,(d\omega/dt)\,(-\sin\theta\mathbf{i}+\cos\theta\mathbf{j})\end{aligned}$$

$$=\mathbf{a}_C+\mathbf{a}_T$$

The quantity $\mathbf{a}_C$, known as the centripetal acceleration is obviously given by

$$\mathbf{a}_C=-\omega^2\,\mathbf{r}(t)$$

Forming the inner product of $\mathbf{r}(t)$, first with $\mathbf{v}(t)$ and then with $\mathbf{a}_T$, known as the tangential acceleration, we have

$$\mathbf{r}\cdot\mathbf{v}=r_0^2\omega(-\cos\theta\sin\theta+\sin\theta\cos\theta)=0$$

$$\mathbf{r}\cdot\mathbf{a}_T=r_0^2\frac{d\omega}{dt}(-\cos\theta\sin\theta+\sin\theta\cos\theta)=0$$

Thus, the velocity and tangential acceleration are perpendicular to $\mathbf{r}(t)$ and hence are tangent to the circle $|\mathbf{r}|=r_0$.

2-3. Differentiation Rules. Let $\varphi(t)$ be a differentiable scalar function of the parameter t, defined by

$$\varphi(t)=\mathbf{u}(t)\cdot\mathbf{v}(t)$$

where $\mathbf{u}(t)$ and $\mathbf{v}(t)$ are non-orthogonal, differentiable vector functions of the parameter t. Then,

$$\varphi(t+\Delta t)-\varphi(t)=\mathbf{u}(t+\Delta t)\cdot\mathbf{v}(t+\Delta t)-\mathbf{u}(t)\cdot\mathbf{v}(t)$$

but,

$$\mathbf{u}(t+\Delta t)=\mathbf{u}(t)+\Delta\mathbf{u}$$
$$\mathbf{v}(t+\Delta t)=\mathbf{v}(t)+\Delta\mathbf{v}$$

so that

$$\frac{\varphi(t+\Delta t)-\varphi(t)}{\Delta t}=\mathbf{u}\cdot(\Delta\mathbf{v}/\Delta t)+\mathbf{v}\cdot(\Delta\mathbf{u}/\Delta t)+(\Delta\mathbf{u}/\Delta t)\cdot\Delta\mathbf{v}+(\Delta\mathbf{v}/\Delta t)\cdot\Delta\mathbf{u}$$

In the lim $\Delta t\to 0$, the last two terms vanish, and we obtain the differentiation formula

$$d(\mathbf{u}\cdot\mathbf{v})/dt=\mathbf{u}\cdot(d\mathbf{v}/dt)+\mathbf{v}\cdot(d\mathbf{u}/dt) \tag{2-3}$$

Let us now consider the vector function of the parameter t which is defined by

$$\mathbf{w}(t)=\mathbf{u}(t)\times\mathbf{v}(t)$$

Then,

$$\frac{\mathbf{w}(t+\Delta t)-\mathbf{w}(t)}{\Delta t}=\mathbf{u}\times(\Delta\mathbf{v}/\Delta t)+(\Delta\mathbf{u}/\Delta t)\times\mathbf{v}+(\Delta\mathbf{u}/\Delta t)\times\Delta\mathbf{v}+\Delta\mathbf{u}\times(\Delta\mathbf{v}/\Delta t)$$

In the lim $\Delta t\to 0$, the last two terms vanish, and we obtain the relation

$$d(\mathbf{u}\times\mathbf{v})/dt=\mathbf{u}\times d\mathbf{v}/dt+(d\mathbf{u}/dt)\times\mathbf{v} \tag{2-4}$$

Finally, if α is any differentiable scalar function of t, and if $\mathbf{u}(t)$ is a differentiable vector function of t, we have the relation

$$d(\alpha\mathbf{u})/dt=\alpha\, d\mathbf{u}/dt+(d\alpha/dt)\,\mathbf{u} \tag{2-5}$$

Let $\mathbf{u}(t)$ be a vector of constant magnitude $|\mathbf{u}|$. It then follows from the definition of the inner product that

$$\mathbf{u}\cdot\mathbf{u}=|\mathbf{u}|^2=\text{constant}$$

Setting $\mathbf{v}=\mathbf{u}$ in Eq. (2-3), we obtain

$$\mathbf{u}\cdot(d\mathbf{u}/dt)+(d\mathbf{u}/dt)\cdot\mathbf{u}=0$$

or

$$\mathbf{u}\cdot(d\mathbf{u}/dt)=0 \tag{2-6}$$

Equation (2-6) implies that either $\mathbf{u}$ is a constant vector, or else it is orthogonal to its derivative. In general, $\mathbf{u}(t)$ is not constant.

If $\mathbf{u}(t)$ is a differentiable vector field which does not have constant magnitude, differentiation of the inner product of u with itself yields

$$\mathbf{u}\cdot(d\mathbf{u}/dt)=\mathbf{u}\; d|\mathbf{u}|/dt$$

This result is not trivial, since $|d\mathbf{u}|\neq d|\mathbf{u}|$ in general.

2-4. The Gradient. One of the most significant vector operations, and which we shall encounter frequently in the applications of vector analysis, is the operation of forming the gradient. Let $\rho(x, y, z)$ be a continuous, differentiable function of the cartesian coordinates (x, y, z). Then,

$$d\rho=(\partial\rho/\partial x)dx+(\partial\rho/\partial y)dy+(\partial\rho/\partial z)dz \tag{2-7}$$

We now define a vector field, known as the gradient of ρ, (written grad ρ, del ρ, or $\nabla\rho$) by the relation

$$\text{grad}\ \rho=(\partial\rho/\partial x)\mathbf{i}+(\partial\rho/\partial y)\mathbf{j}+(\partial\rho/\partial z)\mathbf{k} \tag{2-8}$$

Now, let $\mathbf{r}$ be the radius vector from an arbitrary origin to the point $P(x, y, z)$ at which we have defined the vector field grad ρ. If we move from the point $P(x, y, z)$ to the point $Q(x+dx, y+dy, z+dz)$, the change in the vector $\mathbf{r}$ is

$$d\mathbf{r}=dx\mathbf{i}+dy\mathbf{j}+dz\mathbf{k}$$

so that the inner product of $d\mathbf{r}$ with grad ρ is

$$d\mathbf{r}\cdot\text{grad}\ \rho=\frac{\partial\rho}{\partial x}dx+\frac{\partial\rho}{\partial y}dy+\frac{\partial\rho}{\partial z}dz \tag{2-9}$$

Although in our discussion of the algebra of vectors, we first considered the algebraic operations from the point of view of geometrical definitions, and then formulated computational algorithms in terms of equivalent cartesian representations, it is more convenient to define the differential operators in terms of cartesian representations, and then seek geometrical interpretations. At a particular space point

$P(x_0, y_0, z_0)$ the scalar function $\rho(x, y, z)$ has a particular functional value $\rho(x_0, y_0, z_0)$. Hence, the equation

$$\rho(x, y, z)=\rho(x_0, y_0, z_0)$$

defines a surface which contains the point $P(x_0, y_0, z_0)$. So long as we consider only points on this surface, $\rho(x, y, z)$ has the constant value $\rho(x_0, y_0, z_0)$, and consequently $d\rho=0$. It then follows that

$$d\mathbf{r}\cdot\text{grad }\rho=0 \tag{2-10}$$

Since grad ρ is completely determined by Eq. (2-7) and is generally not identically zero, grad ρ must be orthogonal to $d\mathbf{r}$ so long as $d\mathbf{r}$ represents the change from the space point P to the space point Q, where both P and Q lie in the surface $\rho=\rho(x_0, y_0, z_0)$. Therefore, the vector grad ρ is normal to all possible tangents to the surface which passes through the point P, and hence is normal to the surface ρ=constant. This is illustrated in Fig. 2-2. At any point in space, the vector grad ρ is fixed, so that the differential $d\rho$ will have its maximum value whenever the vector $d\mathbf{r}$ is parallel to the vector grad ρ. Hence, the direction of the vector grad ρ is along the direction of the maximum rate of increase of the scalar function

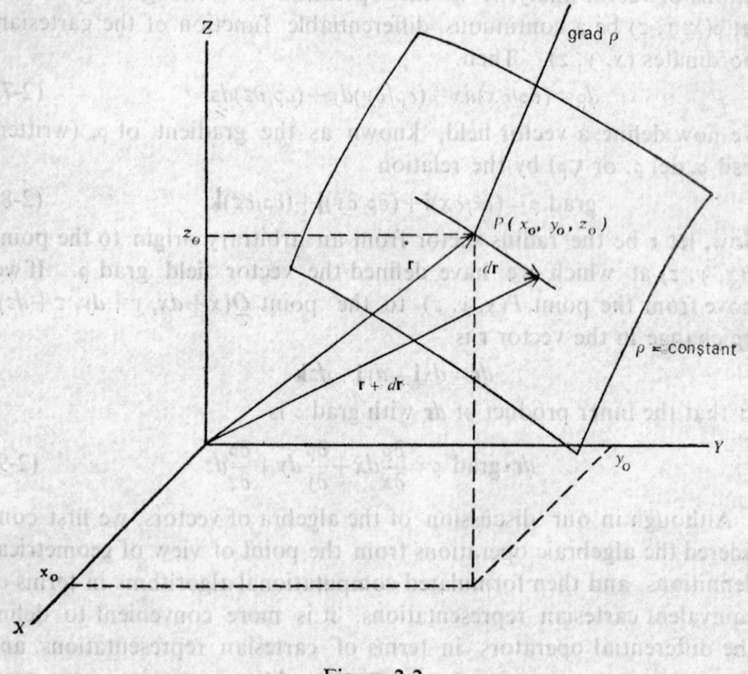

Figure 2-2.

$\rho(x, y, z)$. If we denote the norm of $d\mathbf{r}$ by ds, and let $\mathbf{u}$ be a unit vector in the direction of $d\mathbf{r}$, then

$$d\rho/ds = \mathbf{u} \cdot \text{grad } \rho \tag{2-11}$$

The operation of forming the gradient of a scalar may be interpreted as the application of a vector differential operator, which we shall call the "grad" operator, on the scalar field. Although this interpretation leads to some difficulty whenever we encounter coordinate systems more general than cartesian coordinates, it is useful in obtaining the gradient of products, etc. The application of the vector operator "grad" is similar to the interpretation of the derivative df/dx as the application of the derivative operator $D=d/dx$ to the differentiable function $f(x)$. In terms of a right-handed cartesian coordinate system with the cartesian unit vectors $(\mathbf{i}, \mathbf{j}, \mathbf{k})$, the vector operator "grad" has the representation

$$\text{"grad"} = \mathbf{i}(\partial/\partial x) + \mathbf{j}(\partial/\partial y) + \mathbf{k}(\partial/\partial z) \tag{2-12}$$

Then, the gradient of the scalar field $\rho(x, y, z)$ is given by

$$\begin{aligned} \text{grad } \rho &= [\mathbf{i}(\partial/\partial x)\rho + \mathbf{j}(\partial/\partial y) + \mathbf{k}(\partial/\partial z)]\rho(x, y, z) \\ &= \mathbf{i}(\partial\rho/\partial x) + \mathbf{j}(\partial\rho/\partial y) + \mathbf{k}(\partial\rho/\partial z) \end{aligned}$$

which is, as we have seen, the cartesian representation of the vector grad ρ.

An operator 0 is called linear if

$$0(c_1\rho + c_2\lambda) = c_1 0(\rho) + c_2 0(\lambda)$$

where ρ and λ are any two functions on which the operator 0 acts, and c_1 and c_2 are arbitrary constants with respect to the operator. For example, the differential operator D is linear, since

$$\begin{aligned} D(c_1\rho + c_2\lambda) &= c_1 D(\rho) + c_2 D(\lambda) \\ &= c_1(d\rho/dx) + c_2(d\lambda/dx) \end{aligned}$$

assuming, of course that $d\rho/dx$ and $d\lambda/dx$ are both defined. We shall now show that the operator "grad" is a linear operator with respect to differentiable scalar fields. Let $\rho(x, y, z)$ and $\lambda(x, y, z)$ be two scalar fields which are continuous differentiable functions of their coordinates, such that grad ρ and grad λ are both defined. Then, if c_1 and c_2 are two arbitrary constants,

$$\begin{aligned} \text{grad } (c_1\rho + c_2\lambda) &= \mathbf{i}\frac{\partial}{\partial x}(c_1\rho + c_2\lambda) + \mathbf{j}\frac{\partial}{\partial y}(c_1\rho + c_2\lambda) + \mathbf{k}\frac{\partial}{\partial z}(c_1\rho + c_2\lambda) \\ &= c_1\left[\mathbf{i}\frac{\partial\rho}{\partial x} + \mathbf{j}\frac{\partial\rho}{\partial y} + \mathbf{k}\frac{\partial\rho}{\partial z}\right] + c_2\left[\mathbf{i}\frac{\partial\lambda}{\partial x} + \mathbf{j}\frac{\partial\rho}{\partial y} + \mathbf{k}\frac{\partial\lambda}{\partial z}\right] \\ &= c_1 \text{ grad } \rho + c_2 \text{ grad } \lambda \end{aligned}$$

Again, let $\rho(x, y, z)$ and $\lambda(x, y, z)$ be two scalar fields which are continuous and differentiable functions of the coordinates (x, y, z). Then,

$$\text{grad}\,(\rho\lambda)=\mathbf{i}\frac{\partial\rho\lambda}{\partial x}+\mathbf{j}\frac{\partial\rho\lambda}{\partial y}+\mathbf{k}\frac{\partial\rho\lambda}{\partial z}$$

$$=\rho\,\text{grad}\,\lambda+\lambda\,\text{grad}\,\rho$$

Similarly,

$$\text{grad}\,(\rho/\lambda)=\frac{\lambda\,\text{grad}\,\rho-\rho\,\text{grad}\,\lambda}{\lambda^2}$$

EXAMPLE 3. Consider an ellipse with foci A and B at the points $(-c, 0)$ and $(c, 0)$ respectively. Let P be an arbitrary point on the ellipse, and let $\mathbf{n}$ be the unit inward normal at P as shown in Fig. 2-3. If there is a diverging source of sound or light waves at A which are reflected by the ellipse, we shall show that all of the reflected waves pass through B, i.e, B is the image of A in the elliptic mirror. If R_1 and R_2 denote the distances from A and B to the point P respectively, the definition of an ellipse as the locus of points the sum of whose distances from two fixed points is constant, implies that

$$R_1+R_2=\text{const.}$$

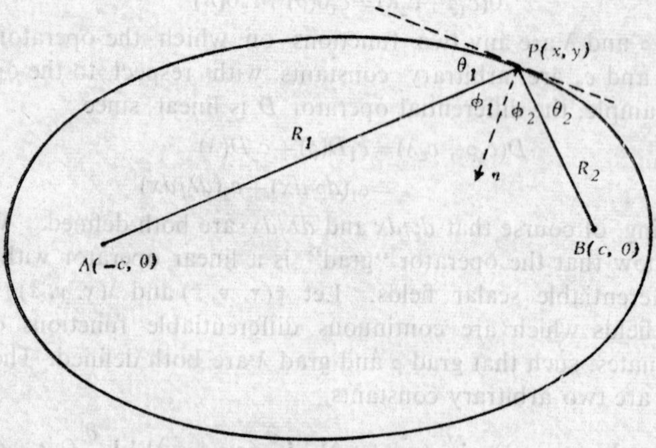

Figure 2-3.

If $\boldsymbol{\tau}$ is the unit tangent to the ellipse at P, the constancy of R_1+R_2 and Eq. (2-10) requires that

$$\text{grad }(R_1+R_2)\cdot\boldsymbol{\tau}=0$$

which implies that

$$\text{grad } R_1\cdot\boldsymbol{\tau}=-\text{grad } R_2\cdot\boldsymbol{\tau}$$

Now

$$R_1=\sqrt{(x+c)^2+y^2}$$

so that

$$\text{grad } R_1=\frac{(x+c)\mathbf{i}+y\mathbf{j}}{R_1}=\frac{\mathbf{R}_1}{R_1}$$

which is a unit vector directed from A toward P. Similarly, grad R_2 is a unit vector in the direction from B to P. Then,

$$\text{grad } R_1\cdot\boldsymbol{\tau}=\cos(\pi-\theta_1)=-\cos\theta_1$$
$$-\text{grad } R_2\cdot\boldsymbol{\tau}=-\cos\theta_2$$

which implies $\theta_1=\theta_2$. Then, $\varphi_1=(\pi/2-\theta_1)=(\pi/2-\theta_2)=\varphi_2$, and the law of reflection is satisfied for rays originating at A if and only if they are reflected through B.

EXAMPLE 4. In one of the formulations of quantum mechanics, the operators corresponding to any cartesian coordinate x and associated momentum P_x are defined to be

$$\hat{x}=x$$
$$\hat{p}_x=-i\hbar(\partial/\partial x)$$

The commutator of the two operators is defined by the relation

$$[\hat{x},\hat{p}_x]\psi=\hat{x}(\hat{p}_x\psi)-\hat{p}_x(\hat{x}\psi)$$

where ψ is the wave function of the quantum system. From the definition of the operators

$$[\hat{x},\hat{p}_x]\psi=-i\hbar[x(\partial\psi/\partial x)-\partial(x\psi)/\partial x]=i\hbar\psi$$

We can immediately generalize to three dimensions and define the operator corresponding to the vector momentum $\mathbf{p}$ as

$$\hat{\mathbf{p}}=-i\hbar\left[\frac{\partial}{\partial x}\mathbf{i}+\frac{\partial}{\partial y}\mathbf{j}+\frac{\partial}{\partial z}\mathbf{k}\right]=-i\hbar\text{ grad}$$

If $f(x, y, z)$ is any function of only the coordinates, the corresponding operator is $\hat{f}=f$. It then follows that

$$[\hat{f},\hat{\mathbf{p}}]=-i\hbar[f\text{ grad }\psi-\text{grad }(f\psi)]$$
$$=-i\hbar[f\text{ grad }\psi-(\text{grad }f)\psi-f\text{ grad }\psi]=i\hbar\text{ grad }f\psi$$

2-5. The Divergence of a Vector Field. Since the operation of forming the gradient of a scalar field can be, in a sense, regarded as the application of a vector differential operator to the scalar field, it is reasonable to expect that the operator "grad" possesses certain vector properties. In particular, we should be able to form the inner product of the operator with an arbitrary vector, and should be able to form the cross product of the operator with a vector. We shall first consider the possibility of forming the inner product of the symbolic operator "grad" with the arbitrary vector **a** which has the cartesian representation

$$\mathbf{a}=a_1\mathbf{i}+a_2\mathbf{j}+a_3\mathbf{k}$$

where we assume that the components a_1, a_2 and a_3 are twice differentiable functions of the cartesian coordinates (x, y, z). This condition is somewhat stronger than is strictly required at this time, but we shall nonetheless impose it in anticipation of later requirements. We shall define the divergence of the vector field **a** in terms of the symbolic operator "grad" by

$$\text{div } \mathbf{a}=\text{"grad"}\cdot\mathbf{a} \tag{2-13}$$

In terms of the cartesian representation of the operator "grad", and the cartesian representation of the vector **a**, the divergence of **a** has the form

$$\text{div } \mathbf{a}=\frac{\partial a_1}{\partial x}+\frac{\partial a_2}{\partial y}+\frac{\partial a_3}{\partial z} \tag{2-14}$$

We shall later see that the representation of the divergence in terms of the operator "grad" is not completely satisfactory. We shall, however, give a definition of the divergence which is completely general and is independent of a particular coordinate system.

There are a number of useful relations concerning the divergence which can be derived immediately from the representation in cartesian coordinates. For example,

$$\begin{aligned}\text{div }(\alpha\mathbf{a})&=\partial(\alpha a_1)/\partial x+\partial(\alpha a_2)/\partial y+\partial(\alpha a_3)/\partial z\\&=\alpha(\partial a_1/\partial x+\partial a_2/\partial y+\partial a_3/\partial z)+\\&\qquad+(a_1\partial\alpha/\partial x+a_2\partial\alpha/\partial y+a_3\partial\alpha/\partial z)\\&=\alpha\text{ div }\mathbf{a}+\mathbf{a}\cdot\text{grad }\alpha\end{aligned}$$

and

$$\begin{aligned}\text{div grad }\rho&=\text{div }(\mathbf{i}\,\partial\rho/\partial x+\mathbf{j}\partial\rho/\partial y+\mathbf{k}\,\partial\rho/\partial z)\\&=\frac{\partial^2\rho}{\partial x^2}+\frac{\partial^2\rho}{\partial y^2}+\frac{\partial^2\rho}{\partial z^2}\end{aligned} \tag{2-15}$$

The second order differential operator whose cartesian representation is given by Eq. (2-15) is of paramount importance in mathematical physics. The operator is known as the Laplacian and is written Lap, ∇^2, or just Δ. Although it is also possible, and sometimes useful to define this operator to operate on vector fields, the definition is not so simple as in the case where the operands are scalar fields. It is important to remember that, although we have considered only the cartesian representation of the divergence and derived some of its properties on the basis of this representation, the divergence is a vector operation, and hence is independent of the coordinate system. In the next chapter, we shall devote considerable effort in obtaining explicit representations of the divergence and the Laplacian in coordinate systems which are not cartesian.

EXAMPLE 5. Consider a fluid with density $\rho(x, y, z)$, which is described by a velocity field whose cartesian representation is

$$\mathbf{u}(x, y, z) = u_1(x, y, z)\,\mathbf{i} + u_2(x, y, z)\,\mathbf{j} + u_3(x, y, z)\,\mathbf{k}$$

If we assume that neither the density nor the velocity field is an explicit function of the time, then the fluid is said to be in steady flow. Let us now consider the fluid flow through the parallelepiped *ABCDEFGH* shown in Fig. 2-4. We first compute the amount of fluid which passes through the face *ABCD* per unit time. The x and

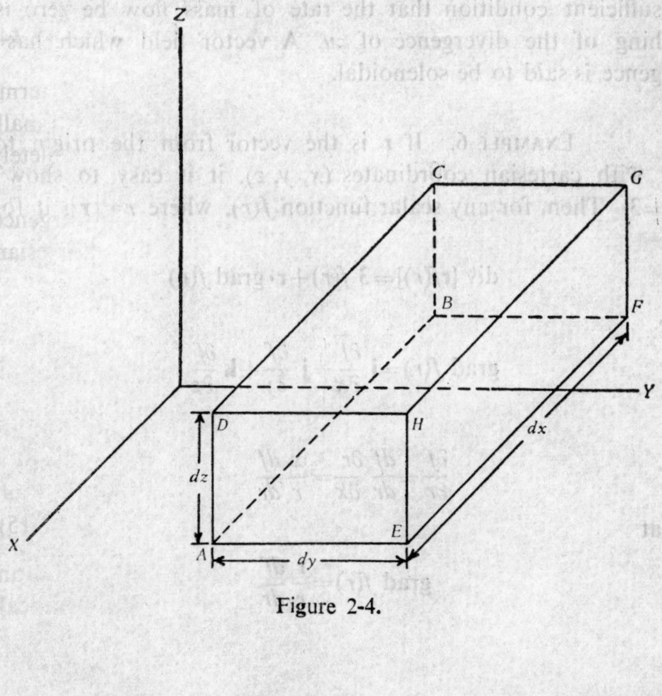

Figure 2-4.

z components of the velocity field contribute nothing to the flow through this face. The mass of fluid which enters the parallelepiped through the face $ABCD$ per unit time is $\rho u_2 dxdz$. The mass of luid leaving the parallelepiped per unit time through the face $EFGH$ is

$$u_2 dxdz + \frac{\partial(\rho u_2)}{\partial y}\, dxdydz$$

Thus, the total rate of mass flow through the two faces normal to the y-axis is

$$\frac{\partial(\rho u_2)}{\partial y}\, dxdydz$$

There are similar terms which represent the net rate of flow through the faces normal to the x- and z-axes. The total rate of mass flow through the parallelopiped is

$$\frac{dM}{dt} = \left[\frac{\partial(\rho u_1)}{\partial x} + \frac{\partial(\rho u_2)}{\partial y} + \frac{\partial(\rho u_3)}{\partial z}\right] dxdydz$$

The bracketed term is the divergence of ρu, and the term $dxdydz$ is volume of the parallelopiped. Hence, the total rate of mass flow is

$$\frac{1}{V}\frac{dM}{dt} = \text{div}\,(\rho u)$$

where V is the volume of fluid under consideration. A necessary and sufficient condition that the rate of mass flow be zero is the vanishing of the divergence of ρu. A vector field which has zero divergence is said to be solenoidal.

EXAMPLE 6. If $\mathbf{r}$ is the vector from the origin to the point with cartesian coordinates (x, y, z), it is easy to show that div $\mathbf{r}=3$. Then, for any scalar function $f(r)$, where $r=|\mathbf{r}|$, it follows that

$$\text{div}\,[\mathbf{r}f(r)] = 3\,f(r) + \mathbf{r}\cdot\text{grad}\,f(r)$$

Now,

$$\text{grad}\,f(r) = \mathbf{i}\,\frac{\partial f}{\partial x} + \mathbf{j}\,\frac{\partial f}{\partial y} + \mathbf{k}\,\frac{\partial f}{\partial z}$$

but

$$\frac{\partial f}{\partial r} = \frac{df}{dr}\frac{\partial r}{\partial x} = \frac{x}{r}\frac{df}{dr}$$

so that

$$\text{grad}\,f(r) = \frac{\mathbf{r}}{r}\frac{df}{dr}$$

and

$$\operatorname{div}\,[\mathbf{r}f(r)]=3\,f(r)+r\,\frac{df}{dr}$$

One particular case of interest arises when $f(r)=r^{n-1}$, n an integer or zero. In this case

$$\operatorname{div}\,\mathbf{r}r^{n-1}=3r^{n-1}+r(n-1)\,r^{n-2}=(n+2)\,r^{n-1}$$

As a concrete example, consider a point charge q at the origin. The electric field at the point $\mathbf{r}$, in *mks* units, is defined by Coulomb's law to be

$$\mathbf{E}=\frac{1}{4\pi\epsilon_0}\,\frac{q\mathbf{r}}{r^3}$$

It then follows from the previous paragraph that

$$\operatorname{div}\,\mathbf{E}=\frac{q}{4\pi\varepsilon_0}\,\operatorname{div}\,(\mathbf{r}r^{-3})=\frac{q}{4\pi\varepsilon_0}\,(-2+2)\,r^{-3}$$

which vanishes at all points in space with the exception of the origin.

EXAMPLE 7. In the Schrödinger formulation of quantum mechanics, the eigenstates ψ_E of a quantum mechanical system are determined from the operator equation

$$\hat{H}\psi_E=E\psi_E$$

where E is the eigen-energy and $\hat{H}$ is the operator corresponding to the Hamiltonian function of classical mechanics. For a classical particle of mass m in a potential field $V(x, y, z)$, the Hamiltonian function is

$$H=\frac{p^2}{2m}+V(x, y, z)$$

Then, using the operators defined in Example 4, we have

$$\begin{aligned}\hat{H}&=\frac{1}{2m}\,(-i\hbar\,\operatorname{grad})\cdot(-i\hbar\,\operatorname{grad})+V(x, y, z)\\ &=-\frac{\hbar^2}{2m}\,\nabla^2+V(x, y, z)\end{aligned}$$

and Schrödinger's equation becomes

$$\nabla^2\psi_E+\frac{2m}{\hbar^2}\,(E-V)\,\psi_E=0$$

EXAMPLE 8. Consider a thin, elastic membrane, such as a drumhead, whose equilibrium configuration lies in the xy-plane.

Define a displacement u which is normal to the equilibrium plane. This displacement varies from point to point on the membrane and from time to time. Hence, we write $u=u(x, y, t)$, and seek an equation of motion for u. Such an equation can be obtained by casting Newton's equation of motion $F=ma$ into a form suitable to the transverse motion of the membrane.

Let the membrane have a mass per unit area μ. Then, the total mass of an infinitesimal area $\Delta x \Delta y$ is $\mu \Delta x \Delta y$ and the right-hand side of Newton's equation of motion is

$$ma=\mu \Delta x \Delta y \frac{\partial^2 u}{\partial t^2}$$

If there is a uniform tension T per unit length of the membrane, and if the displacement u is small, it is not difficult to calculate the total transverse force on the element $\Delta x \Delta y$. In Fig. 2-5, we show the section of the membrane along the line y. From the figure, it is clear that the transverse force at x is

$$-(T\Delta y \sin \theta)_x$$

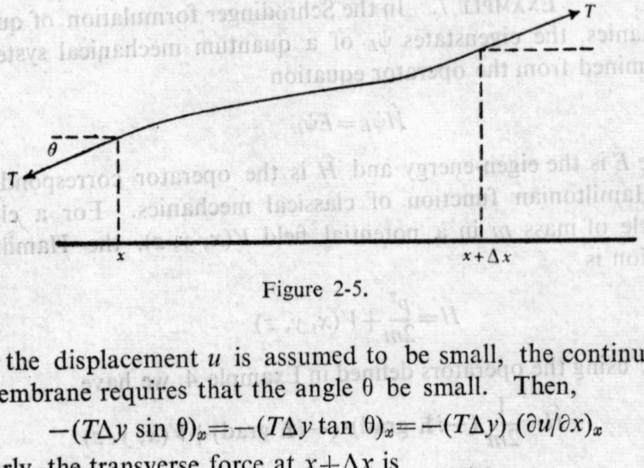

Figure 2-5.

Since the displacement u is assumed to be small, the continuity of the membrane requires that the angle θ be small. Then,

$$-(T\Delta y \sin \theta)_x = -(T\Delta y \tan \theta)_x = -(T\Delta y)\,(\partial u/\partial x)_x$$

Similarly, the transverse force at $x+\Delta x$ is

$$(T\Delta y)\,(\partial u/\partial x)_{x+\Delta x}$$

There are similar expressions for the transverse force at y and $y+\Delta y$. Hence, the total transverse force on the element of the membrane is

$$F_T = T\left\{\left[\left(\frac{\partial u}{\partial x}\right)_{x+\Delta x} - \left(\frac{\partial u}{\partial x}\right)_x\right]\Delta y + \left[\left(\frac{\partial u}{\partial y}\right)_{y+\Delta y} - \left(\frac{\partial u}{\partial y}\right)_y\right]\Delta x\right\}$$

We expand the two square brackets in Taylor series and retain only the first non-vanishing term to obtain

$$F_T=T\left(\frac{\partial^2 u}{\partial x^2}+\frac{\partial^2 u}{\partial y^2}\right)\Delta x\Delta y=T\Delta x\Delta y\nabla^2 u$$

Then, Newton's equation of motion reduces to the two-dimensional wave equation

$$\nabla^2 u-\frac{\mu}{T}\frac{\partial^2 u}{\partial t^2}=0$$

2-6. The Curl of a Vector Field. In defining the cross product of the operator "grad" and an arbitrary vector field **a**, we shall again assume that the cartesian components of **a** are twice differentiable functions of their cartesian coordinates. This is also a stronger condition than is necessary, but will prove convenient in subsequent developments. The curl of the vector field **a** is defined in terms of the operator "grad" by the relation

$$\text{curl }\mathbf{a}=\text{``grad''}\times\mathbf{a} \tag{2-16}$$

The curl is also frequently written as rot **a**, or $\nabla\times\mathbf{a}$. If the vector **a** has the cartesian representation

$$\mathbf{a}=a_1\mathbf{i}+a_2\mathbf{j}+a_3\mathbf{k}$$

then, the cartesian representation of curl **a** is

$$\text{curl }\mathbf{a}=(\partial a_3/\partial y-\partial a_2/\partial z)\,\mathbf{i}+(\partial a_1/\partial z-\partial a_3/\partial x)\,\mathbf{j}+(\partial a_2/\partial x-\partial a_1/\partial y)\,\mathbf{k} \tag{2-17}$$

The cartesian representation of the curl is the formal expansion of the determinant

$$\text{curl }\mathbf{a}=\begin{vmatrix}\mathbf{i} & \mathbf{j} & \mathbf{k}\\ \partial/\partial x & \partial/\partial y & \partial/\partial z\\ a_1 & a_2 & a_3\end{vmatrix} \tag{2-18}$$

Based on the definition and cartesian representation of a vector field, we can immediately establish a number of basic identities:

$$\begin{aligned}\text{curl }(\rho\mathbf{a})&=[\partial(\rho a_3)/\partial y-\partial(\rho a_2)/\partial z]\,\mathbf{i}+\\&\quad[\partial(\rho a_1)/\partial z-\partial(\rho a_3)/\partial x]\,\mathbf{j}+[\partial(\rho a_2)/\partial x-\partial(\rho a_1)/\partial y]\,\mathbf{k}\\&=(\text{grad }\rho)\times\mathbf{a}+\rho\text{ curl }\mathbf{a}\end{aligned}$$

$$\begin{aligned}\text{div }(\mathbf{a}\times\mathbf{b})&=\frac{\partial}{\partial x}(a_2b_3-a_3b_2)+\frac{\partial}{\partial y}(a_3b_1-a_1b_3)+\frac{\partial}{\partial z}(a_1b_2-a_2b_1)\\&=a_1(\partial b_2/\partial z-\partial b_3/\partial y)+a_2(\partial b_3/\partial x-\partial b_1/\partial z)+\\&\quad a_3(\partial b_1/\partial y-\partial b_2/\partial x)-b_1(\partial a_2/\partial z-\partial a_3/\partial y)-\\&\quad b_2(\partial a_3/\partial x-\partial a_1/\partial z)-b_3(\partial a_1/\partial y-\partial a_2/\partial x)\\&=-\mathbf{a}\cdot\text{curl }\mathbf{b}+\mathbf{b}\cdot\text{curl }\mathbf{a}\end{aligned}$$

$$\text{curl }(\mathbf{a}\times\mathbf{b})=(b_1\,\partial/\partial x+b_2\,\partial/\partial y+b_3\,\partial/\partial z)\,\mathbf{a}+(\text{div }\mathbf{b})\,\mathbf{a}-(a_1\,\partial/\partial x+a_2\,\partial/\partial y+a_3\,\partial/\partial yz)\,b-(\text{div }\mathbf{a})\,\mathbf{b}$$

The first and third terms may be symbolically written as **b**·grad and **a**·grad, so that

$$\text{curl }(\mathbf{a}\times\mathbf{b})=(\mathbf{b}\cdot\text{grad})\,\mathbf{a}-(\mathbf{a}\cdot\text{grad})\,\mathbf{b}+(\text{div }\mathbf{b})\,\mathbf{a}-(\text{div }\mathbf{a})\,\mathbf{b}$$

The final identity we shall obtain here is

$$\begin{aligned}\text{curl curl }\mathbf{a}=&[(\partial^2a_1/\partial x^2+\partial^2a_2/\partial y\partial x+\partial^2a_3/\partial z\partial x)-\\&(\partial^2a_1/\partial x^2+\partial^2a_1/\partial y^2+\partial^2a_1/\partial z^2)]\,\mathbf{i}+\\&[(\partial^2a_1/\partial x\partial y+\partial^2a_2/\partial y^2+\partial^2a_3/\partial z\partial y)-\\&(\partial^2a_2/\partial x^2+\partial^2a_2/\partial y^2+\partial^2a_2/\partial z^2)]\,\mathbf{j}+\\&[(\partial^2a_1/\partial x\partial z+\partial^2a_2/\partial y\partial z+\partial^2a_3/\partial z^2)-\\&(\partial^2a_3/\partial x^2+\partial^2a_3/\partial y^2+\partial^2a_3/\partial z^2)]\,\mathbf{k}\\=&\,\mathbf{i}\,\frac{\partial}{\partial x}(\text{div }\mathbf{a})+\mathbf{j}\,\frac{\partial}{\partial y}(\text{div }\mathbf{a})+\mathbf{k}\,\frac{\partial}{\partial z}(\text{div }\mathbf{a})-\\&\mathbf{i}\,\nabla^2a_1-\mathbf{j}\nabla^2a_2-\mathbf{k}\nabla^2a_3\end{aligned}\tag{2-19}$$

Since each of the last three terms of Eq. (2-19) is in the form of the Laplacian operator operating on one of the cartesian components of **a**, we shall take the cartesian representation of the Laplacian operating on a vector field to be of this form, and write, in general

$$\text{curl curl }\mathbf{a}=\text{grad div }\mathbf{a}-\nabla^2\mathbf{a}\tag{2-20}$$

It is not difficult to show that if $\mathbf{a}_i$ is any cartesian component of the vector **a**, then

$$\nabla^2a_i=\partial^2a_i/\partial x^2+\partial^2a_i/\partial y^2+\partial^2a_i/\partial z^2\tag{2-21}$$

The simple relation (2-21) holds only for cartesian coordinates, and in the more general case, the Laplacian of a vector field must be calculated from Eq. (2-20).

Although we were able to obtain physical interpretation of the gradient of a scalar, and the divergence of a vector, it is not possible to obtain a physical interpretation of the curl until we have at our disposal certain integration theorems.

EXAMPLE 9. In a region of free space (vacuum) which contains no free charges or currents, the electromagnetic field can be described by two vector fields **E** (electric intensity) and **B** (magnetic induction). These two vector fields are related by the vacuum Maxwell equations

$$\text{curl }\mathbf{E}=-\frac{\partial\mathbf{B}}{\partial t}\,,\quad \text{div }\mathbf{E}=0$$

$$\text{curl }\mathbf{B}=\frac{1}{c^2}\frac{\partial\mathbf{E}}{\partial t}\,,\quad \text{div }\mathbf{B}=0$$

where c is the speed of light in free space. Take the curl of the first of these equations to obtain

$$\text{curl curl } \mathbf{E} = -\text{curl } (\partial \mathbf{B}/\partial t) = -\frac{\partial}{\partial t} (\text{curl } \mathbf{B}) = -\frac{1}{c^2}\frac{\partial^2 \mathbf{E}}{\partial t^2}$$

However, from Eq. (2-20),

$$\text{curl curl } \mathbf{E} = \text{grad div } \mathbf{E} - \nabla^2 \mathbf{E}$$

so that the electric intensity field satisfies the vector wave equation

$$\nabla^2 \mathbf{E} = \frac{1}{c^2}\frac{\partial^2 \mathbf{E}}{\partial t^2}$$

In exactly the same way, we can show that

$$\nabla^2 \mathbf{B} = \frac{1}{c^2}\frac{\partial^2 \mathbf{B}}{\partial t^2}$$

EXAMPLE 10. The quantum mechanical model of the electron due to Pauli involves the operator expression

$$(-i\hbar \text{ grad} - e\mathbf{A}) \times (-i\hbar \text{ grad} - e\mathbf{A})\, \psi$$

where e is the electron charge, $\mathbf{A}$ is the vector potential for the magnetic field, i.e. $\mathbf{B} = \text{curl } \mathbf{A}$, and $-i\hbar$ grad is the quantum mechanical operator corresponding to the linear momentum $\mathbf{p}$. Expanding the cross product, we have

$$\begin{aligned}
&(-i\hbar \text{ grad} - e\mathbf{A}) \times (-i\hbar \text{ grad} - e\mathbf{A})\, \psi \\
&= -\hbar^2 \text{ curl grad } \psi + i\hbar\, e\mathbf{A} \times \text{grad } \psi + i\hbar\, e \text{ curl } (\psi \mathbf{A}) + e^2\, \mathbf{A} \times \mathbf{A}\, \psi \\
&= i\hbar e\, [\mathbf{A} \times \text{grad } \psi + \text{grad } \psi \times \mathbf{A} + (\text{curl } A)\, \psi] \\
&= i\hbar e\, (\text{curl } \mathbf{A})\psi = i\hbar e \mathbf{B}\psi
\end{aligned}$$

EXAMPLE 11. For an ideal fluid in isentropic motion, the Euler equation has the form

$$\frac{\partial \mathbf{v}}{\partial t} + (\mathbf{v} \text{ grad})\, \mathbf{v} = -\text{grad } w$$

where $\mathbf{v}$ is the velocity of the moving fluid and w is the enthalpy (heat function per unit mass). It is frequently desirable to obtain a form of Euler's equation which involves only the velocity of the fluid. To do this, consider

$$\begin{aligned}
\text{grad } v^2 = \text{grad } (\mathbf{v} \cdot \mathbf{v}) = \mathbf{i} \frac{\partial}{\partial x} (v_x^2 + v_y^2 + v_z^2) + \mathbf{j} \frac{\partial}{\partial y} (v_x^2 + v_y^2 + v_z^2) \\
+ \mathbf{k} \frac{\partial}{\partial z} (v_x^2 + v_y^2 + v_z^2)
\end{aligned}$$

$$=2\left[\mathbf{i}\left(v_x\frac{\partial v_x}{\partial x}+v_y\frac{\partial v_y}{\partial x}+v_z\frac{\partial v_z}{\partial x}\right)+\mathbf{j}\left(v_x\frac{\partial v_x}{\partial y}+v_y\frac{\partial v_y}{\partial y}+v_z\frac{\partial v_z}{\partial y}\right)+\mathbf{k}\left(v_x\frac{\partial v_x}{\partial z}+v_y\frac{\partial v_y}{\partial z}+v_z\frac{\partial v_z}{\partial z}\right)\right]$$

$$=2\mathbf{i}\left[v_x\frac{\partial v_x}{\partial x}+v_y\frac{\partial v_x}{\partial y}+v_z\frac{\partial v_x}{\partial z}+v_y\left(\frac{\partial v_y}{\partial x}-\frac{\partial v_x}{\partial y}\right)-v_z\left(\frac{\partial v_x}{\partial z}-\frac{\partial v_z}{\partial x}\right)\right]+$$
$$2\mathbf{j}\left[v_x\frac{\partial v_y}{\partial x}+v_y\frac{\partial v_y}{\partial y}+v_z\frac{\partial v_y}{\partial z}+v_z\left(\frac{\partial v_z}{\partial y}-\frac{\partial v_y}{\partial z}\right)-v_x\left(\frac{\partial v_y}{\partial x}-\frac{\partial v_x}{\partial y}\right)\right]+$$
$$2\mathbf{k}\left[v_x\frac{\partial v_z}{\partial x}+v_y\frac{\partial v_z}{\partial y}+v_z\frac{\partial v_z}{\partial z}+v_x\left(\frac{\partial v_x}{\partial z}-\frac{\partial v_z}{\partial x}\right)-v_y\left(\frac{\partial v_z}{\partial y}-\frac{\partial v_y}{\partial z}\right)\right]$$
$$=2\,[(\mathbf{v}\cdot\text{grad})\,\mathbf{v}+\mathbf{v}\times\text{curl}\,\mathbf{v}]$$

This result is a special case of the identity

$$\text{grad}\,(\mathbf{a}\cdot\mathbf{b})=(\mathbf{a}\cdot\text{grad})\,\mathbf{b}+(\mathbf{b}\cdot\text{grad})\,\mathbf{a}+\mathbf{a}\times\text{curl}\,\mathbf{b}+\mathbf{b}\times\text{curl}\,\mathbf{a}$$

Then,

$$(\mathbf{v}\cdot\text{grad})\,\mathbf{v}=\tfrac{1}{2}\,\text{grad}\,v^2-\mathbf{v}\times\text{curl}\,\mathbf{v}$$

and the Euler equation can be written in the form

$$\frac{\partial\mathbf{v}}{\partial t}+\tfrac{1}{2}\,\text{grad}\,v^2-\mathbf{v}\times\text{curl}\,\mathbf{v}=-\text{grad}\,w$$

Now, taking the curl of both sides,

$$\frac{\partial}{\partial t}(\text{curl}\,\mathbf{v})=\text{curl}\,(\mathbf{v}\times\text{curl}\,\mathbf{v})$$

which involves only the velocity $\mathbf{v}$.

2-7. The Line Integral over a Vector Field. In the previous sections, we have considered most of the operations of the differential calculus of vectors. In the present and succeeding sections, we shall consider the theory of integration over vector fields. It is assumed in all cases that we are considering the Riemann integral and that such an integral exists. Although it is possible to define the Stieltjes and Lebesque integrals over vector fields, we are usually considering the Riemann integral in the applications.

Let $\mathbf{a}(x, y, z)$ be a continuous vector field which has the cartesian representation

$$\mathbf{a}(x, y, z)=a_1\,(x, y, z)\,\mathbf{i}+a_2\,(x, y, z)\,\mathbf{j}+a_3\,(x, y, z)\,\mathbf{k}$$

and consider the line integral

$$\int_a^b \mathbf{a}\cdot d\mathbf{r}=\int_a^b (\mathbf{a}\cdot d\mathbf{r}/ds)\,ds$$

along any piecewise continuous curve, to which can be assigned a definite length S, and along which $\mathbf{r}=\mathbf{r}(s)$ (see Fig. 2-6). Since

$$(d\mathbf{r}/ds)\, ds = dx\ \mathbf{i} + dy\ \mathbf{j} + dz\ \mathbf{k}$$

the line integral may be written

$$\int_S (\mathbf{a}\cdot d\mathbf{r}/ds)\, ds = \int_S (a_1\, dx + a_2\, dy + a_3\, dz) \tag{2-22}$$

If we represent the space curve in parametric form, $x=x(t)$, $y=y(t)$, $z=z(t)$ we may write

$$\int_S (\mathbf{a}\cdot d\mathbf{r}/ds)\ ds = \int_{t_1}^{t_2} \left[a_1(t)\frac{dx}{dt} + a_2(t)\frac{dy}{dt} + a_3(t)\frac{dz}{dt} \right] dt \tag{2-23}$$

where t_1 and t_2 are the values of the parameter t which correspond to end points of the contour.

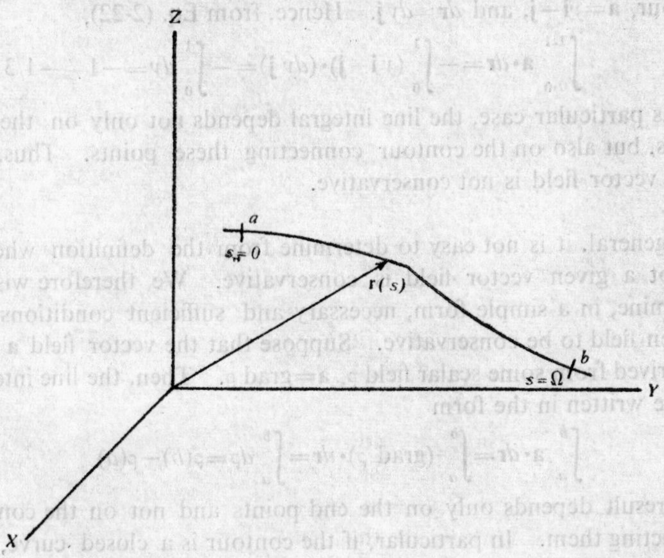

Figure 2-6.

In general, the value of the line integral depends not only on the end points of the space curve, but also on the particular space curve which connects these end points. In the event that the integral depends only on the end points, and not on the connecting contour, we say that **a** is a conservative vector field. Otherwise, *a* is said to be a non-conservative vector field.

EXAMPLE 12. Consider the vector field

$$\mathbf{a} = y\ \mathbf{i} - x\ \mathbf{j}$$

integrated between the points (0, 0) and (1, 1). We shall first take the contour connecting these points to be the parabola $y=x^2$. In parametric form, the equations of the contour are $y=t^2$, $x=t$. The end points of the contour correspond to the values $t=0$, and $t=1$, respectively. Applying Eq. (2-23), we obtain

$$\int_{0,0}^{1,1} \mathbf{a}\cdot d\mathbf{r}=\int_0^1 (t^2-2t^2)\, dt=-\int_0^1 t^2\, dt=-1/3$$

Next, consider the same vector field and the same end points, but choose the contour to lie along the x-axis from $x=0$ to $x=1$, and then along the line $x=1$ from $y=0$ to $y=1$. Along the first part of the contour, $\mathbf{a}=-x\,\mathbf{j}$, and $d\mathbf{r}=dx\,\mathbf{i}$. Along the second part of the contour, $\mathbf{a}=y\mathbf{i}-\mathbf{j}$, and $d\mathbf{r}=dy\,\mathbf{j}$. Hence, from Eq. (2-22),

$$\int_{0,0}^{1,1} \mathbf{a}\cdot d\mathbf{r}=-\int_0^1 (y\,\mathbf{i}-\mathbf{j})\cdot(dy\,\mathbf{j})=-\int_0^1 dy=-1\neq-1/3$$

In this particular case, the line integral depends not only on the end points, but also on the contour connecting these points. Thus, the given vector field is not conservative.

In general, it is not easy to determine from the definition whether or not a given vector field is conservative. We therefore wish to determine, in a simple form, necessary and sufficient conditions for a given field to be conservative. Suppose that the vector field $\mathbf{a}$ can be derived from some scalar field ρ, $\mathbf{a}=\text{grad}\,\rho$. Then, the line integral can be written in the form

$$\int_a^b \mathbf{a}\cdot d\mathbf{r}=\int_a^b (\text{grad}\,\rho)\cdot d\mathbf{r}=\int_a^b d\rho=\rho(b)-\rho(a)$$

This result depends only on the end points and not on the contour connecting them. In particular, if the contour is a closed curve, the line integral vanishes identically. Hence, a sufficient condition for the vector field $\mathbf{a}$ to be conservative is that $\mathbf{a}=\text{grad}\,\rho$ for some scalar function ρ. Since the curl of the gradient of any scalar field is a null vector (see Prob. 2-11), the sufficient condition may be stated as follows: If curl $\mathbf{a}=\mathbf{0}$, then $\mathbf{a}$ is a conservative vector field.

We next obtain a necessary condition for the arbitrary vector field $\mathbf{a}$ to be conservative. Assume that the given field is conservative, and define a scalar field ρ by

$$\rho(x, y, z)=\int_{P_0}^{P} \mathbf{a}\cdot d\mathbf{r}$$

where $P_0=P_0(x_0, y_0, z_0)$ is some fixed point, and $P(x, y, z)$ is a variable point. It then follows that

$$\frac{\rho(x+\Delta x, y, z)-\rho(x, y, z)}{\Delta x}=\frac{1}{\Delta x}\int_{P(x,y,z)}^{P(x+\Delta x,y,z)}(\mathbf{a}\cdot d\mathbf{r}/ds)\,ds$$

Since we are assuming that **a** is a conservative field, the line integral is independent of the path between $P(x, y, z)$ and $P(x+\Delta x, y, z)$. We shall choose this path to be the straight line between the two end points, i.e. $d\mathbf{r}=dx\,\mathbf{i}$. Then,

$$\frac{\partial\rho}{\partial x}=\lim_{\Delta x\to 0}\frac{1}{\Delta x}\int_{P(x,y,z)}^{P(x+\Delta x,y,z)} a_1(x, y, z)\,dx$$
$$=a_1(x, y, z)$$

where we have assumed that **a** is a continuous vector field. Similarly,

$$\frac{\partial\rho}{\partial y}=a_2(x, y, z), \quad \frac{\partial\rho}{\partial z}=a_3(x, y, z)$$

Combining these expressions with the cartesian representation of **a,** we see that a necessary condition for **a** to be conservative is

$$\mathbf{a}=a_1(x, y, z)\,\mathbf{i}+a_2(x, y, z)\,\mathbf{j}+a_3(x, y, z)\mathbf{k}$$
$$=\frac{\partial\rho}{\partial x}\mathbf{i}+\frac{\partial\rho}{\partial y}\mathbf{j}+\frac{\partial\rho}{\partial z}\mathbf{k}=\text{grad }\rho$$

As we have already indicated, $\mathbf{a}=\text{grad }\rho$ implies that curl $\mathbf{a}=\mathbf{0}$. We now wish to investigate the converse. In order for curl **a** to be a null vector, we must, in general, require that

$$\frac{\partial a_1}{\partial y}=\frac{\partial a_2}{\partial x}, \quad \frac{\partial a_2}{\partial z}=\frac{\partial a_3}{\partial y}, \quad \frac{\partial a_3}{\partial x}=\frac{\partial a_1}{\partial z}$$

Define a scalar function

$$\rho(x, y, z)=\int_{x_0}^{x} a_1(x, y, z)dx+\int_{y_0}^{y} a_2(x_0, y, z)dy+\int_{z_0}^{z} a_3(x_0, y_0, z)dz \quad (2\text{-}24)$$

where (x_0, y_0, z_0) is some fixed point. Then,

$$\frac{\partial\rho}{\partial x}=a_1(x, y, z)$$

$$\frac{\partial\rho}{\partial y}=\int_{x_0}^{x}\partial a_1/\partial y\,dx+a_2(x_0, y, z)$$
$$=\int_{x_0}^{x}\partial a_2/\partial x\,dx+a_2(x_0, y, z)$$
$$=a_2(x, y, z)-a_2(x_0, y, z)+a_2(x_0, y, z)$$
$$=a_2(x, y, z)$$

and

$$\frac{\partial\rho}{\partial z}=\int_{x_0}^{x}\partial a_1/\partial z\,dx+\int_{y_0}^{y}\partial a_2/\partial z\,dy+a_3(x_0, y_0, z)$$

$$=\int_{x_0}^{x} \partial a_3/\partial x\, dx+\int_{y_0}^{y} \partial a_3/\partial y\, dy+a_3(x_0, y_0, z)$$
$$=a_3(x, y, z)-a_3(x_0, y, z)+a_3(x_0, y, z)-a_3(x_0, y_0, z)+a_3(x_0, y_0, z)$$
$$=a_3(x, y, z)$$

and hence

$$\mathbf{a}=\frac{\partial \rho}{\partial x}\mathbf{i}+\frac{\partial \rho}{\partial y}\mathbf{j}+\frac{\partial \rho}{\partial z}\mathbf{k}$$

Thus, the requirement that curl $\mathbf{a}=\mathbf{0}$ implies that $\mathbf{a}=\text{grad}\ \rho$, where ρ is given by Eq. (2-24).

We may summarize the results of this analysis as follows: A vector field $\mathbf{a}$ is conservative if and only if it satisfies either of the equivalent conditions, $\mathbf{a}=\text{grad}\ \rho$, or curl $\mathbf{a}=\mathbf{0}$.

EXAMPLE 13. A particle with charge q, moving with velocity $\mathbf{v}$ in an electromagnetic field $\mathbf{E}$ and $\mathbf{B}$ is subject to the Lorentz force

$$\mathbf{F}=q(\mathbf{E}+\mathbf{v}\times\mathbf{B})$$

In order to determine whether or not this force is conservative, we form the curl and obtain

$$\text{curl}\ \mathbf{F}=q\ \text{curl}\ \mathbf{E}+\text{curl}\ (\mathbf{v}\times\mathbf{B})$$
$$=q[\text{curl}\ \mathbf{E}+v\ \text{div}\ \mathbf{B}-\mathbf{B}\ \text{div}\ \mathbf{v}+(\mathbf{v}\cdot\text{grad})\ \mathbf{B}-(\mathbf{B}\cdot\text{grad})\ \mathbf{v}]$$
$$=q[-(\partial\mathbf{B}/\partial t-\mathbf{B}\ \text{div}\ \mathbf{v}+(\mathbf{v}\cdot\text{grad})\ \mathbf{B}-(\mathbf{B}\cdot\text{grad})\ \mathbf{v}]$$

since, from Maxwell's equations

$$\text{curl}\ \mathbf{E}=-\frac{\partial\mathbf{B}}{\partial t};\ \text{div}\ \mathbf{B}=0$$

The expression for curl $\mathbf{F}$ is not, in general, zero, so that the Lorentz force is not conservative.

Now, if the fields are static, i.e. independent of the time, $\partial\mathbf{B}/\partial t=0$ and curl $\mathbf{E}$ vanishes. Further, if $\mathbf{B}$ is a homogeneous field (independent of the coordinates), the term $(\mathbf{v}\cdot\text{grad})\ \mathbf{B}$ is zero and we have

$$\text{curl}\ \mathbf{F}=-\mathbf{B}\ \text{div}\ \mathbf{v}-(\mathbf{B}\cdot\text{grad})\ \mathbf{v}$$

Let the coordinate system be defined so that $\mathbf{B}=\mathbf{B}_z\mathbf{k}$. Then,

$$\text{curl}\ \mathbf{F}=-B_z\left[\mathbf{i}\frac{\partial v_x}{\partial z}+\mathbf{j}\frac{\partial v_y}{\partial z}+\mathbf{k}\left(\frac{\partial v_x}{\partial x}+\frac{\partial v_y}{\partial y}+2\frac{\partial v_z}{\partial z}\right)\right]$$

Thus, a final condition for the force field to be conservative is that

$$\frac{\partial v_x}{\partial z}=\frac{\partial v_y}{\partial z}=\frac{\partial v_x}{\partial x}+\frac{\partial v_y}{\partial y}+2\frac{\partial v_z}{\partial z}=0$$

One particular case of interest obtains when the particle motion is restricted to lie in a plane normal to the magnetic field. In this case, the magnitude of the velocity v is constant, since the force acting on the particle is normal to $\mathbf{v}$ and can change only its direction. Then, we have

$$\mathbf{v}=v_0 \cos \theta \, \mathbf{i}+v_0 \sin \theta \, \mathbf{j}$$

where θ is the instantaneous angle between v and the positive x-axis. In terms of this angle,

$$x=-\frac{r}{\omega_0} \sin \theta, \; y=\frac{r}{\omega_0} \cos \theta$$

where $\omega_0=d\theta/dt$. Then,

$$\frac{\partial v_x}{\partial z}=0, \quad \frac{\partial v_y}{\partial z}=0, \quad \frac{\partial v_z}{\partial z}=0,$$

$$\frac{\partial v_x}{\partial x}=\frac{\partial}{\partial x}(\omega_0 v_0 y/r)=0, \quad \frac{\partial v_y}{\partial y}=\frac{\partial}{y}(-\omega_0 v_0 x/r)=0$$

Thus, in this case, the Lorentz force is conservative.

EXAMPLE 14. The result that curl $\mathbf{F}=0$ if, and only if, $\mathbf{F}=\text{grad}\,\phi$ is the basis of much of potential theory. As an example, consider the electro-static field E produced by a continuous distribution of charge with density ρ. From Maxwell's equations, the electric intensity is defined by the two vector equations

$$\text{curl}\,\mathbf{E}=0, \quad \text{div}\,\mathbf{E}=\frac{1}{\epsilon}\rho$$

where ϵ is a constant which specifies the electrical properties of the medium in which the field exists. Since curl $\mathbf{E}=0$, we define a scalar potential ϕ by

$$\mathbf{E}=-\text{grad}\,\phi$$

Note that ϕ is defined only to within an arbitrary additive constant. This is unimportant, since we can observe only changes in potential or the electric field $\mathbf{E}$. It immediately follows from the divergence condition on $\mathbf{E}$ that the scalar potential satisfies the Poisson equation

$$\nabla^2\phi=\frac{1}{\epsilon}\rho$$

2-8. The Divergence Theorem. In the study of volume and surface integrals over vector fields, the analysis can be frequently simplified by the application of certain general theorems. The first of these general theorems we shall consider is the so-called divergence theorem which is originally due to Gauss. We shall first investigate this in a rather intuitive fashion, and then give a rigorous mathematical proof. The rigorous proof will be in terms of a more general definition of the divergence than we have used so far in our discussion.

In our physical interpretation of the divergence, we found it useful to consider the motion of an incompressible fluid under steady flow. If we are given an arbitrary vector field **a**, it is possible to treat the field as the product of a mass density and a velocity field for some hypothetical fluid. Now suppose that **a** and div **a** are continuous and single valued in a bounded region of space V, which is contained within the closed surface S. These continuity conditions imply that our hypothetical fluid is incompressible and in steady flow. As we have already seen, for such a fluid the net loss of fluid per unit volume per unit time is given by div **a**. Then, the total loss of fluid per unit time from the bounded volume V is given by the integral

$$\int_V \operatorname{div} \mathbf{a}\, dv$$

where dv is the differential element of volume.

Since **a** and div **a** have been assumed to be continuous in V and on the bounding surface S, there can be no point in V or on S at which fluid is either being created or destroyed. Hence, the loss of fluid per unit time from the volume V can be a result only of a flow of fluid through the bounding surface S. Consider some point on this bounding surface, with differential area ds, and let **n** denote the unit normal to S at this point, in a direction which is out of V. If we let **u** denote the velocity field of the fluid, then the component of the velocity field which is normal to the surface element ds is $\mathbf{u}\cdot\mathbf{n}$. The rate of mass flow per unit area of surface is given by

$$\rho \mathbf{u}\cdot\mathbf{n}\, ds = \mathbf{a}\cdot n\, ds$$

where ρ is the mass density of the fluid. The mass flow per unit time through the surface S is

$$\int_S \mathbf{a}\cdot\mathbf{n}\, ds$$

From conservation arguments, this must be equal to the rate at which fluid is lost from the volume V, and we obtain the divergence theorem

$$\int_V \operatorname{div} \mathbf{a}\, dv = \int_S \mathbf{a}\cdot\mathbf{n}\, ds \tag{2-25}$$

Before proceeding with a rigorous proof of (2-25), we shall define the divergence of a vector field in a quite general way. This definition will be independent of any coordinate system. Let **a** be a vector field which is differentiable in a bounded volume V, and on the closed surface S which bounds V. Enclose any point P within V by a small volume Δv, which is bounded by the closed surface Δs. Denote the unit outward normal to Δs by **n**, and form the integral

$$\int_{\Delta s} \mathbf{a}\cdot\mathbf{n}\, ds$$

We shall then define

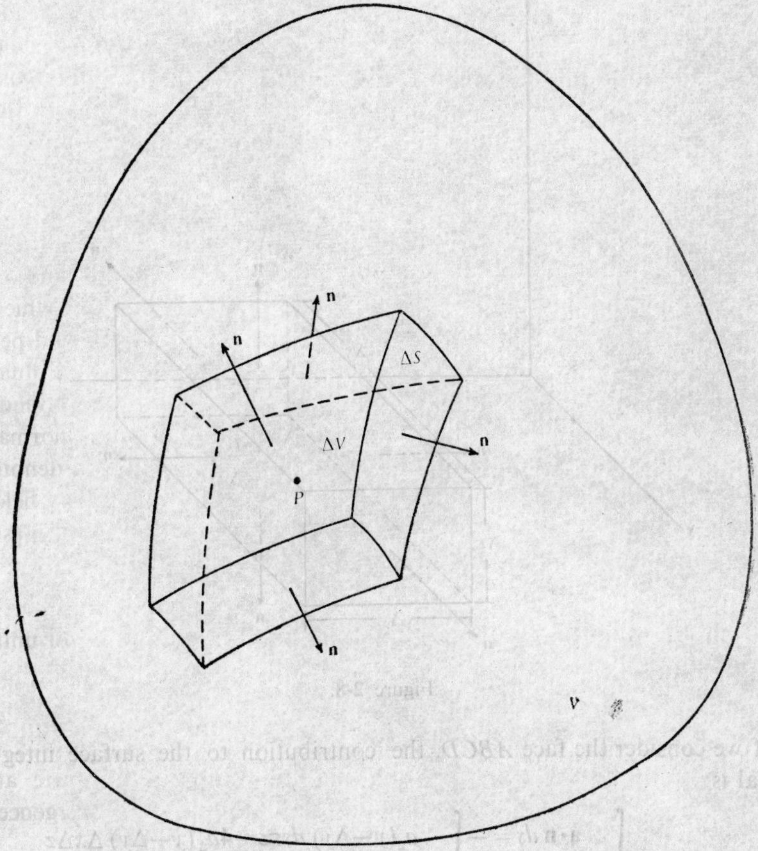

Figure 2-7.

$$\operatorname{div} \mathbf{a} = \lim_{\Delta v \to 0} \frac{\int_{\Delta s} \mathbf{a} \cdot \mathbf{n} \, ds}{\Delta v} \tag{2-26}$$

if the limit exists and is independent of the manner in which Δv tends to zero. The geometry of this definition is shown in Fig. 2-7.

We shall now show that the general definition (2-26) gives rise to the cartesian representation (2-14). Let the region Δv consist of the parallelepiped *ABCDEFGH* shown in Fig. 2-8, and let the cartesian representation of the vector field **a** be given by

$$\mathbf{a} = a_1 \mathbf{i} + a_2 \mathbf{j} + a_2 \mathbf{k}$$

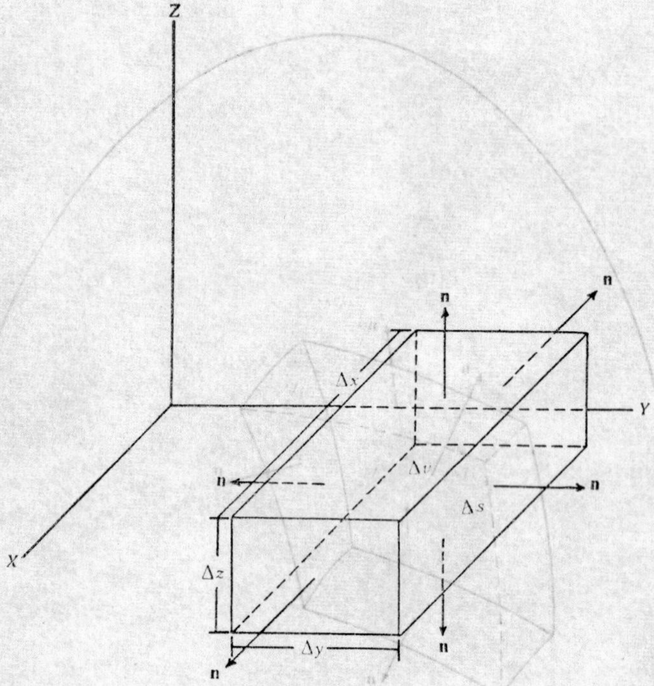

Figure 2-8.

If we consider the face *ABCD*, the contribution to the surface integral is

$$\int_{\Delta s_1} \mathbf{a} \cdot \mathbf{n} \, ds = -\int_{\Delta s_1} a_2(y - \Delta y) \, ds \approx -4a_2 (y - \Delta y) \, \Delta x \Delta z$$

Similarly, the contribution to the surface integral from the face *EFGH* is

$$\int_{\Delta s_2} \mathbf{a}\cdot\mathbf{n}\, ds = \int_{\Delta s_2} a_2(y+\Delta y)\, ds \approx 4a_2\,(y+\Delta y)\Delta x\Delta z$$

Including the contributions to the surface integral from the remaining faces of the parallelepiped, we have from Eq. (2-26)

$$\text{div}\,\mathbf{a} = \lim_{\Delta x\to 0} \frac{a_1(x+\Delta x)-a_1(x-\Delta x)}{2\Delta x} + \lim_{\Delta y\to 0} \frac{a_2(y+\Delta y)-a_2(y-\Delta y)}{2\Delta y} + \lim_{\Delta z\to 0} \frac{a_3(z+\Delta z)-a_3(z-\Delta z)}{2\Delta z}$$

where we have set $\Delta v = 8\,\Delta x\Delta y\Delta z$. Hence,

$$\text{div}\,\mathbf{a} = \frac{\partial a_1}{\partial x} + \frac{\partial a_2}{\partial y} + \frac{\partial a_3}{\partial z} \tag{2-27}$$

which is the cartesian representation of div **a**. We have previously obtained this representation by considering the inner product of the operator "grad" with the vector field **a** in terms of their cartesian representations.

Now, let **a** and div **a** be continuous in the bounded region V and on the closed surface S which bounds V. Again, denote the unit outward normal to S by **n**. Then, at any point in V, we have from (2-26)

$$\text{div}\,\mathbf{a}\,\Delta v = \int_{\Delta s} a\cdot\mathbf{n}\, ds + \epsilon\,\Delta v \tag{2-28}$$

where $\epsilon\to 0$ as $\Delta v\to 0$. If we subdivide the region V into many elemental volumes, we obtain an expression similar to (2-28) for each elemental volume. Forming the sum of the expressions, as we let Δv become an infinitesimal, we obtain the divergence theorem

$$\int_V \text{div}\,\mathbf{a}\, dv = \int_S \mathbf{a}\cdot\mathbf{n}\, ds \tag{2-29}$$

In obtaining (2-29), it is clear that the surfaces interior to V contribute nothing to the surface integral, since for each positive **n** ds, there is a corresponding negative **n** ds. This is shown in Fig. 2-9. The sum

$$\Sigma\epsilon_i\Delta v_i$$

vanishes is the limit, since

$$\left|\sum_i \epsilon_i\,\Delta v_i\right| \leqslant \sum_i |\epsilon|_{\max}\,\Delta v_i \leqslant |\epsilon|_{\max} V$$

and $\quad |\epsilon|_{\max}\to 0 \quad \text{as} \quad \Delta v\to 0.$

Since we are considering vector fields which are, in general, assumed to be continuous, except possibly along isolated lines, or at isolated points, it is possible to define lines of flow for the vector field as being lines which are everywhere tangent to the vectors of the field. In order to indicate the magnitude of the vector field at the points along the flow lines, we shall in some arbitrary fashion associate the number of lines in the region of a point with the magnitude of the vector field at that point, i.e. the larger the magnitude of the vector field the greater the number of flow lines. One property of these

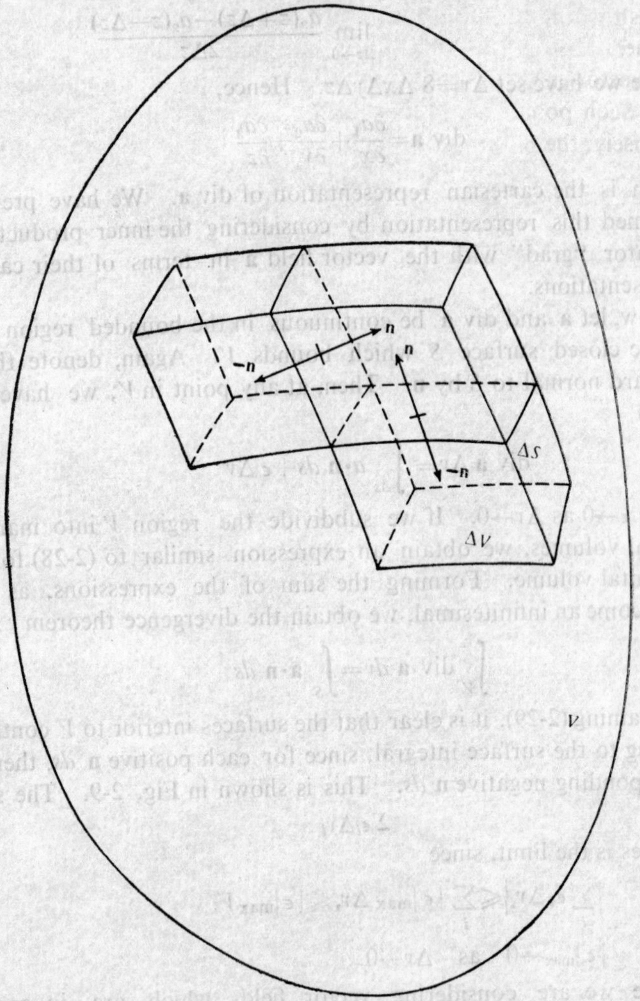

Figure 2-9.

flow lines which is of great help in interpreting the divergence of the vector field is the net outflow of lines from a closed region of space. The surface integral which appears in (2-29) is just the number of flow lines of the vector field which cross the surface S bounding the volume V. If no lines of flow originate within the volume V, the total number of lines crossing the surface S must be zero, since for every line which crosses in the outward direction, there must be a line which crosses in the inward direction. If this is the case, then it is clear from the divergence theorem that the divergence of **a** must vanish within the volume V, since S is any surface which bounds V. If there are lines of flow which originate or disappear within V, then the vector field must have a point, or points of discontinuity within V. Such points are called source points for the vector field. More precisely, the points at which lines of flow originate are called sources,

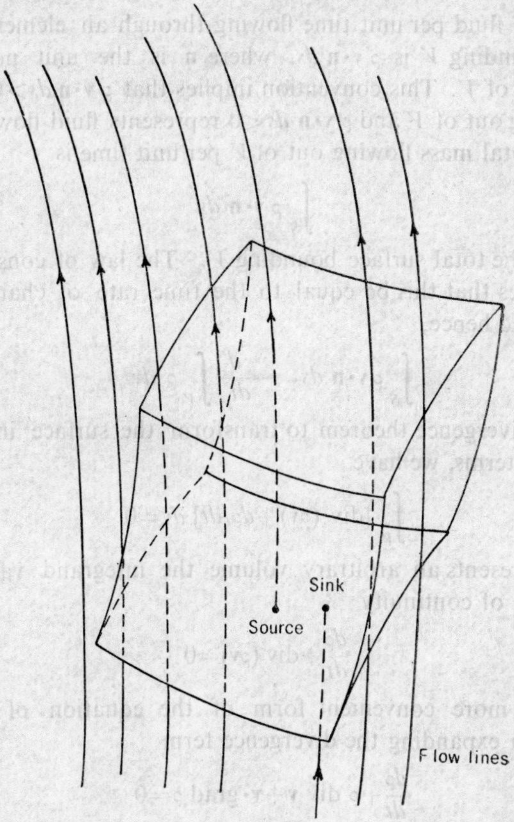

Figure 2-10.

and points at which lines of flow disappear are called sinks. It is thus seen that the divergence of the vector field **a** is a measure of strength of the sources of the field within a given region of space. This is illustrated in Fig. 2-10.

EXAMPLE 15. The most directed application of the divergence theorem is in the derivation of continuity equations, which are the mathematical expression of physical conservation laws. To illustrate this point, we consider a fluid of density ρ flowing with a velocity **v**, and look for a mathematical statement of the law of conservation of mass. Consider some volume in space V. The mass of fluid in this volume is

$$M = \int_V \rho dv$$

The mass of fluid per unit time flowing through an element ds of the surface bounding V is $\rho\, \mathbf{v}\cdot\mathbf{n}\, ds$, where **n** is the unit normal to ds directed out of V. This convention implies that $\rho\, \mathbf{v}\cdot\mathbf{n}\, ds > 0$ represents fluid flowing out of V and $\rho\, \mathbf{v}\cdot\mathbf{n}\, ds < 0$ represents fluid flowing into V. Then, the total mass flowing out of V per unit time is

$$\int_S \rho\, \mathbf{v}\cdot\mathbf{n}\, ds$$

where S is the total surface bounding V. The law of conservation of mass requires that this be equal to the time rate of change of mass within V, and hence,

$$\int_S \rho\, \mathbf{v}\cdot\mathbf{n}\, ds = -\frac{d}{dt}\int_V \rho\, dv$$

Using the divergence theorem to transform the surface integral, and rearranging terms, we have

$$\int_V [\mathrm{div}\, (\rho\mathbf{v}) + d\rho/dt]\, dv = 0$$

Since V represents an arbitrary volume the integrand vanishes, and the equation of continuity

$$\frac{d\rho}{dt} + \mathrm{div}\, (\rho\mathbf{v}) = 0$$

follows. A more convenient form of the equation of continuity obtains from expanding the divergence term

$$\frac{d\rho}{dt} + \rho\, \mathrm{div}\, \mathbf{v} + \mathbf{v}\cdot \mathrm{grad}\, \rho = 0$$

EXAMPLE 16. The electric field intensity **E**, in vacuum, due to a point charge q is defined by Coulomb's Law

$$E=\frac{1}{4\pi\epsilon_0}\frac{q}{r^3}\,\mathbf{r}$$

where **r** is the vector from q to the point at which **E** is to be determined. Let S be a closed surface with unit outward normal **n** which contains the point charge q. Forming the inner product of **n** with both sides of the defining equation for **E**, and integrating over S

$$\int_S \mathbf{E}\cdot\mathbf{n}\,ds=\frac{q}{4\pi\varepsilon_0}\int_S \frac{1}{r^3}\mathbf{r}\cdot\mathbf{n}\,ds$$

Consider the term $(1/r^3)\,\mathbf{r}\cdot\mathbf{n}\,ds$. Let $d\Omega$ be the portion of a sphere of radius r and centre at the charge q which is cut out by the cone with vertex at q connecting all of the points on ds. Then, the solid angle subtended by ds at q is defined to be

$$d\omega=\frac{d\Omega}{r^2}$$

If θ is the angle between **r and n**,

$$d\omega=\frac{ds\cos\theta}{r^2}=ds\,\frac{\mathbf{r}\cdot\mathbf{n}}{r^3}$$

and

$$\int_S \mathbf{E}\cdot\mathbf{n}\,ds=\frac{q}{4\pi\epsilon_0}\int_S d\omega$$

The solid angle subtended at q by any closed surface S is just 4π, and Gauss' Law follows

$$\int_S \mathbf{E}\cdot\mathbf{n}\;ds=\frac{q}{\epsilon_0}$$

It is experimentally known that the principle of superposition holds for the electrostatic field, i.e. if the point cha ge q_1 at the point $\mathbf{r}_1$ produces an electric field intensity $\mathbf{E}_1$ at the point **r** and the point charge q_2 at $\mathbf{r}_2$ produces a field E_2 at **r**, then the total field at **r** is $\mathbf{E}_1+\mathbf{E}_2$. Applying the principle of linear superposition to a continuous charge distribution with density ρ, we write Gauss' Law for this case as

$$\int_S \mathbf{E}\cdot\mathbf{n}\,ds=\frac{1}{\epsilon_0}\int_V \rho\,dv$$

where V is an arbitrary volume containing all of the charge, which is bounded by the closed surface S. Transforming the surface integral by the divergence theorem and rearranging terms,

$$\int_V\left[\operatorname{div}\mathbf{E}-\frac{1}{\epsilon_0}\rho\right]dv=0$$

Since V is an arbitrary volume, the integrand must vanish and we obtain the Maxwell equation

$$\text{div } \mathbf{E} = \frac{1}{\epsilon_0} \rho$$

2-9. Green's Theorem. A technique which is frequently used in the solution of many of the partial differential equations of mathematical physics is based on a vector theorem due to Green. The theorem is a direct consequence of the divergence theorem. Let V be a bounded region of space, bounded by the closed surface S, and let the outward unit normal to S at any point be denoted by $\mathbf{n}$. Further, let ϕ and ψ be any two scalar fields which have continuous second derivatives in V and on S. It follows from the divergence theorem that

$$\int_V \text{div } (\phi \text{ grad } \psi) \, dv = \int_S (\phi \text{ grad } \psi) \cdot \mathbf{n} \, ds \tag{2-30}$$

If we expand the integrand of the term on the left hand side of (2-30), we immediately obtain Green's first identity

$$\int_V (\text{grad } \phi \cdot \text{grad } \psi + \phi \nabla^2 \psi) \, dv = \int_S (\phi \text{ grad } \psi) \cdot \mathbf{n} \, ds \tag{2-31}$$

In particular, if we set $\phi = \psi$, and require that ψ be a solution of of Laplace's differential equation

$$\nabla^2 \psi = \frac{\partial^2 \psi}{\partial x^2} + \frac{\partial^2 \psi}{\partial y^2} + \frac{\partial^2 \psi}{\partial z^2} = 0$$

then, Green's first identity reduces to the form

$$\int_V (\text{grad } \psi)^2 \, dv = \int_S (\psi \text{ grad } \psi) \cdot \mathbf{n} \, ds$$

If we interchange ϕ and ψ in Green's first identity, we obtain

$$\int_V (\text{grad } \psi \cdot \text{grad } \phi + \psi \nabla^2 \phi) \, dv = \int_S (\psi \text{ grad } \phi) \cdot \mathbf{n} \, ds \tag{2-32}$$

Subtracting (2-32) from (2-31), we obtain Green's theorem or Green's second identity

$$\int_V (\phi \nabla^2 \psi - \phi \nabla^2 \phi) \, dv = \int_S (\phi \text{ grad } \psi - \psi \text{ grad } \phi) \cdot \mathbf{n} \, ds \tag{2-33}$$

EXAMPLE 17. One of the common problems of electrostatics is the determination of the electrostatic scalar potential due to a charge distribution of density ρ, contained in a finite region of space in the presence of one or more conducting surfaces. In order to be more specific, suppose that the charge distribution is contained within the closed surface S, and that on the surface S the scalar potential ϕ has the constant value ϕ_0. Then, at all points interior to S, the scalar potential satisfies the Poisson equation

$$\nabla^2\phi = -\frac{1}{\epsilon_0}\rho$$

and the boundary condition $\phi=\phi_0$ for points on S. We define a Green's function $G(x, y, z; x', y', z')$ by the differential equation

$$\nabla'^2 G = -\delta(x-x')\,\delta(y-y')\,\delta(z-z')$$

where ∇'^2 is the Laplacian operator with respect to the primed coordinates, and $\delta(x-x')$, known as the Dirac-delta, has the property

$$\int_{x-\epsilon}^{x+\epsilon} f(x')\,\delta(x-x')\,dx' = f(x)$$

for all continuous functions $f(x)$ and all $\epsilon>0$. In addition to the differential equation, we also require the Green's function to satisfy the boundary condition

$$G(x, y, z; x', y', z') = 0$$

for $(x'\ y'\ z')$ on S. It is convenient to change the Poisson equation for ϕ to one in terms of the primed coordinates. We therefore write

$$\nabla'^2\phi(x', y', z') = -\frac{1}{\epsilon_0}\rho(x', y', z')$$

Multiply the equation for the Green's function by ϕ, the equation for ϕ by G and subtract to obtain

$$\phi(x', y'\ z')\,\nabla'^2 G - G(x, y, z;\ x'\ y', z')\,\nabla'^2\phi$$
$$= \frac{1}{\epsilon_0}\rho(x', y', z')\,G(x, y, z; x', y', z') - \phi(x', y', z')\,\delta(x-x')\,\delta(y-y')\,\delta(z-z')$$

Integrating over the volume V,

$$\int_V \phi\delta(x-x')\,\delta(y-y')\,\delta(z-z')\,dv' = \frac{1}{\epsilon_0}\int_V \rho G\,dv' + \int_V [G\nabla'^2\phi - \phi\nabla'^2 G]\,dv'$$

where $dv' = dx'\ dy'\ dz'$. Then, from the properties of the Dirac-delta,

$$\phi(x, y, z) = \frac{1}{\epsilon_0}\int_V \rho G\ dv' + \int_S [G \text{ grad}'\ \phi - \phi \text{ grad}'\ G]\cdot n\ ds'$$

where we have transformed the second volume integral on the right-hand side by Green's theorem. We now impose the boundary conditions, $\phi=\phi_0$, $G=0$ on S, to obtain the final form

$$\phi(x, y, z) = \frac{1}{\epsilon_0}\int_V \rho G\ dv' - \phi_0 \int_S \text{grad}'\ G\cdot\mathbf{n}\ ds'$$

Although it is almost as difficult to determine the Green's function as it is to determine the scalar potential for a given problem, once the Green's function is known for the given geometry, the scalar potential is determined for all possible boundary values ϕ_0, and all charge distributions ρ.

2-10. Stokes' Theorem. In a manner analogous to our definition of the divergence, we define the gradient of a scalar field as the limit

$$\text{grad } \rho = \lim_{\Delta v \to 0} \frac{\int_{\Delta S} \rho \mathbf{n} \, ds}{\Delta v} \tag{2-34}$$

if the limit exists and is independent of the way in which $\Delta v \to 0$. Before making use of this definition, we shall show that it reduces to the proper representation in terms of a set of orthogonal cartesian coordinates. If we form the inner product of the cartesian unit vector **i** with (2-34), we obtain

$$\mathbf{i} \cdot \text{grad } \rho = \lim_{\Delta v \to 0} \frac{\int_{\Delta S} \mathbf{i} \cdot \mathbf{n} \, \rho ds}{\Delta v}$$

where **i** can be taken inside the integral and the limit since it is a constant vector. We shall evaluate the surface integral over a region of cylindrical shape, as shown in Fig. 2-11, whose axis is parallel to the unit vector **i**. For such a region, $\mathbf{i} \cdot \mathbf{n}$ vanishes over the sides of the cylinder, and is unity over the ends. If the ends of the cylinder are at $x=a$ and $x=b$, the surface integral reduces to an integration over the end faces, and the increment of volume is $(b-a)$ times the area of one of the end faces,

$$\mathbf{i} \cdot \text{grad } \rho = \lim_{\Delta v \to 0} \frac{\int_{\Delta S} \rho(b) \, ds - \int_{\Delta S} \rho(a) \, ds}{(b-a) \int_{\Delta S} ds}$$

In taking the limit $\Delta v \to 0$, we shall first shrink the cylinder radius, and then let $b \to a$. Assuming that ρ is continuous, we obtain in the limit

$$\mathbf{i} \cdot \text{grad } \rho = \lim_{b \to a} \frac{\rho(b) - \rho(a)}{b-a} = \frac{\partial \rho}{\partial x}$$

If we repeat this process with the cartesian unit vectors **j** and **k**, we obtain the cartesian representation of grad ρ,

$$\text{grad } \rho = \frac{\partial \rho}{\partial x} \mathbf{i} + \frac{\partial \rho}{\partial y} \mathbf{j} + \frac{\partial \rho}{\partial z} \mathbf{k} \tag{2-35}$$

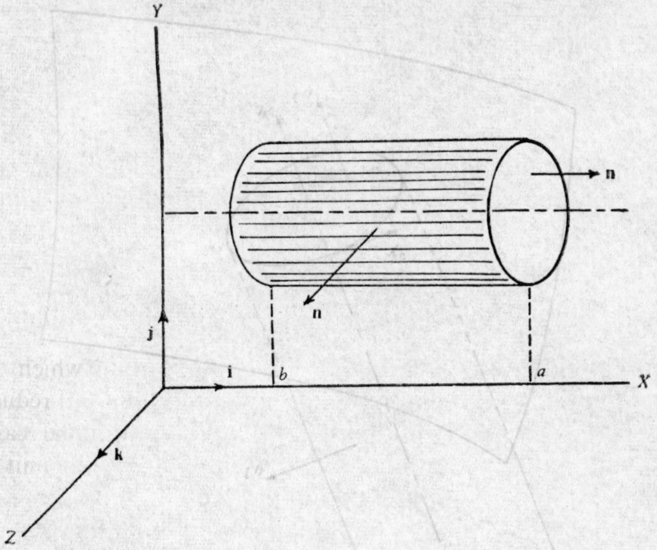

Figure 2-11.

Finally, it is reasonable to take as our general definition of the curl of the vector field **a**, the limit

$$\text{curl } \mathbf{a} = \lim_{\Delta v \to 0} \frac{\int_{\Delta S} \mathbf{n} \times \mathbf{a} \, ds}{\Delta v} \tag{2-36}$$

where, as before, the limit is assumed to exist and be independent of the way in which $\Delta v \to 0$. It can be shown, rather easily, that the general definition (2-36) reduces to the cartesian representation in cartesian coordinates. This demonstration is very similar to that used in obtaining the cartesian representation of gradient from the general definition.

Let us now consider an unclosed surface bounded by the closed contour C which has line element $d\mathbf{l}$. Since there is no inside and outside associated with the surface, we define the positive normal to the surface as follows: The positive sense of the contour C is clockwise whenever we look through the surface S in the direction of the positive normal. Consider a cylinder of volume Δv, whose axis is parallel to the positive normal n to the open surface S, and which intersects this surface in the element ds with contour element $d\mathbf{l}$ (see Fig. 2-12). If we evaluate the normal component of curl **a** on the open surface, we have from (2-36)

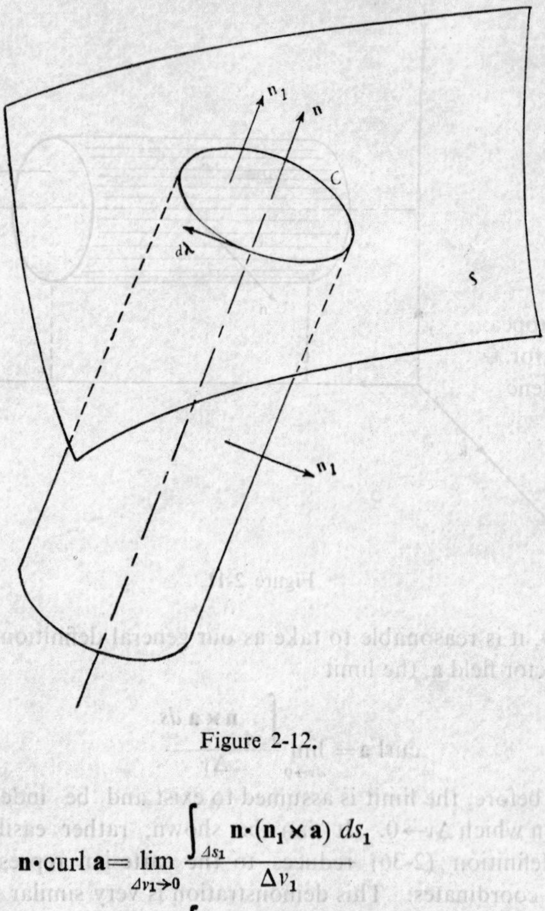

Figure 2-12.

$$\mathbf{n}\cdot\text{curl }\mathbf{a} = \lim_{\Delta v_1 \to 0} \frac{\int_{\Delta s_1} \mathbf{n}\cdot(\mathbf{n}_1 \times \mathbf{a})\, ds_1}{\Delta v_1}$$

$$= \lim_{\Delta v_1 \to 0} \frac{\int_{\Delta s_1} (\mathbf{n}\times\mathbf{n}_1)\cdot\mathbf{a}\, ds_1}{\Delta v_1}$$

where the subscript 1 refers to the closed cylinder. Now, on the ends of the cylinder, $\mathbf{n}\times\mathbf{n}_1=\mathbf{0}$, and, on the sides, $\mathbf{n}\times\mathbf{n}_1$ is parallel to $d\mathbf{l}$. Further, if the length of the cylinder is dx, the element of area along the side is $ds=dx\,|\,d\mathbf{l}\,|$, so that $\mathbf{n}\times\mathbf{n}_1\ ds_1=d\mathbf{l}$, and $\Delta v_1=dx\ \Delta s_1$. Hence, we have

$$\mathbf{n}\cdot\text{curl }\mathbf{a} = \lim_{\Delta s \to 0} \frac{\oint_C \mathbf{a}\cdot d\mathbf{l}}{\Delta s}$$

or

$$\int_S (\text{curl } \mathbf{a}) \cdot \mathbf{n}\, ds = \oint_C \mathbf{a} \cdot d\mathbf{l} \tag{2-37}$$

which is known as Stokes' theorem.

Now that we have proved Stokes' theorem, we are in a position to give a physical interpretation of the curl of a vector field. In making this interpretation, it is again convenient to regard the given vector field as the product of a mass density ρ and a velocity field $\mathbf{v}$ for a suitably defined hypothetical fluid. The line integral

$$\oint_C \mathbf{a} \cdot d\mathbf{l}$$

which appears in Stokes' theorem is called the net circulation integral for the vector field $\mathbf{a}$ around the contour C, and measures the tendency of the flow lines to circulate. This terminology is by analogy with the physical fluid flow field, since, if we interpet $\mathbf{a}$ as a fluid velocity, then the line integral gives the actual circulation of the fluid around the contour C. If the line integral vanishes, we say that the flow is irrotational. It is clear that a condition for a vector field to be irrotational in a region is the vanishing of the curl of the vector field in that region.

In those cases where there is a net circulation, there must be a source which produces the circulation. If we again make use of the fluid flow analogy, we may call these regions which produce a net circulation "vortex" regions. These vortex regions must be tubular shaped without beginning or end. That is to say, such regions must either extend to infinity in both directions, or else must close on themselves to produce toroidal shaped regions.

EXAMPLE 18. In order to obtain the equation of motion for an ideal fluid, we apply Newton's second law to an arbitrary volume V in the fluid. If the fluid pressure is p, then the total force acting on the element of fluid is

$$-\int_S p \cdot \mathbf{n}\, ds$$

where S is the closed surface bounding V with unit outward normal $\mathbf{n}$. The surface integral can be transformed into a volume integral by Eq. (2-34),

$$-\int_S p\mathbf{n}\, ds = -\int_V \text{grad } p\, dv$$

Now, the product of the mass and acceleration of the fluid element can be written as

$$\int_V \rho \frac{d\mathbf{u}}{dt} dv$$

where **u** is the fluid velocity. Since the fluid velocity at a given point varies from time to time, and at a given time varies from point to point

$$\frac{d\mathbf{u}}{dt} = \frac{\partial \mathbf{u}}{\partial t} + \frac{\partial \mathbf{u}}{\partial x}\frac{dx}{dt} + \frac{\partial \mathbf{u}}{\partial y}\frac{dy}{dt} + \frac{\partial \mathbf{u}}{\partial z}\frac{dz}{dt}$$
$$= \frac{\partial \mathbf{u}}{\partial t} + (\mathbf{u} \cdot \text{grad})\,\mathbf{u}$$

Then from Newton's second law,

$$\int_V \left\{ \rho \left[\frac{\partial \mathbf{u}}{\partial t} + (\mathbf{u} \cdot \text{grad})\,\mathbf{u} \right] + \text{grad } p \right\} dv = 0$$

However, this must hold for any volume V, and hence the integrand must vanish and we obtain Euler's equation

$$\frac{\partial \mathbf{u}}{\partial t} + (\mathbf{u} \cdot \text{grad})\,\mathbf{u} = -\frac{1}{\rho}\,\text{grad } p$$

EXAMPLE 19. In order to produce stationary currents in an electrical circuit, it is experimentally found that it is necessary to have electric fields which are not conservative. This situation is described by the Faraday induction law. Mathematically, the most general form of this law can be written

$$\oint_C \mathbf{E} \cdot d\mathbf{l} = -\frac{d}{dt} \int_S \mathbf{B} \cdot \mathbf{n}\, ds$$

where S is an open surface bounded by the closed curve C. In the special case of an electric circuit of resistance R carrying a current J, the contour C is the actual physical contour of the circuit; the integral

$$\int_S \mathbf{B} \cdot \mathbf{n}\, ds = \Phi_m$$

is the magnetic flux linking the circuit; and

$$\oint_C \mathbf{E} \cdot d\mathbf{l} = JR - \mathcal{E}$$

where $\mathcal{E}$ is the electromotive force driving the current J. In this particular case, the Faraday induction law states that the current in the circuit differs from that predicted by Ohm's law by an amount equal to the negative time rate of change of the magnetic flux linking the circuit.

Applying Stokes' theorem to the left hand side of the general form of Faraday's law, and rearranging terms

$$\int_S (\text{curl } \mathbf{E}+\partial\mathbf{B}/\partial t)\cdot\mathbf{n}\, ds=0$$

Since this relation must hold for an arbitrary open surface S, the integrand must vanish, and we obtain the Maxwell equation

$$\text{curl } \mathbf{E}=-\partial\mathbf{B}/\partial t$$

2-11. Helmholtz' Theorem. We have proved that the curl of a vector field vanishes if and only if the vector field is the gradient of a scalar field. We now wish to investigate whether or not a similar requirement holds for the vanishing of the divergence of a vector field. Let the vector field **a** be derived as the curl of a "vector potential" field **A**. It then follows that div **a**=0. Hence, it is sufficient that **a**=curl **A**, for div **a**=0. We note at this point, that the vector A is not unique, since curl (A+grad ρ)=curl **A**, for any reasonably well behaved scalar function ρ. We shall later make use of this lack of uniqueness in discussing the electromagnetic field potentials.

Suppose that div **a**=0, and define a vector field

$$\mathbf{A}=A_1\mathbf{i}+A_2\mathbf{j}+A_3\mathbf{k}$$

where A_1, A_2, and A_3 are as yet undetermined. We shall determine these functions so that the vector **A** is a vector potential for the vector field **a**. If **a**=curl **A**, then the components of a are given by

$$a_1=\frac{\partial A_3}{\partial y}-\frac{\partial A_2}{\partial z},\quad a_2=\frac{\partial A_1}{\partial z}-\frac{\partial A_3}{\partial x},\quad a_3=\frac{\partial A_2}{\partial x}-\frac{\partial A_1}{\partial y} \tag{2-38}$$

If we assume that $A_1=0$, it follows from (2-38) that

$$a_1=\frac{\partial A_3}{\partial y}-\frac{\partial A_2}{\partial z},\quad a_2=-\frac{\partial A_3}{\partial x},\quad a_3=\frac{\partial A_2}{\partial x}$$

Hence, if there is a solution of the system of Eqs. (2-38) with $A_1=0$, then such a solution must be of the form

$$A_2=\int_{x_0}^{x} a_3\, dx+f(y,z),\quad A_3=-\int_{x_0}^{x} a_2\, dx+g(y,z)$$

where $f(y, z)$ and $g(y, z)$ are arbitrary functions. The assumption div **a**=0 implies that

$$\frac{\partial a_1}{\partial x}=-\left[\frac{\partial a_2}{\partial y}+\frac{\partial a_3}{\partial x}\right]$$

Hence,

$$\frac{\partial A_3}{\partial y}-\frac{\partial A_2}{\partial z}=-\int_{x_0}^{x}\left[\frac{\partial a_2}{\partial y}+\frac{\partial a_3}{\partial z}\right]dx+\frac{\partial g}{\partial y}-\frac{\partial f}{\partial z}$$

$$=\int_{x_0}^{x}\frac{\partial a_1}{\partial x}dx+\frac{\partial g}{\partial y}-\frac{\partial f}{\partial z}$$

$$=a_1(x, y, z)-a_1(x_0, y, z)+\frac{\partial g}{\partial y}-\frac{\partial f}{\partial z}=a_1(x, y, z)$$

This latter system of equations can be satisfied by setting

$$g(y, z)=0, f(y, z)=-\int_{x_0}^{x} a_1(x_0, y, z)\, dz$$

Thus, the system of Eqs. (2-38) has the solution

$$A_1=0,\ A_2=\int_{x_0}^{x} a_3dx+f(y, z),\ A_3=-\int_{x_0}^{x} a_2dx$$

We have thus shown that div $\mathbf{a}=0$ implies that $\mathbf{a}=$curl $\mathbf{A}$, where $\mathbf{A}$ is given by

$$\mathbf{A}(x,y,z)=\left[\int_{x_0}^{x} a_3\, dx+f(y,z)\right]\mathbf{j}-\int_{x_0}^{x} a_2\, dx\, \mathbf{k}+\text{grad }\rho \qquad (2\text{-}39)$$

In summary, div $\mathbf{a}=0$, if and only if $\mathbf{a}=$curl $\mathbf{A}$, where $\mathbf{A}$ is given by Eq. (2-39)

Before we proceed to the principal result of this section, it is convenient to introduce the Dirac-delta, which we shall define by the properties

$$\delta(x-x')=0,\ x\neq x'$$

$$\int_{x-\epsilon}^{x+\epsilon} f(x')\ \delta(x-x')\ dx'=f(x) \qquad (2\text{-}40)$$

where the integral relation holds for any continuous function $f(x')$, and any $\epsilon>0$. Although there is some objection to the definition (2-40) from a rigorous mathematical point of view, the two properties can be rather well justified, and constitute an adequate description of the Dirac-delta for our purposes. We may immediately generalize the one-dimensional Dirac-delta to three dimensions. Let the vectors $\mathbf{r}$ and $\mathbf{r}'$ have the cartesian components, (x,y,z) and (x',y',z') respectively. We then define the three-dimensional Dirac-delta as

$$\delta(\mathbf{r}-\mathbf{r}')=\delta(x-x')\ \delta(y-y')\ \delta(z-z') \qquad (2\text{-}41)$$

We immediately see from (2-41), that if V' is any volume with volume element $dx'dy'dz'$, which contains the point (x,y,z), then

$$\int_{V'} f(x',y',z')\ \delta(\mathbf{r}-\mathbf{r}')\ dv'=f(x,y,z)$$

for any function $f(x,y,z)$ which is continuous in V'.

Now, consider two points which are specified by the radius vectors $\mathbf{r}$ and $\mathbf{r}'$, as shown in Fig. 2-13. Define $\mathbf{R}=\mathbf{r}-\mathbf{r}'$. Then,

$$|\mathbf{R}|=R=\sqrt{(x-x')^2+(y-y')^2+(z-z')^2}$$

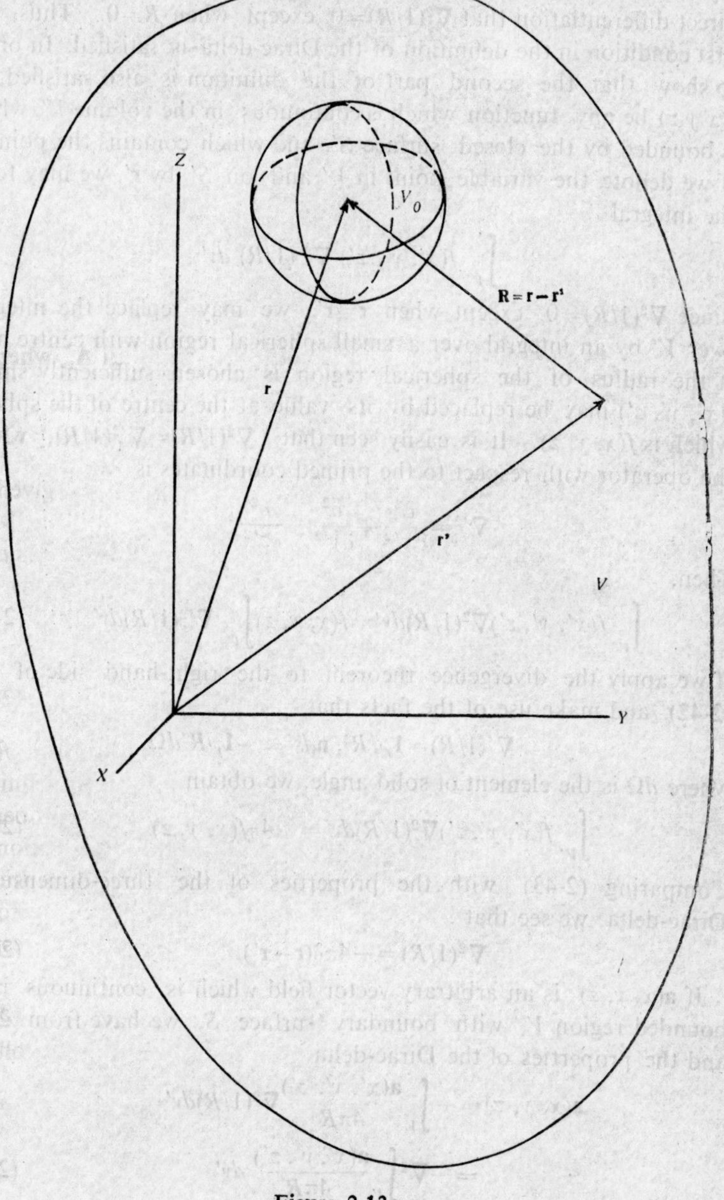

Figure 2-13.

We shall now show that $\nabla^2(1/R)=-4\pi\,\delta(r-r')$. It can be verified by direct differentiation that $\nabla^2(1/R)=0$, except when $R=0$. Thus, the first condition in the definition of the Dirac-delta is satisfied. In order to show that the second part of the difinition is also satisfied, let $f(x,y,z)$ be any function which is continuous in the volume V', which is bounded by the closed surface S', and which contains the point $\mathbf{r}$. If we denote the variable point in V' and on S' by $\mathbf{r}'$, we may form the integral

$$\int_{V'} f(x', y', z')\,\nabla^2\,(1/R)\,dv'$$

Since $\nabla^2(1/R)=0$, except when $\mathbf{r}=\mathbf{r}'$, we may replace the integral over V' by an integral over a small spherical region with centre at $\mathbf{r}$. If the radius of the spherical region is chosen sufficiently small, $f(x', y', z')$ may be replaced by its value at the centre of the sphere, which is $f(x, y, z)$. It is easily seen that $\nabla^2(1/R)=\nabla'^2(1/R)$, where the operator with respect to the primed coordinates is

$$\nabla'^2=\frac{\partial^2}{\partial x'^2}+\frac{\partial^2}{\partial y'^2}+\frac{\partial^2}{\partial z'^2}$$

Then,

$$\int_{V'} f(x', y', z')\nabla^2(1/R)dv'=f(x, y, z)\int_{V_0} \nabla'^2(1/R)dv' \qquad (2\text{-}42)$$

If we apply the divergence theorem to the right-hand side of Eq. (2-42), and make use of the facts that

$$\nabla'(1/R)=\mathbf{1}_{r'}/R^2,\ \mathbf{n}ds'=-\mathbf{1}_{r'}R^2d\Omega$$

where $d\Omega$ is the element of solid angle, we obtain

$$\int_{V'} f(x', y', z')\nabla^2(1/R)dv'=-4\pi f(x, y, z) \qquad (2\text{-}43)$$

Comparing (2-43) with the properties of the three-dimensional Dirac-delta, we see that

$$\nabla^2(1/R)=-4\pi\delta(\mathbf{r}-\mathbf{r}') \qquad (2\text{-}44)$$

If $\mathbf{a}(x, y, z)$ is an arbitrary vector field which is continuous in a bounded region V, with boundary surface S, we have from (2-44) and the properties of the Dirac-delta

$$\mathbf{a}(x, y, z)=-\int_V \frac{\mathbf{a}(x', y', z')}{4\pi R}\nabla^2(1/R)dv'$$

$$=-\nabla^2\int_V \frac{\mathbf{a}(x', y', z')}{4\pi R}\,dv' \qquad (2\text{-}45)$$

Making use of the vector identity $\nabla^2\mathbf{a}=\text{curl curl }\mathbf{a}-\text{grad div }\mathbf{a}$, we may write

$$\mathbf{a}(x, y, z)=\text{curl curl}\int_V \mathbf{a}(x', y', z')/4\pi R\, dv' - \text{grad div}\int_V \mathbf{a}(x', y', z')/4\pi R\, dv'$$

Let us consider the divergence term first. Since the divergence operator is with respect to the unprimed coordinates only,

$$\text{div}\int_V \mathbf{a}(x', y' z')/4\pi R\, dv' = \frac{1}{4\pi}\int_V \mathbf{a}(x', y', z')\ \text{grad}\,(1/R)\, dv'$$

But,

$$\mathbf{a}(x', y', z')\ \text{grad}\,(1/R) = -\mathbf{a}(x', y', z')\ \text{grad}'\,(1/R) = -\text{div}'\frac{\mathbf{a}(x', y', z')}{R} + \text{div}'\frac{\mathbf{a}(x', y', z')}{R}$$

and hence

$$\text{div}\int_V \frac{\mathbf{a}(x', y', z')dv'}{4\pi R} = -\int_S \frac{\mathbf{a}(x',y', z')\cdot\mathbf{n}\, ds'}{4\pi R} + \int_V \frac{\text{div}'\,\mathbf{a}(x', y' z')dv'}{4\pi R} = \rho(x, y, z) \qquad (2\text{-}46)$$

We next consider the curl term

$$\text{curl}\int_V \frac{\mathbf{a}(x', y', z')\, dv'}{4\pi R} = \frac{1}{4\pi}\int_V \mathbf{a}(x', y', z')\times\text{grad}'\,(1/R)\, dv'$$

Now,

$$\text{curl}'\frac{\mathbf{a}(x', y', z')}{R} = -\mathbf{a}(x', y', z')\times\text{grad}'\,(1/R) + \frac{\text{curl}'\,\mathbf{a}(x', y', z')}{R}$$

and hence,

$$\text{curl}\int_V \frac{\mathbf{a}(x', y', z')dv'}{4\pi R} = \int_V \frac{\text{curl}'\,\mathbf{a}(x', y', z')dv'}{4\pi R} - \frac{1}{4\pi}\int_V \text{curl}'\frac{\mathbf{a}(x', y', z')}{R}dv'$$

It can be shown from (2-36) that

$$\int_V \text{curl}'\frac{\mathbf{a}(x', y', z')}{R}dv' = -\int_S \frac{\mathbf{a}(x', y', z')\times\mathbf{n}\, ds'}{R}$$

and hence,

$$\text{curl}\int_V \frac{\mathbf{a}(x', y', z')dv'}{4\pi R} = \int_V \frac{\text{curl}'\,\mathbf{a}(x', y', z')dv'}{4\pi R} + \int_S \frac{\mathbf{a}(x', y', z')\times\mathbf{n}\, ds'}{4\pi R} = \mathbf{A}(x, y, z) \qquad (2\text{-}47)$$

Thus, (2-45) can be written in the form

$$\mathbf{a}(x, y, z)=\mathbf{a}_1(x, y, z)+\mathbf{a}_2(x, y, z) \tag{2-48}$$

where

$$\mathbf{a}_1=-\text{grad } \rho, \qquad \mathbf{a}_2=\text{curl } \mathbf{A}$$

with (x, y, z) and $A(x, y, z)$ given by (2-46) and (2-47) respectively. The result (2-48) which shows that any continuous vector field can be decomposed into the sum of two vector fields, one of which is irrotational, and the other of which is solenoidal, is known as the Helmholtz' theorem.

EXAMPLE 20. In any region of free space, the complete electromagnetic field is determined by the set of four Maxwell equations

$$\text{curl } \mathbf{E}=-\frac{\partial \mathbf{B}}{\partial t}, \quad \text{div } \mathbf{E}=\frac{1}{\epsilon_0}\rho$$

$$\text{curl } \mathbf{B}=\mathbf{J}+\mu_0\epsilon_0\frac{\partial \mathbf{E}}{\partial t}, \quad \text{div } \mathbf{B}=0$$

Since the divergence of **B** vanishes, it follows that the field **B** can be derived from a vector potential

$$\mathbf{B}=\text{curl } \mathbf{A}$$

On the other hand, we require both a scalar and vector potential to describe **E**. Now,

$$\text{curl } \mathbf{E}=-\frac{\partial \mathbf{B}}{\partial t}=-\frac{\partial}{\partial t}(\text{curl } \mathbf{A})=-\text{curl } (\partial \mathbf{A}/\partial t)$$

which implies that $(\mathbf{E}+\partial \mathbf{A}/\partial t)$ is irrotational and can be derived from a scalar potential ϕ. Hence,

$$\mathbf{E}=-\text{grad } \phi-\partial \mathbf{A}/\partial t$$

2-12. Summary of Identities.

The Gradient:

$$\text{grad } \rho=\lim_{\Delta v\to 0}\frac{\int_{\Delta s}\rho\mathbf{n}ds}{\Delta v} \tag{2-49}$$

$$\text{grad } \rho=\frac{\partial \rho}{\partial x}\mathbf{i}+\frac{\partial \rho}{\partial y}\mathbf{j}+\frac{\partial \rho}{\partial z}\mathbf{k} \qquad \text{(cartesian representation)}$$

$$\text{grad } (\lambda\rho)=\lambda \text{ grad } \rho+\rho \text{ grad } \lambda \tag{2-50}$$

$$\text{grad } (\mathbf{a}\cdot\mathbf{b})=(\mathbf{a}\cdot\text{grad})\,\mathbf{b}+(\mathbf{b}\cdot\text{grad})\,\mathbf{a}+\mathbf{a}\times\text{curl } \mathbf{b}+\mathbf{b}\times\text{curl } \mathbf{a} \tag{2-51}$$

The Divergence:

$$\operatorname{div} \mathbf{a} = \lim_{\Delta v \to 0} \frac{\int_{\Delta s} \mathbf{a} \cdot \mathbf{n}\, ds}{\Delta v} \tag{2-52}$$

$$= \frac{\partial a_2}{\partial x} + \frac{\partial a_2}{\partial y} + \frac{\partial a_3}{\partial z} \quad \text{(cartesian representation)}$$

$$\operatorname{div}(\mathbf{a}+\mathbf{b}) = \operatorname{div} \mathbf{a} + \operatorname{div} \mathbf{b} \tag{2-53}$$

$$\operatorname{div}(\rho \mathbf{a}) = \mathbf{a} \operatorname{grad} \rho + \rho \operatorname{div} \mathbf{a} \tag{2-54}$$

$$\operatorname{div}(\mathbf{a} \times \mathbf{b}) = \mathbf{b} \operatorname{curl} \mathbf{a} - \mathbf{a} \operatorname{curl} \mathbf{b} \tag{2-55}$$

$$\operatorname{div} \operatorname{curl} \mathbf{a} = 0 \tag{2-56}$$

$$\operatorname{div} \mathbf{r} = 3, \; \mathbf{r} = x\,\mathbf{i} + y\,\mathbf{j} + z\,\mathbf{k} \tag{2-57}$$

$$\nabla^2 \rho = \operatorname{div} \operatorname{grad} \rho \tag{2-58}$$

$$\int_V \operatorname{div} \mathbf{a}\, dv = \oint_S \mathbf{a} \cdot \mathbf{n}\, ds, \quad \text{(divergence theorem)} \tag{2-59}$$

The Curl:

$$\operatorname{curl} \mathbf{a} = \lim_{\Delta v \to 0} \frac{\int_{\Delta s} \mathbf{n} \times \mathbf{a}\, ds}{\Delta v} \tag{2-60}$$

$$\operatorname{curl} \mathbf{a} = \left(\frac{\partial a_3}{\partial y} - \frac{\partial a_2}{\partial z}\right)\mathbf{i} + \left(\frac{\partial a_1}{\partial z} - \frac{\partial a_3}{\partial x}\right)\mathbf{j} + \left(\frac{\partial a_2}{\partial x} - \frac{\partial a_1}{\partial y}\right)\mathbf{k} \quad \text{(cartesian representation)}$$

$$\operatorname{curl}(\mathbf{a}+\mathbf{b}) = \operatorname{curl} \mathbf{a} + \operatorname{curl} \mathbf{b} \tag{2-61}$$

$$\operatorname{curl}(\rho \mathbf{a}) = (\operatorname{grad} \rho) \times \mathbf{a} + \rho \operatorname{curl} \mathbf{a} \tag{2-62}$$

$$\operatorname{curl}(\mathbf{a} \times \mathbf{b}) = \mathbf{a} \operatorname{div} \mathbf{b} - \mathbf{b} \operatorname{div} \mathbf{a} + (\mathbf{b} \cdot \operatorname{grad})\, \mathbf{a} - (\mathbf{a} \cdot \operatorname{grad})\, \mathbf{b} \tag{2-63}$$

$$\operatorname{curl} \operatorname{grad} \rho = 0 \tag{2-64}$$

$$\operatorname{curl} \mathbf{r} = 0, \; \mathbf{r} = x\mathbf{i} + y\mathbf{j} + z\mathbf{k} \tag{2-65}$$

$$\operatorname{curl} \operatorname{curl} \mathbf{a} = \operatorname{grad} \operatorname{div} \mathbf{a} - \nabla^2 \mathbf{a} \tag{2-66}$$

$$\int_S (\operatorname{curl} \mathbf{a}) \cdot \mathbf{n}\, ds = \oint_C \mathbf{a} \cdot d\mathbf{l} \quad \text{(Stokes' theorem)} \tag{2-67}$$

PROBLEMS

2-1. Determine the derivative, with respect to t, of the following vector fields:

(i) $\mathbf{r}(x, y, z, t) = (x_0 + v_0 t + 1/2\, at^2)\,\mathbf{i}$

(ii) $\mathbf{r}(x, y, z, t) = xt\,\mathbf{i} + xyt^2\,\mathbf{j} + xyz\,\mathbf{k}$

2-2. For a particle moving in a circular orbit, $\mathbf{r} = r \cos \omega t\,\mathbf{i} + r \sin \omega t\,\mathbf{j}$, where $r = |\mathbf{r}| =$ constant and ω is constant.

(i) Evaluate $\mathbf{r}\times\dot{\mathbf{r}}$.

(ii) Show that $\ddot{\mathbf{r}}+\omega^2\mathbf{r}=0$.

2-3. If $\mathbf{u}=x\mathbf{i}+y\mathbf{j}+z\mathbf{k}$, show by direct calculation that $|d\mathbf{u}|\neq d|\mathbf{u}|$.

2-4. Determine the unit normals to the surface $2xz-3xy-4x=1$, at the point (1, 1, 4).

2-5. If the vector field $\mathbf{A}$ is an explicit function of both the space coordinates (x, y, z) and the time t, find an expression for $d\mathbf{A}$.

2-6. The electrostatic potential due to a distribution of electric charge of volume density ρ is

$$\phi(\mathbf{r})=\frac{1}{4\pi\epsilon_0}\int_V \frac{\rho(\mathbf{r}')}{|\mathbf{r}-\mathbf{r}'|}dv'$$

where $\mathbf{r}'$ is the position of a charge element, $\mathbf{r}$ the point at which the potential is determined, $dv'=dx'\,dy'\,dz'$, and V is any volume which contains all of the charge. The electric intensity is determined by $\mathbf{E}=-\text{grad}\,\phi$. Calculate $\mathbf{E}$.

2-7. Determine a representation of the gradient of a scalar field in terms of the left-handed cartesian coordinate system defined in Problem 1-12. Compare this representation with the representation in a right-handed cartesian system.

2-8. Let $u(x, y, z)$ and $v(x, y, z)$ be two scalar functions which are related by some function $f(u, v)=0$. Show that it is both necessary and sufficient that

$$(\text{grad}\,\mathbf{u})\times(\text{grad}\,\mathbf{v})=0.$$

2-9. If $\mathbf{r}=x\mathbf{i}+y\mathbf{j}+z\mathbf{k}$, show by direct calculation in cartesian coordinates that div $\mathbf{r}=3$.

2-10. The Yukawa potential has the form

$$\phi(x, y, z)=\phi_0\,\frac{e^{-|\mathbf{r}|}}{|\mathbf{r}|}$$

where $\mathbf{r}=x\mathbf{i}+y\mathbf{j}+z\mathbf{k}$.

(i) Determine the electrostatic field E corresponding to this potential.

(ii) What charge density ρ is required to generate such a field ?

2-11. Use cartesian representations to prove.

(i) div curl $\mathbf{a}=0$ for all twice differentiable vector fields a.

(ii) curl grad $\rho=0$ for all twice differentiable scalar fields.

(iii) curl $\mathbf{r}=0$, $\mathbf{r}=x\mathbf{i}+y\mathbf{j}+z\mathbf{k}$.

2-12. A rigid body in the xy-plane rotates about the z-axis with constant angular velocity $\boldsymbol{\omega}=\omega_z\,\mathbf{k}$. The linear velocity at any point $\mathbf{r}=x\mathbf{i}+y\mathbf{j}$ is $\mathbf{v}=\boldsymbol{\omega}\times\mathbf{r}$. Show that curl $\mathbf{v}=2\omega$.

2-13. Show that

$$\text{grad}\,(\mathbf{A}\cdot\mathbf{B})=(\mathbf{B}\cdot\text{grad})\,\mathbf{A}+(\mathbf{A}\cdot\text{grad})\,\mathbf{B}+\mathbf{B}\times\text{curl}\,\mathbf{A}+\mathbf{A}\times\text{curl}\,\mathbf{B}$$

2-14. Determine the representation of the curl of a vector field in terms of a left-handed cartesian coordinate system. What conclusions do you draw concerning the curl of a vector field?

2-15. In quantum mechanics, the operator corresponding to the angular momentum vector $\mathbf{L}$ is $\mathbf{r}\times\text{grad}$. If ψ is any scalar wave function, show that

$$\hat{\mathbf{L}}^2\psi=\hat{\mathbf{L}}\cdot\hat{\mathbf{L}}\psi=(\mathbf{r}\times\text{grad})\cdot(\mathbf{r}\times\text{grad})\,\psi$$

$$\hat{\mathbf{L}}^2\psi=r^2\nabla^2\psi-r^2\,\frac{\partial^2\psi}{\partial r^2}-2r\,\frac{\partial\psi}{\partial r}$$

where $r=|\mathbf{r}|=(x^2+y^2+z^2)^{1/2}$.

2-16. If a_i is any cartesian component of the vector **a**, show that

$$\nabla^2 a_i = \frac{\partial^2 a_i}{\partial x^2} + \frac{\partial^2 a_i}{\partial y^2} + \frac{\partial^2 a_i}{\partial z^2}$$

2-17. If the vector **A** satisfies the vector Helmholtz equation

$$\nabla^2 \mathbf{A} + k^2 \mathbf{A} = 0$$

and the auxilliary condition

$$\text{div } \mathbf{A} = 0$$

show that

$$\text{curl curl } \mathbf{A} - k^2 \mathbf{A} = 0$$

2-18. Calculate $\nabla^2 \psi$ in terms of a left-handed cartesian coordinate system.

2-19. Let $\mathbf{f} = x^2 \mathbf{i} + y^2 \mathbf{j}$ and consider the line integral

$$\int \mathbf{f} \cdot d\mathbf{r}$$

from the point (0, 0) to the point (1, 1) along the following contours:

(i) the parabola $y = x^2$

(ii) the x-axis from (0, 0) to (1, 0) and then from (1, 0) to (1, 1) along the line $x = 1$.

2-20. According to Newton's law of gravitation, the force exerted on a point mass m_1 by a point mass m_2 is

$$\mathbf{F}_{12} = G \frac{m_1 m_2}{r^3} \mathbf{r}$$

where G is a universal constant and **r** is the vector from m_1 to m_2.

(i) Show that $\mathbf{F}_{12}$ is a conservative field.

(ii) Determine the gravitational potential.

2-21. The magnetic field at a distance **r** from a straight wire carrying an electric current density **J** is

$$\mathbf{B} = \frac{\mu_0}{4\pi} \frac{\mathbf{J} \times \mathbf{r}}{r^3}$$

Determine whether or not the magnetic field **B** is conservative.

2-22. Show that $\mathbf{f} = x^2 \mathbf{i} + y^3 \mathbf{j}$ is conservative and determine the scalar potential for **f**.

2-23. A force field **F** is given by

$$\mathbf{F} = \frac{\mathbf{r}}{r^n}, \quad \mathbf{r} = x\mathbf{i} + y\mathbf{j} + z\mathbf{k}$$

(i) Show that **F** is conservative.

(ii) Assuming $n \neq 1$, determine the scalar potential for F

(iii) What happens when $n = 1$?

2-24. If $\mathbf{a} = 4xz\,\mathbf{i} + y^2\,\mathbf{j} + yz\,\mathbf{k}$, evaluate the volume integral of the divergence of **a** within the cube bounded by the planes $x = 0$, $x = 1$, $y = 0$, $y = 1$, $z = 0$, $z = 1$.

2-25. Consider a fluid of density ρ and velocity **v** in some region V_0. Show that

$$\frac{\partial \rho}{\partial t} + \rho \text{ div } \mathbf{v} + \mathbf{v} \cdot \text{grad } \rho = 0$$

which is one form of the continuity equation in hydrodynamics.

2-26. The electric displacement field **D** is defined by the Maxwell equation

$$\text{div } \mathbf{D} = \rho$$

where ρ is the volume electric charge density. At the interface between two different media (medium 1 and medium 2), show that

$$(\mathbf{D}_2-\mathbf{D}_1)\cdot\mathbf{n}=\sigma$$

where σ is the area charge density on the interface, and $\mathbf{n}$ is a unit normal from medium 1 to medium 2. (Hint: Construct a thin cylinder with one face in medium 1, the other in medium 2 and make use of the Gauss law.)

2-27. If V is any volume bounded by the closed surface S with unit outward normal $\mathbf{n}$, and if the vector field $\mathbf{a}$ has continuous derivatives within and on S, show that

$$\int_V \text{curl } \mathbf{a}\, dv=-\int_S \mathbf{a}\times\mathbf{n}\, ds$$

2-28. If $\mathbf{F}=(x-2y)\,\mathbf{i}+yz^3\,\mathbf{j}+xyz^2\,\mathbf{k}$, evaluate

$$\int_S (\text{curl } \mathbf{F})\cdot\mathbf{n}\, ds$$

where S is the upper half of the spherical surface $x^2+y^2+z^2=1$.

2-29. Evaluate

$$\int_S (\text{curl } \mathbf{F})\cdot\mathbf{n}\, ds$$

over the upper surface of the ellipsoid $(x^2/4)+(y^2/9)+z^2=1$, when

$$\mathbf{F}=(x+y)\,\mathbf{i}+xy\,\mathbf{j}+(x^2z-yz^2)\,\mathbf{k}.$$

2-30. If $\mathbf{a}=z\,\mathbf{i}+x\mathbf{j}+y\mathbf{k}$, show that $\mathbf{a}=\text{curl } \mathbf{A}$ and determine $\mathbf{A}$.

2-31. Determine which of the following vector fields can be derived from a scalar potential; which from a vector potential; and which must be derived from both. Determine the potentials in each case.

(i) $\mathbf{a}=yz\,\mathbf{i}+zx\,\mathbf{j}+yz\,\mathbf{k}$

(ii) $\mathbf{a}=x\,\mathbf{i}+y\,\mathbf{j}+z\,\mathbf{k}$

(iii) $\mathbf{a}=(x+y)\,\mathbf{i}+(y+z)\,\mathbf{j}+(z+x)\,\mathbf{k}$

CHAPTER 3

Curvilinear Coordinates

The great utility of vector analysis in the formulation and solution of physical problems lies in the fact that results which are expressed in terms of vector equations are independent of the coordinate system used to define the points of space. So far, in our discussion of vectors, we have considered only descriptions in terms of cartesian coordinates. In the present chapter, we shall derive descriptions of vectors in very general coordinate systems. In fact, the analysis is too general to be of maximum utility in the applications. In the last section of this chapter, we shall specialize the general results to the case of orthogonal systems. However, the more general discussion will be of great value in our later discussion of the tensor formalism.

3-1. Unitary and Reciprocal Unitary Vectors. In a given region of space, we may associate with each point of the region the cartesian coordinates (x,y,z). This description of the points of space is unique so long as we restrict ourselves to the given cartesian system. However, in the region of space, we can define three independent, single valued functions of the cartesian set

$$u=f_1(x,y,z),\quad v=f_2(x,y,z),\quad w=f_3(x,y,z) \tag{3-1}$$

If we consider a particular point in the region, say $P(x_0,y_0,z_0)$, we can associate with each such point the three functional values u_0,v_0, and w_0 which are obtained by setting x, y, and z equal to x_0, y_0, and z_0 respectively.

Under very general conditions, we can solve the set of Eqs. (3-1) to obtain

$$x=g_1(u, v, w),\quad y=g_2(u, v, w),\quad z=g_3(u, v, w) \tag{3-2}$$

where the functions g_1, g_2, and g_3 are also independent and continuous functions. Generally, these functions are not single valued for the entire range of u, v, and w. At those points where there are multiple values, some care must be exercised in the application of the results we obtain here. Thus, for each triplet of numbers u, v, and w, there will generally correspond one and only one point, $P(x, y, z)$, in the given region of space. There is, therefore, a one-to-one correspondence between the triplet (u, v, w) and the points of a region of space. The set of functions (u, v, w) can be termed a set of coordinates for the points in space. These coordinates are generally known as generalized or curvilinear coordinates.

Through each point P in the given region of space, there will pass the three surfaces

$$u=\text{constant},\quad v=\text{constant},\quad w=\text{constant}$$

which are known as the coordinate surfaces. Any two of these constant surfaces intersect in a space curve. The set of three curves through the point P are known as the coordinate curves of the point P. As a matter of nomenclature, we shall adopt the following convention to distinguish between the coordinate curves:

$$u\text{-coordinate curve: } v=\text{constant},\quad w=\text{constant}$$
$$v\text{-coordinate curve: } w=\text{constant},\quad u=\text{constant}$$
$$w\text{-coordinate curve: } u=\text{constant},\quad v=\text{constant}$$

The coordinate surfaces and coordinate curves associated with the point P are shown in Fig. 3-1.

If $\mathbf{r}$ is the radius vector from an arbitrary origin to the point P, it can be regarded as a function of the generalized coordinates of the point,

$$\mathbf{r}=\mathbf{r}(u, v, w)$$

The change in the radius vector r due to infinitesimal displacements along the three coordinate curves is

$$d\mathbf{r}=\frac{\partial \mathbf{r}}{\partial u}\,du+\frac{\partial \mathbf{r}}{\partial v}\,dv+\frac{\partial \mathbf{r}}{\partial w}\,dw$$

Thus, the change in $\mathbf{r}$ due to a unit displacement along the u-coordinate curve is $d\mathbf{r}/du$. Similarly, unit displacements along the v- and w-coordinate curves induce the changes

$$d\mathbf{r}=\frac{\partial \mathbf{r}}{\partial v},\quad d\mathbf{r}=\frac{\partial \mathbf{r}}{\partial w}$$

respectively. We define the set of vectors

$$\mathbf{e}_1=\frac{\partial \mathbf{r}}{\partial u},\quad \mathbf{e}_2=\frac{\partial \mathbf{r}}{\partial v},\quad \mathbf{e}_3=\frac{\partial \mathbf{r}}{\partial w} \tag{3-3}$$

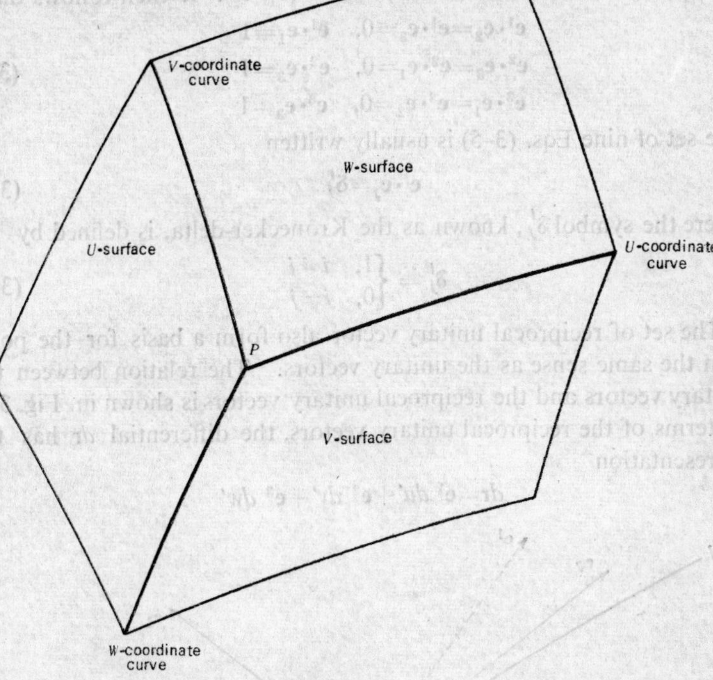

Figure 3-1.

which are known as the unitary vectors associated with the point P. We note that these vectors are not generally of unit length, and that their dimensions depend on the nature of the generalized coordinates. They do, however, serve as a base of reference in the sense that any vector whose initial point is at P can be expressed as a linear combination of the set of unitary vectors. In particular,

$$d\mathbf{r}=\mathbf{e}_1\, du+\mathbf{e}_2\, dv+\mathbf{e}_3\, dw$$

Since the set of unitary vectors are non-coplanar, they define a parallelepiped whose volume is given by

$$V=\mathbf{e}_1\cdot(\mathbf{e}_2\times\mathbf{e}_3)=\mathbf{e}_2\cdot(\mathbf{e}_3\times\mathbf{e}_1)=\mathbf{e}_3\cdot(\mathbf{e}_1\times\mathbf{e}_2)$$

We may define a triplet of vectors, known as the reciprocal unitary vectors by the relations

$$\mathbf{e}^1=\frac{\mathbf{e}_2\times\mathbf{e}_3}{V},\quad \mathbf{e}^2=\frac{\mathbf{e}_3\times\mathbf{e}_1}{V},\quad \mathbf{e}^3=\frac{\mathbf{e}_1\times\mathbf{e}_2}{V} \tag{3-4}$$

It is clear from the definition that $\mathbf{e}^1$ is normal to the plane defined by the unitary vectors $(\mathbf{e}_2, \mathbf{e}_3)$. Similarly, $\mathbf{e}^2$ is normal to the plane of $(\mathbf{e}_3, \mathbf{e}_1)$, and $\mathbf{e}^3$ is normal to the plane of $(\mathbf{e}_1, \mathbf{e}_2)$. It then follows that

$$\begin{aligned} \mathbf{e}^1 \cdot \mathbf{e}_2 = \mathbf{e}^1 \cdot \mathbf{e}_3 = 0, \quad \mathbf{e}^1 \cdot \mathbf{e}_1 = 1 \\ \mathbf{e}^2 \cdot \mathbf{e}_3 = \mathbf{e}^2 \cdot \mathbf{e}_1 = 0, \quad \mathbf{e}^2 \cdot \mathbf{e}_3 = 1 \\ \mathbf{e}^3 \cdot \mathbf{e}_1 = \mathbf{e}^3 \cdot \mathbf{e}_2 = 0, \quad \mathbf{e}^3 \cdot \mathbf{e}_3 = 1 \end{aligned} \tag{3-5}$$

The set of nine Eqs. (3–5) is usually written

$$\mathbf{e}^i \cdot \mathbf{e}_j = \delta^i_j \tag{3-6}$$

where the symbol δ^i_j, known as the Kronecker-delta, is defined by

$$\delta^i_j = \begin{cases} 1, & i=j \\ 0, & i \neq j \end{cases} \tag{3-7}$$

The set of reciprocal unitary vector also form a basis for the point P in the same sense as the unitary vectors. The relation between the unitary vectors and the reciprocal unitary vectors is shown in Fig. 3-2. In terms of the reciprocal unitary vectors, the differential $d\mathbf{r}$ has the representation

$$d\mathbf{r} = \mathbf{e}^1\, du' + \mathbf{e}^2\, dv' + \mathbf{e}^3\, dw'$$

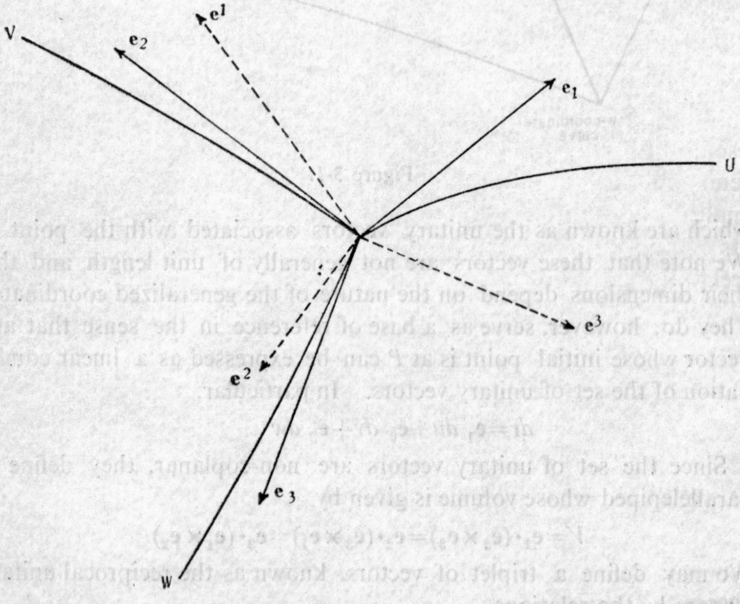

Figure 3-2.

The differentials du', dv', and dw' are obviously the components of $d\mathbf{r}$ in the directions specified by the reciprocal unitary vectors, and are functions of the generalized coordinate (u, v, w). However, these differentials are related to the differentials of the generalized coordinates by a set of linear equations which are not, in general integrable. If we equate the representations of $d\mathbf{r}$ in terms of unitary and reciprocal unitary vectors, we have

$$d\mathbf{r} = \mathbf{e}_1\, du + \mathbf{e}_2\, dv + \mathbf{e}_3\, dw = \mathbf{e}^1\, du' + \mathbf{e}^2\, dv' + \mathbf{e}^3\, dw'$$

Now, form the inner product of $d\mathbf{r}$ successively with $\mathbf{e}^1$, $\mathbf{e}^2$, and $\mathbf{e}^3$, and use the orthogonality relations to obtain the set of equations

$$du = \mathbf{e}^1\cdot\mathbf{e}^1\, du' + \mathbf{e}^1\cdot\mathbf{e}^2\, dv' + \mathbf{e}^1\cdot\mathbf{e}^3\, dw'$$
$$dv = \mathbf{e}^2\cdot\mathbf{e}^1\, du' + \mathbf{e}^2\cdot\mathbf{e}^2\, dv' + \mathbf{e}^2\cdot\mathbf{e}^3\, dw'$$
$$dw = \mathbf{e}^3\cdot\mathbf{e}^1\, du' + \mathbf{e}^3\cdot\mathbf{e}^2\, dv' + \mathbf{e}^3\cdot\mathbf{e}^3\, dw'$$

Similarly, if we take the inner product of $d\mathbf{r}$ with $\mathbf{e}_1$, $\mathbf{e}_2$, and $\mathbf{e}_3$ we obtain the differential equations

$$du' = \mathbf{e}_1\cdot\mathbf{e}_1\, du + \mathbf{e}_1\cdot\mathbf{e}_2\, dv + \mathbf{e}_1\cdot\mathbf{e}_3\, dw$$
$$dv' = \mathbf{e}_2\cdot\mathbf{e}_1\, du + \mathbf{e}_2\cdot\mathbf{e}_2\, dv + \mathbf{e}_2\cdot\mathbf{e}_3\, dw$$
$$dw' = \mathbf{e}_3\cdot\mathbf{e}_1\, du + \mathbf{e}_3\cdot\mathbf{e}_2\, dv + \mathbf{e}_3\cdot\mathbf{e}_3\, dw$$

It is convenient to represent the inner products of the unitary vectors, and the reciprocal unitary vectors by the sets of quantities

$$g^{ij} = \mathbf{e}^i\cdot\mathbf{e}^j = g^{ji}$$
$$g_{ij} = \mathbf{e}_i\cdot\mathbf{e}_j = g_{ji} \tag{3-8}$$

These quantities are fundamental in the representation of the vector differential operators in terms of generalized coordinates, and in the formalism of tensor analysis. In terms of this notation, the components of $d\mathbf{r}$ with respect to the unitary and reciprocal unitary vectors are related by

$$du = g^{11}\, du' + g^{12}\, dv' + g^{13}\, dw'$$
$$dv = g^{21}\, du' + g^{22}\, dv' + g^{23}\, dw'$$
$$dw = g^{31}\, du' + g^{32}\, dv' + g^{33}\, dw'$$

and

$$du' = g_{11}\, du + g_{12}\, dv + g_{13}\, dw$$
$$dv' = g_{21}\, du + g_{22}\, dv + g_{23}\, dw$$
$$dw' = g_{31}\, du + g_{32}\, dv + g_{33}\, dw$$

Now, consider an arbitrary vector $\mathbf{a}$ which has its initial point at the point P. This vector may be represented in terms of either the set of unitary or reciprocal unitary vectors

$$\mathbf{a} = \sum_{i=1}^{3} a^i \mathbf{e}_i = \sum_{i=1}^{3} a_i \mathbf{e}^i$$

where a^i and a_i are the projections of a in the directions of $\mathbf{e}_i$ and $\mathbf{e}^i$ respectively. If we form the inner product of $\mathbf{a}$ with $\mathbf{e}^i$, we obtain as a result of the orthogonality relations

$$a^i = \mathbf{a}\cdot\mathbf{e}^i$$

Similarly,

$$a_i = \mathbf{a}\cdot\mathbf{e}_i$$

The two sets of components of the vector $\mathbf{a}$ are related to one another by

$$a^j = \sum_{i=1}^{3} g^{ji}a_i, \quad a_j = \sum_{i=1}^{3} g_{ji}a^i$$

The arbitrary vector $\mathbf{a}$ thus has either of the representations

$$\mathbf{a} = \sum_{i=1}^{3} (\mathbf{a}\cdot\mathbf{e}^i)\mathbf{e}_i = \sum_{i=1}^{3} (\mathbf{a}\cdot\mathbf{e}_i)\mathbf{e}^i \tag{3-9}$$

in the generalized coordinate system (u, v, w). The quantities a^i are known as the contravariant components of the vector $\mathbf{a}$ relative to the coordinate system (u, v, w). Similarly, the set of quantities a_i are known as the covariant components of the vector $\mathbf{a}$ relative to the coordinate system (u, v, w).

Since the lengths and dimensions of the unitary and reciprocal unitary vectors depend on the nature of the set of generalized coordinates, it is clear that the covariant and controvariant components of a given vector do not necessarily have the same dimensions as the given vector. In order to avoid difficulties which may be introduced by this fact, it is frequently desirable to define a set of unit vectors which are parallel to the set of unitary vectors, but which are dimensionless

$$\mathbf{i}_1 = \frac{\mathbf{e}_1}{\sqrt{\mathbf{e}_1\cdot\mathbf{e}_1}} = \frac{\mathbf{e}_1}{\sqrt{g_{11}}}, \quad \mathbf{i}_2 = \frac{\mathbf{e}_2}{\sqrt{\mathbf{e}_2\cdot\mathbf{e}_2}} = \frac{\mathbf{e}_2}{\sqrt{g_{22}}}$$

$$\mathbf{i}_3 = \frac{\mathbf{e}_3}{\sqrt{\mathbf{e}_3\cdot\mathbf{e}_3}} = \frac{\mathbf{e}_3}{\sqrt{g_{33}}} \tag{3-10}$$

In terms of this set of dimensionless unit vectors, the arbitrary vector $\mathbf{a}$ has the representation

$$\mathbf{a} = A_1\mathbf{i}_1 + A_2\mathbf{i}_2 + A_3\mathbf{i}_3$$

The set of components A_i have the same dimensionality as the given vector $\mathbf{a}$. They are known as the physical components of the vector $\mathbf{a}$ relative to the coordinate system (u, v, w). Whenever we are dealing with a single coordinate system, and are not considering transformations from one system to another, it is generally preferable to use the physical components to describe a given vector. On the other hand, in discussing the behaviour of vectors as we transform from one coordinate system to another, there is some advantage in using the covariant or contravariant components, since the resulting equations appear in a more symmetric form. The physical components of a given vector **a** are related to the contravariant components by

$$A_i = \sqrt{g_{ii}}\, a^i \tag{3-11}$$

EXAMPLE 1. A system of generalized coordinates, known as spherical polar coordinates, is defined by the transformation equations,

$$u = r = (x^2+y^2+z^2)^{1/2}$$
$$v = \theta = \tan^{-1}(x^2+y^2)^{1/2/z} \qquad 0 < v < \pi$$
$$w = \varphi = \tan^{-1}(y/x) \qquad 0 \leqslant w < 2\pi$$

It is clear that the functions (u, v, w) are continuous, single valued functions of the cartesian coordinates (x, y, z), so long as x and y are not both zero. The set of generalized coordinates can be made complete by requiring that points on the positive z-axis to be defined as $u = |z|$, $v = 0$, $w = 0$; and those on the negative z-axis to be defined by $u = |z|$, $v = \pi$, $w = 0$. However, this convention implies that w is discontinuous in the limit $x \to 0$, $y \to 0$.

It follows from the transformation equations that

$$x^2+y^2+z^2 = u^2$$
$$x^2+y^2 = z^2 \tan^2 v$$
$$y^2 = x^2 \tan^2 w$$

From the second and third of these equations

$$x^2+y^2+z^2 = z^2 \sec^2 v = u^2$$
$$x^2+y^2 = x^2 \sec^2 w = z^2 \tan^2 v$$

or

$$z = u \cos v$$
$$x = u \sin v \cos w$$

It then follows that the inverse transformation is

$$x = u \sin v \cos w$$
$$y = u \sin v \sin w$$
$$z = u \cos v$$

The coordinate surfaces are most easily determined by setting u, v, and w equal to constants in the direct transformation equations. These coordinate surfaces are

u=constant;	the spheres	$x^2+y^2+z^2=u^2$
v=constant;	right circular cones	$x^2+y^2=z^2 \tan^2 w$
w=constant:	azimuthal planes	$y=x \tan w$

The determination of the unitary and reciprocal unitary vectors is straightforward, but somewhat tedious. We have

$$d\mathbf{r}=dx\mathbf{i}+dy\mathbf{j}+dz\mathbf{k}=\mathbf{e}_1du+\mathbf{e}_2dv+\mathbf{e}_3dw$$

but,

$$dx=\frac{\partial x}{\partial u}\,du+\frac{\partial x}{\partial v}\,dv+\frac{\partial x}{\partial w}\,dw=\sin v\cos w du+u\cos v\cos w\,dv-u\sin v\sin w\,dw$$

$$dy=\frac{\partial y}{\partial u}\,du+\frac{\partial y}{\partial v}\,dv+\frac{\partial y}{\partial w}\,dw=\sin v\sin w du+u\cos v\sin w\,dv+u\sin v\cos w\,dw$$

$$dz=\frac{\partial z}{\partial u}\,du+\frac{\partial z}{\partial v}\,dv+\frac{\partial z}{\partial w}\,dw=\cos v du-u\sin v\,dv$$

Then,

$$(\sin v\cos w\mathbf{i}+\sin v\sin w\mathbf{j}+\cos v\mathbf{k})\,du+(u\cos v\cos w\mathbf{i}+u\cos v\sin w\mathbf{j}-u\sin v\mathbf{k})\,dv+(-u\sin v\sin w\mathbf{i}+u\sin v\cos w\mathbf{j})\,dw$$
$$=\mathbf{e}_1du+\mathbf{e}_2dv+\mathbf{e}_3dw$$

Since du, dv, and dw are independent variables,

$$\mathbf{e}_1=\sin v\cos w\mathbf{i}+\sin v\sin w\mathbf{j}+\cos v\mathbf{k}$$
$$\mathbf{e}_2=u\,(\cos v\cos w\mathbf{i}+\cos v\sin w\mathbf{j}-\sin v\mathbf{k})$$
$$\mathbf{e}_3=u\,(-\sin v\sin w\mathbf{i}+\sin v\cos w\mathbf{j})$$

The volume of the parallelepiped defined by the unitary vectors is

$$V=\mathbf{e}_1\cdot\mathbf{e}_2\times\mathbf{e}_3=u^2\sin v$$

and the reciprocal unitary vectors are

$$\mathbf{e}^1=\frac{\mathbf{e}_2\times\mathbf{e}_3}{V}=\sin v\cos w\mathbf{i}+\sin v\sin w\mathbf{j}+\cos v\mathbf{k}=\mathbf{e}_1$$

$$\mathbf{e}^2=\frac{\mathbf{e}_3\times\mathbf{e}_1}{V}=\frac{1}{u}\,(\cos v\cos w\mathbf{i}+\cos v\sin w\mathbf{j}-\sin v\mathbf{k})=\frac{1}{u^2}\,\mathbf{e}_2$$

$$\mathbf{e}^3=\frac{\mathbf{e}_1\times\mathbf{e}_2}{V}=\frac{1}{u\sin v}\,(-\sin w\mathbf{i}+\cos w\mathbf{j})=\frac{1}{u^2\sin^2 v}\,\mathbf{e}_3$$

The quantities g_{ij} and g^{ij} are given by

$$g_{11}=\mathbf{e}_1\cdot\mathbf{e}_1=1\quad;\quad g_{12}=g_{21}=\mathbf{e}_1\cdot\mathbf{e}_2=0\quad;\quad g_{13}=g_{31}=\mathbf{e}_1\cdot\mathbf{e}_3=0$$

$$g_{22}=\mathbf{e}_2\cdot\mathbf{e}_2=u^2 \quad ; \quad g_{23}=g_{32}=\mathbf{e}_2\cdot\mathbf{e}_3=0$$
$$g_{33}=\mathbf{e}_3\cdot\mathbf{e}_3=u^2 \sin^2 v$$
$$g^{11}=\mathbf{e}^1\cdot\mathbf{e}^1=1 \quad ; \quad g^{12}=g^{21}=\mathbf{e}^1\cdot\mathbf{e}^2=0 \quad ; \quad g^{13}=g^{31}=\mathbf{e}^1\cdot\mathbf{e}^3=0$$
$$g^{22}=\mathbf{e}^2\cdot\mathbf{e}^2=1/u^2; \quad g^{23}=g^{32}=\mathbf{e}^2\cdot\mathbf{e}^3=0$$
$$g^{33}=\mathbf{e}^3\cdot\mathbf{e}^3=1/(u^2 \sin^2 v)$$

We see that the unitary vectors are a mutually orthogonal triplet, and hence define an orthogonal curvilinear system, which is a rather special case of generalized coordinates.

EXAMPLE 2. A set of generalized coordinates, which is sometimes used in certain problems in classical mechanics, is defined by the transformations

$$x=a\,u+bv$$
$$y=a\sqrt{1-u^2}+b\sqrt{1-v^2}$$
$$z=w$$

where a and b are constant parameters. Although this set of transformation equations can be inverted to find the constant coordinate surfaces, these surfaces turn out to be fourth-degree surfaces, and cannot be expressed in any particularly simple form.

It is easy enough, however, to analyze the other properties of this coordinate system. Now,

$$dx=a\,du+b\,dv$$
$$dy=-au\,(1-u^2)^{-1/2}\,du-bv\,(1-v^2)^{-1/2}$$
$$dz=dw$$

so that

$$d\mathbf{r}=a\,[\mathbf{i}-u(1-u^2)^{-1/2}\,\mathbf{j}]\,du+b\,[\mathbf{i}-v(1-v^2)^{-1/2}\,\mathbf{j}]\,dv+\mathbf{k}\,dw$$
$$=\mathbf{e}_1\,du+\mathbf{e}_2\,dv+\mathbf{e}_3\,dw$$

We immediately identify the unitary vectors as

$$\mathbf{e}_1=a\,\mathbf{i}-\frac{au}{\sqrt{1-u^2}}\,\mathbf{j}$$
$$\mathbf{e}_2=b\,\mathbf{i}-\frac{bv}{\sqrt{1-v^2}}\,\mathbf{j}$$
$$\mathbf{e}_3=\mathbf{k}$$

It then follow that the g_{ij} are

$$g_{11}=\frac{a^2}{1-u^2};\ g_{12}=g_{21}=ab\left(1+\frac{uv}{\sqrt{(1-u^2)\,(1-v^2)}}\right);$$
$$g_{13}=g_{31}=0;$$

$$g_{22}=\frac{b^2}{1-v^2};\ g_{23}=g_{32}=0;\ g_{33}=1$$

We immediately see from the form of the g_{ij} that the coordinate system is not orthogonal. However, the plane defined by the unitary vectors $\mathbf{e}_1$ and $\mathbf{e}_2$ is orthogonal to $\mathbf{e}_3$.

The volume of the parallelopiped defined by the unitary vectors is

$$V=\mathbf{e}_1\cdot\mathbf{e}_2\times\mathbf{e}_3=\begin{vmatrix} a & -au(1-u^2)^{-1/2} & 0 \\ b & -bv(1-v^2)^{-1/2} & 0 \\ 0 & 0 & 1 \end{vmatrix}$$

$$=ab\left(\frac{u\sqrt{1-v^2}-v\sqrt{1-u^2}}{\sqrt{(1-u^2)(1-v^2)}}\right)$$

Then, the reciprocal unitary vectors are

$$\mathbf{e}^1=\frac{1}{V}\,\mathbf{e}_2\times\mathbf{e}_3=-\frac{v\sqrt{1-u^2}}{a(u\sqrt{1-v^2}-v\sqrt{1-u^2})}\,\mathbf{i}-\frac{\sqrt{(1-u^2)(1-v^2)}}{a(u\sqrt{1-v^2}-v\sqrt{1-u^2})}\,\mathbf{j}$$

$$\mathbf{e}^2=\frac{1}{V}\,\mathbf{e}_3\times\mathbf{e}_1=\frac{u\sqrt{1-v^2}}{b(u\sqrt{1-v^2}-v\sqrt{1-u^2})}\,\mathbf{i}+\frac{\sqrt{(1-u^2)(1-v^2)}}{b(u\sqrt{1-v^2}-v\sqrt{1-u^2})}\,\mathbf{j}$$

$$\mathbf{e}^3=\frac{1}{V}\,\mathbf{e}_1\times\mathbf{e}_2=\mathbf{k}$$

The calculation of the second metric coefficients yields

$$g^{11}=\frac{1-u^2}{a^2\,[u^2\,(1-v^2)+v^2\,(1-u^2)-2uv\sqrt{(1-u^2)\,(1-v^2)}]^2}$$

$$g^{12}=g^{21}=\frac{-uv\sqrt{(1-u^2)\,(1-v^2)}-(1-u^2)\,(1-v^2)}{ab\,[u^2\,(1-v^2)+v^2(1-u^2)-2uv\sqrt{(1-u^2)(1-v^2)}]^2}$$

$$g^{13}=g^{31}=0$$

$$g^{22}=\frac{1-v^2}{b^2\,[u^2\,(1-v^2)+v^2\,(1-u^2)-2uv\sqrt{(1-u^2)(1-v^2)}]^2}$$

$$g^{23}=g^{32}=0$$

$$g^{33}=1$$

3-2. Line, Surface and Volume Elements. The vector $d\mathbf{r}$ represents a displacement from the point $P(u, v, w)$ to the point $Q(u+du, v+dv, w+dw)$. We shall define

$$\begin{aligned} ds^2=d\mathbf{r}\cdot d\mathbf{r}=&\mathbf{e}_1\cdot\mathbf{e}_1\,du\,du+\mathbf{e}_1\cdot\mathbf{e}_2\,du\,dv+\mathbf{e}_1\cdot\mathbf{e}_3\,du\,dw+\\ &\mathbf{e}_2\cdot\mathbf{e}_1\,dv\,du+\mathbf{e}_2\cdot\mathbf{e}_2\,dv\,dv+\mathbf{e}_2\cdot\mathbf{e}_3\,dv\,dw+\\ &\mathbf{e}_3\cdot\mathbf{e}_1\,dw\,du+\mathbf{e}_3\cdot\mathbf{e}_2\,dw\,dv+\mathbf{e}_3\cdot\mathbf{e}_3\,dw\,dw \end{aligned}$$

In terms of the coefficients g_{ij}, this may be written

$$ds^2 = g_{11}\,du\,du + g_{12}\,du\,dv + g_{13}\,du\,dw + g_{21}\,dv\,du + g_{22}\,dv\,dv + g_{23}\,dv\,dw + g_{31}\,dw\,du + g_{32}\,dw\,dv + g_{33}\,dw\,dw$$

We may also express the quantity ds^2 in terms of the reciprocal unitary vectors by

$$ds^2 = g^{11}\,du'\,du' + g^{12}\,du'\,dv' + g^{13}\,du'\,dw' + g^{21}\,dv'\,du' + g^{22}\,dv'\,dv' + g^{23}\,dv'\,dw' + g^{31}\,dw'\,du' + g^{32}\,dw'\,dv' + g^{33}\,dw'\,dw'$$

The g_{ij} and g^{ij} appear as the coefficients of two differential quadratic forms which express the square of the arc length in the space of the generalized coordinates (u, v, w) or the related coordinates (u', v', w'). As we shall see later in this section, all of the measurable properties of the coordinate system are expressed in terms of these coefficients. For this reason, the g_{ij} and g^{ij} are known as the first and second metric coefficients respectively.

Now, let ds_1 represent an infinitesimal displacement along the u-coordinate curve from the point $P(u, v, w)$. Then

$$d\mathbf{s}_1 = \mathbf{e}_1\,du, \quad ds_1 = |\,d\mathbf{s}_1\,| = \sqrt{g_{11}}\,du$$

Similarly, the length of arc along the v- and w-coordinate curves are given by

$$ds_2 = \sqrt{g_{22}}\,dv, \quad ds_3 = \sqrt{g_{33}}\,dw$$

respectively.

Consider an infinitesimal element of the u-coordinate surface, which is bounded by intersecting v- and w-coordinate curves as shown in Fig. 3-3. The area of this element is

$$da_1 = |\,d\mathbf{s}_2 \times d\mathbf{s}_3\,| = |\,\mathbf{e}_2 \times \mathbf{e}_3\,|\,dv\,dw$$

$$= \sqrt{(\mathbf{e}_2 \times \mathbf{e}_3)\cdot(\mathbf{e}_2 \times \mathbf{e}_3)}\,dv\,dw \qquad (3\text{-}12)$$

If we make use of the identity $(\mathbf{a}\times\mathbf{b})\cdot(\mathbf{c}\times\mathbf{d}) = (\mathbf{a}\cdot\mathbf{c})\,(\mathbf{b}\cdot\mathbf{d}) = (\mathbf{a}\cdot\mathbf{d})\,(\mathbf{b}\cdot\mathbf{c})$, we may reduce the quantity under the radical in (3-12) to

$$(\mathbf{e}_2 \times \mathbf{e}_3)\cdot(\mathbf{e}_2 \times \mathbf{e}_3) = (\mathbf{e}_2\cdot\mathbf{e}_2)\,(\mathbf{e}_3\cdot\mathbf{e}_3) - (\mathbf{e}_2\cdot\mathbf{e}_3)\,(\mathbf{e}_2\cdot\mathbf{e}_3)$$

$$= g_{22}\,g_{33} - g_{23}^2$$

Hence, the elemental area in the u-coordinate surface is

$$da_1 = \sqrt{g_{22}\,g_{33} - g_{23}^2}\,dv\,dw$$

Similarly, the elemental areas in the v- and w-coordinate surfaces are given by

$$da_2 = \sqrt{g_{11}\,g_{33} - g_{31}^2}\,du\,dw$$

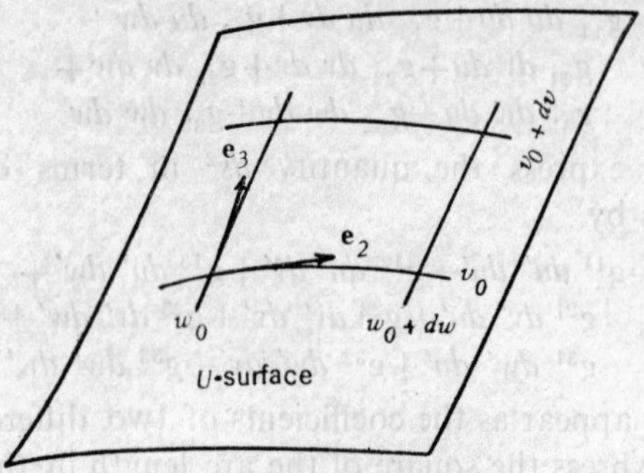

Figure 3-3.

and

$$da_3 = \sqrt{g_{11}g_{22} - g_{12}^2}\; du\, dv$$

respectively.

The infinitesimal region of space bounded by the three coordinate surfaces has a volume given by

$$dV = d\mathbf{s}_1 \cdot (d\mathbf{s}_2 \times d\mathbf{s}_3) = \mathbf{e}_1 \cdot \mathbf{e}_2 \times \mathbf{e}_3\; du\, dv\, dw$$

It follows from Eq. (3-9) that the cross product $\mathbf{e}_2 \times \mathbf{e}_3$ can be written as

$$\mathbf{e}_2 \times \mathbf{e}_3 = (\mathbf{e}^1 \cdot \mathbf{e}_2 \times \mathbf{e}_3)\, \mathbf{e}_1 + (\mathbf{e}^2 \cdot \mathbf{e}_2 \times \mathbf{e}_3)\, \mathbf{e}_2 + (\mathbf{e}^3 \cdot \mathbf{e}_2 \times \mathbf{e}_3)\, \mathbf{e}_3 \tag{3-13}$$

It follows from the definition of the reciprocal unitary vectors that we can rewrite (3-13) in the form

$$\mathbf{e}_1 \cdot \mathbf{e}_2 \times \mathbf{e}_3 = \frac{\mathbf{e}_1}{\mathbf{e}_1 \cdot \mathbf{e}_2 \times \mathbf{e}_3} \cdot [(\mathbf{e}_2 \times \mathbf{e}_3) \cdot (\mathbf{e}_2 \times \mathbf{e}_2)\, \mathbf{e}_1 + (\mathbf{e}_3 \times \mathbf{e}_1) \cdot (\mathbf{e}_2 \times \mathbf{e}_3)\, \mathbf{e}_2 + (\mathbf{e}_1 \times \mathbf{e}_2) \cdot (\mathbf{e}_2 \times \mathbf{e}_3)\, \mathbf{e}_3]$$

which is equivalent to

$$\begin{aligned}(\mathbf{e}_1 \cdot \mathbf{e}_2 \times \mathbf{e}_3)^2 = {} & \mathbf{e}_1 \cdot \mathbf{e}_1\, [(\mathbf{e}_2 \cdot \mathbf{e}_2)\, (\mathbf{e}_3 \cdot \mathbf{e}_3) - (\mathbf{e}_2 \cdot \mathbf{e}_3)\, (\mathbf{e}_3 \cdot \mathbf{e}_2)] + \\ & \mathbf{e}_1 \cdot \mathbf{e}_2\, [(\mathbf{e}_2 \cdot \mathbf{e}_3)\, (\mathbf{e}_3 \cdot \mathbf{e}_1) - (\mathbf{e}_2 \cdot \mathbf{e}_1)\, (\mathbf{e}_3 \cdot \mathbf{e}_3)] + \\ & \mathbf{e}_1 \cdot \mathbf{e}_3\, [(\mathbf{e}_2 \cdot \mathbf{e}_1)\, (\mathbf{e}_3 \cdot \mathbf{e}_2) - (\mathbf{e}_2 \cdot \mathbf{e}_2)\, (\mathbf{e}_3 \cdot \mathbf{e}_1)] \\ = {} & g_{11}\, (g_{22}\, g_{33} - g_{23}\, g_{32}) + g_{22}\, (g_{23}\, g_{31} - g_{21}\, g_{33}) \\ & g_{33}\, (g_{21}\, g_{32} - g_{22}\, g_{31})\end{aligned} \tag{3-14}$$

The right-hand side of (3-14) is the expansion of the determinant

$$g = \begin{vmatrix} g_{11} & g_{12} & g_{13} \\ g_{21} & g_{22} & g_{23} \\ g_{31} & g_{32} & g_{33} \end{vmatrix}$$

Hence, the incremental volume is expressed in terms of the generalized coordinates (u, v, w) by

$$dV=\sqrt{g}\,du\,dv\,dw$$

In the balance of this chapter, and in the following chapters, it will be convenient to make certain changes in notation, and to introduce the so-called Einstein summation convention. The change in notation which we require is the following:

$$u=u^1, \qquad v=u^2, \qquad w=u^3$$

$$x=x^1, \qquad y=x^2, \qquad z=x^3$$

The summation convention which we shall adopt is the following: Whenever a Latin index appears both as a subscript and a superscript in the same expression, it is summed from 1 to 3. For example,

$$g_{ij}\,du^j=g_{i1}\,du^1+g_{i2}\,du^2+g_{i3}\,du^3=du_i$$

and similarly,

$$g^{ij}\,du_j=g^{i1}\,du_1+g^{i2}\,du_2+g^{i3}\,du_3=du^i$$

In order to complete the convention, we shall adopt the rule that whenever an index appears as a subscript in the denominator of a fraction, it is regarded as equivalent to a superscript in the numerator for purposes of summation, and vice versa. For example,

$$\frac{\partial u^k}{\partial x^k}=\frac{\partial u^1}{\partial x^1}+\frac{\partial u^2}{\partial x^2}+\frac{\partial u^3}{\partial x^3}$$

and similarly,

$$\frac{\partial u_i}{\partial x_i}=\frac{\partial u_1}{\partial x_1}+\frac{\partial u_2}{\partial x_2}+\frac{\partial u_3}{\partial x_3}$$

Since the metric coefficients g_{ij} are required in order for us to characterize the geometrical properties of space in terms of an arbitrary generalized coordinate system (u^1, u^2, u^3), it is essential to know how these coefficients may be calculated. In terms of an orthogonal cartesian system (x^1, x^2, x^3), the Euclidean distance function is

$$ds^2=dx^i\,dx^i \qquad \text{(sum on } i\text{)}$$

This Euclidean distance is also expressed in terms of a fundamental quadratic form whose coefficients are the metric coefficients. In terms of the orthogonal cartesian system, we can immediately identify the metric coefficients to be

$$g_{ij}=\delta_{ij} \tag{3-15}$$

where δ_{ij} is another form of the Kronecker-delta. Since the coordinate surfaces for cartesian coordinates are a set of mutually orthogonal planes, it is evident that the unitary and reciprocal unitary vectors for this simple coordinate system are the same, are of unit length, and are the base vectors (**i, j, k**) which we have previously considered. A further fundamental property of cartesian coordinates is that there is no distinction between the covariant, contravariant and physical components of a given vector when represented in this coordinate system.

Suppose that the cartesian coordinates (x^1, x^2, x^3) are related to a given set of generalized coordinates (u^1, u^2, u^3) by the set of independent functional equations

$$x^1 = x^1 (u^1, u^2, u^3)$$
$$x^2 = x^2 (u^1, u^2, u^3)$$
$$x^3 = x^3 (u^1, u^2, u^3)$$

The differentials of the cartesian coordinates are expressed in terms of the differentials of the generalized coordinates by

$$dx^i = \frac{\partial x^i}{\partial u^j} du^j \tag{3-16}$$

Now,

$$ds^2 = g_{ij} du^i du^j = \delta_{mn} dx^m dx^n$$

If we substitute the differentials (3-16) and add, we obtain

$$ds^2 = \left[\frac{\partial x^1}{\partial u^i}\frac{\partial x^1}{\partial u^j} + \frac{\partial x^2}{\partial u^i}\frac{\partial x^2}{\partial u^j} + \frac{\partial x^3}{\partial u^i}\frac{\partial x^3}{\partial u^j}\right] du^i du^j$$

From this, we immediately identify the first metric coefficients

$$g_{ij} = \frac{\partial x^1}{\partial u^i}\frac{\partial x^1}{\partial u^j} + \frac{\partial x^2}{\partial u^i}\frac{\partial x^2}{\partial u^j} + \frac{\partial x^3}{\partial u^i}\frac{\partial x^3}{\partial u^j} \tag{3-17}$$

The calculation of the second metric coefficients is somewhat more complicated, and will be postponed until we consider the tensor formalism.

EXAMPLE 3. The calculation of the differential elements of length, area and volume is relatively simple if the unitary vectors $(\mathbf{e}_1, \mathbf{e}_2, \mathbf{e}_3)$ are mutually orthogonal. We shall illustrate this for the case of the spherical coordinate system considered in Example 1. For ready references, we again write the metric coefficients which were calculated in the example

$$g_{11} = 1,\ g_{22} = u^2,\ g_{33} = u^2 \sin^2 v,\ g_{12} = g_{21} = g_{13} = g_{31} = g_{23} = g_{32} = 0$$

It then the follows that the differential elements of length along the coordinate lines are

$$ds_1 = \sqrt{g_{11}}\, du = du$$
$$ds_2 = \sqrt{g_{22}}\, dv = u\, dv$$
$$ds_3 = \sqrt{g_{33}}\, dw = u \sin v\, dw$$

From Eq. (3-12) we have

$$da_1=\sqrt{g_{22}\,g_{33}-g_{23}^2}\;dv\;dw=u^2\sin v\;dv\;dw$$

$$da_2=\sqrt{g_{11}\,g_{33}-g_{13}^2}\;du\;dw=u\sin v\;du\;dw$$

$$da_3=\sqrt{g_{11}\,g_{22}-g_{12}^2}\;du\;dv=u\;du\;dv$$

The determinant of the metric coefficients is

$$g=\begin{vmatrix}1 & 0 & 0\\ 0 & u^2 & 0\\ 0 & 0 & u^2\sin^2 v\end{vmatrix}=u^4\sin^2 v$$

It then follows that

$$dV=\sqrt{g}\;du dv dw=u^2\sin v\;du\;dv\;dw$$

EXAMPLE 4. Although the computation of line, area and volume elements is somewhat more complicated when the unitary vectors are not orthogonal, it is still straightforward. Consider the coordinate system defined in Example 2 of Section 3-1,

$$x=a\,u+b\,v,\quad 0\leqslant u\leqslant 1,\ 0\leqslant v\leqslant 1$$

$$y=a\sqrt{1-u^2}+b\sqrt{1-v^2}$$

$$z=w$$

The set of metric coefficients are

$$g_{11}=\frac{a}{1-u^2},\ g_{12}=g_{21}=ab\left(1+\frac{uv}{\sqrt{(1-u^2)\,(1-v^2)}}\right),$$

$$g_{22}=\frac{b^2}{1-v^2},\ g_{33}=1,\ g_{13}=g_{31}=g_{23}=g_{32}=0$$

Then,

$$ds_1=\sqrt{g_{11}}\;du=a(1-u^2)^{-1/2}\;du$$

$$ds_2=\sqrt{g_{22}}\;dv=b(1-v^2)^{-1/2}\;dv$$

$$ds_3=\sqrt{g_{33}}\;dw=dw$$

The differential elements of area in the coordinate surfaces are

$$da_1=\sqrt{g_{22}\,g_{33}-g_{23}^2}\;dv\;dw=b(1-v^2)^{-1/2}\;dvdw$$

$$da_2=\sqrt{g_{11}\,g_{33}-g_{13}^2}\;du\;dw=a(1-u^2)^{-1/2}\;dudw$$

$$da_3=\sqrt{g_{11}\,g_{22}-g_{12}^2}\;dudv$$

$$=ab\left[\frac{(u^2+v^2)-2u^2v^2-2uv\,(1-u^2)\,(1-v^2)}{(1-u^2)\,(1-v^2)}\right]^{1/2}dudv$$

In order to obtain the differential volume element, we first calculate the determinant of the metric coefficients

$$g=\begin{vmatrix} \dfrac{a^2}{1-u^2} & ab\left(1+\dfrac{uv}{\sqrt{(1-u^2)(1-v^2)}}\right) & 0 \\ ab\left(1+\dfrac{uv}{\sqrt{(1-u^2)(1-v^2)}}\right) & \dfrac{b^2}{1-v^2} & 0 \\ 0 & 0 & 1 \end{vmatrix}$$

$$=a^2b^2\left[\frac{(u^2+v^2)-2u^2v^2-2uv\sqrt{(1-u^2)(1-v^2)}}{(1-u^2)(1-v^2)}\right]$$

It then follows immediately that

$$dV=ab\left[\frac{(u^2+v^2)-2u^2v^2-2uv\sqrt{(1-u^2)(1-v^2)}}{(1-u^2)(1-v^2)}\right]^{1/2} du\,dv\,dw$$

3-3. The Differential Operators in Generalized Coordinates. Now that we have obtained a basis for the representation of vectors in terms of the generalized coordinates (u^1, u^2, u^3), we may turn our attention to finding representations of the differential operators in terms of these generalized coordinates. We shall make use of the general definitions of these operators which we formulated in Chapter 2. These definitions are explicitly independent of the coordinate system used to define the points of space.

The gradient of a scalar function $\rho(u^1, u^2, u^3)$ is a fixed vector which is defined to have the direction and magnitude of the maximum rate of change of ρ with respect to the coordinates. The variation in ρ corresponding to the infinitesimal displacement $d\mathbf{r}$ is

$$d\rho=d\mathbf{r}\cdot\text{grad }\rho=\frac{\partial\rho}{\partial u^i}\,du^i$$

Now, the du^i are the contravariant components of the infinitesimal displacement $d\mathbf{r}$, and hence,

$$du^i=d\mathbf{r}\cdot\mathbf{e}^i$$

Then,

$$\left(\text{grad }\rho-\frac{\partial\rho}{\partial u^i}\,\mathbf{e}^i\right)\cdot d\mathbf{r}=0 \tag{3-18}$$

Since the displacement $d\mathbf{r}$ is arbitrary, it follows that the bracketed term in (3-18) must vanish identically. Thus, the representation of the gradient of a scalar field in terms of the generalized coordinates (u^1, u^2, u^3) is

$$\text{grad }\rho=\frac{\partial\rho}{\partial u^i}\,\mathbf{e}^i \tag{3-19}$$

The representation (3-19) is in terms of the reciprocal unitary vectors, and hence, the quantities $(\partial\rho/\partial u^i)$ represent the covariant components of the gradient of ρ. We may obtain a representation in terms of the unitary vectors from the relations

$$\mathbf{e}^i = g^{ij}\,\mathbf{e}_j$$

If we substitute these relations in (3-19), we obtain the representation

$$\text{grad}\,\rho = g^{ij}\frac{\partial\rho}{\partial u^j}$$

We then identify the contravariant components of grad ρ as

$$(\text{grad}\,\rho)^i = g^{ij}\frac{\partial\rho}{\partial u^j} \tag{3-20}$$

The representation of the divergence of the vector field $\mathbf{a}(u^1, u^2, u^3)$ in terms of the generalized coordinates is easily obtained from the definition

$$\text{div}\,\mathbf{a}(u^1, u^2, u^3) = \lim_{\Delta v \to 0} \frac{\int_{\Delta s} \mathbf{a}\cdot\mathbf{n}\,ds}{\Delta v}$$

We shall evaluate the surface integral over the surface bounding the volume shown in Fig. 3-4. The volume is bounded by the coordinate surfaces $u^1 - du^1$, $u^1 + du^1$, $u^2 - du^2$, $u^2 + du^2$, $u^3 - du^3$, $u^3 + du^3$. Consider the contribution to the surface integral from the two ends of the volume element which lie in u^1-surfaces. At the face $u^1 + du^1$, we have

$$\mathbf{n}\,ds = 4\,(\mathbf{e}_2 \times e_3)\,du^2\,du^3$$

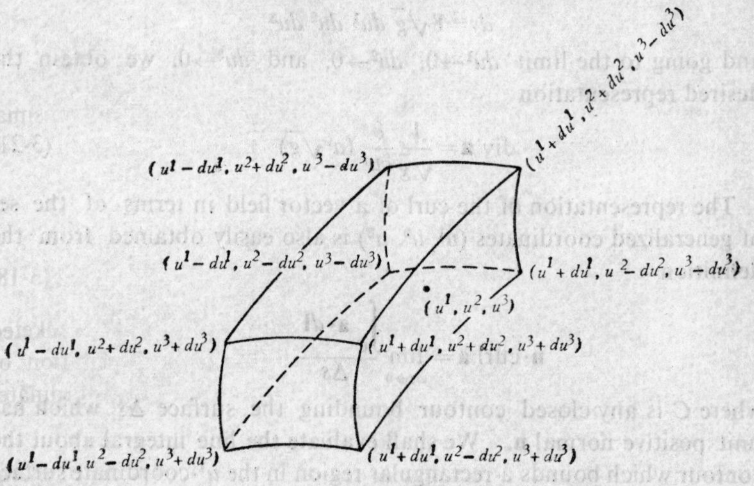

Figure 3-4.

while at the face $u^1 - du^1$,

$$\mathbf{n}\, ds = 4\,(\mathbf{e}_3 \times \mathbf{e}_2)\, du^2\, du^3$$

The contribution to the surface integral from these two faces is

$$4\,[\mathbf{a}\cdot(\mathbf{e}_2 \times \mathbf{e}_3)\, du^2\, du^3]_{u^1+du^1} + 4\,[\mathbf{a}\cdot(\mathbf{e}_3 \times \mathbf{e}_2)\, du^3\, du^2]_{u^1-du^1}$$

where the subscripts indicate that the quantity in brackets is to be evaluated at the point indicated. We expand both terms in Taylor's series about u^1 and retain only the linear terms to obtain

$$4\,[\mathbf{a}\cdot(\mathbf{e}_2 \times \mathbf{e}_3)]_{u^1}\, du^2\, du^3 + 4\left\{\frac{\partial}{\partial u^1}[\mathbf{a}\cdot(\mathbf{e}_2 \times \mathbf{e}_3)]\right\}_{u^1} du^1 du^2 du^3 -$$

$$4\,[\mathbf{a}\cdot(\mathbf{e}_2 \times \mathbf{e}_3)]_{u^1}\, du^2 du^3 + 4\left\{\frac{\partial}{\partial u^1}\,[\mathbf{a}\cdot(\mathbf{e}_2 \times \mathbf{e}_3)]\right\}_{u^1} du^1 du^2 du^3$$

We note that

$$\mathbf{a}\cdot(\mathbf{e}_2 \times \mathbf{e}_3) = (\mathbf{a}\cdot\mathbf{e}^i)\,(\mathbf{e}_i\cdot\mathbf{e}_2 \times \mathbf{e}_3) = (\mathbf{a}\cdot\mathbf{e}^1)\,(\mathbf{e}_1\cdot\mathbf{e}_2 \times \mathbf{e}_3)$$

But, $a\cdot\mathbf{e}^1 = a^1$, and $\mathbf{e}_1\cdot\mathbf{e}_2 \times \mathbf{e}_3 = \sqrt{g}$, and hence

$$\mathbf{a}\cdot(\mathbf{e}_2 \times \mathbf{e}_3) = a^1\sqrt{g}$$

The contribution to the surface integral from the faces of the volume element at $u^1 + du^1$ and $u^1 - du^1$ is

$$8\,\frac{\partial}{\partial u^1}\,(a^1\sqrt{g})\, du^1 du^2 du^3$$

There are analogous contributions from the remaining two pairs of faces. Adding all of the contributions to the surface integral dividing by

$$dv = 8\sqrt{g}\, du^1\, du^2\, du^3$$

and going to the limit $du^1 \to 0$, $du^2 \to 0$, and $du^3 \to 0$, we obtain the desired representation

$$\operatorname{div} \mathbf{a} = \frac{1}{\sqrt{g}}\frac{\partial}{\partial u^i}\,(a^i\sqrt{g}) \tag{3-21}$$

The representation of the curl of a vector field in terms of the set of generalized coordinates (u^1, u^2, u^3) is also easily obtained from the definition

$$\mathbf{n}\cdot\operatorname{curl} \mathbf{a} = \lim_{\Delta s \to 0} \frac{\int_C \mathbf{a}\cdot d\mathbf{l}}{\Delta s}$$

where C is any closed contour bounding the surface Δs, which has unit positive normal $\mathbf{n}$. We shall evaluate the line integral about the contour which bounds a rectangular region in the u^1-coordinate surface as shown in Fig. 3-5. The direction of the contour has been chosen so that the positive normal is in the direction of the positive u^1-coordinate curve. The contribution to the line integral from the sides which are parallel to the u^2-coordinate curve is

$$-[(\mathbf{a}\cdot\mathbf{e}_2)\,du^2]_{u^3+du^3} + [(\mathbf{a}\cdot\mathbf{e}_2)\,du^2]_{u^3-du^3} \qquad (3\text{-}22)$$

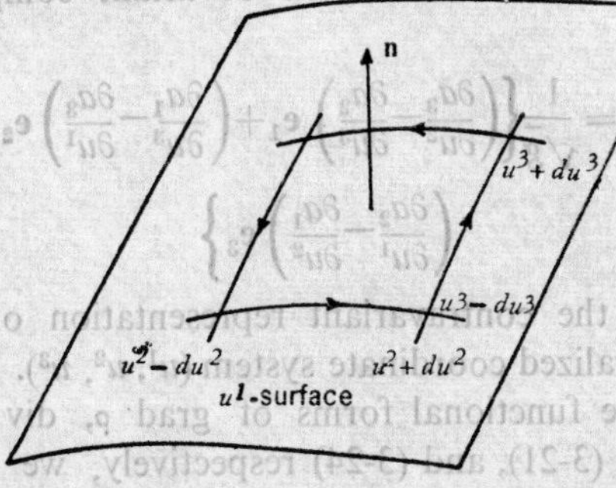

Figure 3-5.

Similarly, the sides parallel to the u^3-coordinate curves provide a contribution

$$[(\mathbf{a}\cdot\mathbf{e}_3)\,du^3]_{u^2+du^2} + [(\mathbf{a}\cdot\mathbf{e}_3)\,du^3]_{u^2-du^2} \qquad (3\text{-}23)$$

Approximating (3-22) and (3-23) by the linear terms of their Taylor series we find that the line integral is approximated by

$$4\left(\frac{\partial}{\partial u^2}(a\cdot e_3) - \frac{\partial}{\partial u^3}(a\cdot e_2)\right) du^2 du^3$$

This quantity must now be divided by the area of the rectangular region which is

$$\Delta s = 4\sqrt{(\mathbf{e}_2\times\mathbf{e}_3)\cdot(\mathbf{e}_2\times\mathbf{e}_3)}\;du^2\,du^3$$

The reciprocal unitary vector $\mathbf{e}^1$ is always orthogonal to the u^1-coordinate surface, so that the unit normal to the open surface is

$$\mathbf{n} = \frac{\mathbf{e}^1}{\sqrt{\mathbf{e}^1\cdot\mathbf{e}^1}}$$

Then,

$$(\text{curl}\,\mathbf{a})\cdot\mathbf{n} = (\text{curl}\,\mathbf{a})\cdot(\mathbf{e}^1/\sqrt{\mathbf{e}^1\cdot\mathbf{e}^1})$$

$$= \frac{1}{\sqrt{(\mathbf{e}_2\times\mathbf{e}_3)\cdot(\mathbf{e}_2\times\mathbf{e}_3)}}\left[\frac{\partial}{\partial u^2}(\mathbf{a}\cdot\mathbf{e}_3) - \frac{\partial}{\partial u^3}(\mathbf{a}\cdot\mathbf{e}_2)\right]$$

As we have previously seen,

$$\mathbf{e}_2\times\mathbf{e}_3 = \sqrt{g}\,\mathbf{e}^1$$

and thus,

$$(\text{curl } \mathbf{a})^1 = (\text{curl } \mathbf{a}) \cdot \mathbf{e}^1$$
$$= \frac{1}{\sqrt{g}} \left[\frac{\partial}{\partial u^2} (\mathbf{a} \cdot \mathbf{e}_3) - \frac{\partial}{\partial u^3} (\mathbf{a} \cdot \mathbf{e}_2) \right]$$

The other components of curl **a** may be obtained by a cyclic permutation of indices. Since $\mathbf{a} \cdot \mathbf{e}^i$ are the covariant components of the vector **a**, we have

$$\text{curl } \mathbf{a} = \frac{1}{\sqrt{g}} \left\{ \left(\frac{\partial a_3}{\partial u^2} - \frac{\partial a_2}{\partial u^3} \right) \mathbf{e}_1 + \left(\frac{\partial a_1}{\partial u^3} - \frac{\partial a_3}{\partial u^1} \right) \mathbf{e}_2 + \left(\frac{\partial a_2}{\partial u^1} - \frac{\partial a_1}{\partial u^2} \right) \mathbf{e}_3 \right\} \quad (3\text{-}24)$$

Thus, Eq. (3-24) is the contravariant representation of curl **a** with respect to the generalized coordinate system (u^1, u^2, u^3).

If we compare the functional forms of grad ρ, div **a**, and curl **a**, that is, Eqs. (3-19), (3-21), and (3-24) respectively, we see one of the difficulties in representing the differential operators in terms of the symbolic operator "grad". We note that a different form is required for the operator, depending on whether it is being used to form the gradient of a scalar, the divergence of a vector, or the curl of a vector. The use of the symbolic operator works well in obtaining representations of the differential operators only in terms of cartesian coordinates. This is a result of the particularly simple structure of cartesian coordinates, and the fact that there is no distinction between contravariant, covariant and physical components of a given vector in this system.

The final result which we shall obtain in this section, is a representation of the Laplacian of a scalar field in terms of the generalized coordinates (u^1, u^2, u^3). This representation is obtained by setting $\mathbf{a} = \text{grad } \rho$ in the representation of the divergence. The representation of the divergence requires the contravariant components of the vector field, which are, in the case of grad ρ,

$$(\text{grad } \rho)^i = g^{ij} \frac{\partial \rho}{\partial u^j}$$

If we substitute this result in Eq. (3–21), we obtain the representation

$$\nabla^2 \rho = \frac{1}{\sqrt{g}} \frac{\partial}{\partial u^i} \left(g^{ij} \sqrt{g} \frac{\partial \rho}{\partial u^j} \right) \quad (3\text{-}25)$$

EXAMPLE 5. To illustrate the calculation of the differential operators, we shall again use the spherical coordinate system

$$x = u \sin v \cos w, \; y = u \sin v \sin w, \; z = u \cos v$$

with metric coefficients

$$g_{11}=1,\ g_{22}=u^2,\ g_{33}=u^2 \sin^2 v,\ g_{12}=g_{21}=g_{13}=g_{31}=0$$
$$g_{23}=g_{32}=0,\ g=u^4 \sin^2 v$$

From Eq. (3-19), if ρ is any scalar function of (u, v, w), then

$$\text{grad } \rho=\frac{\partial \rho}{\partial u}\,\mathbf{e}^1+\frac{\partial \rho}{\partial v}\,\mathbf{e}^2+\frac{\partial \rho}{\partial w}\,\mathbf{e}^3$$

or, in terms of the unitary vectors,

$$\text{grad } \rho=\frac{\partial \rho}{\partial u}\,\mathbf{e}_1+\frac{1}{u^2}\frac{\partial \rho}{\partial v}\,\mathbf{e}_2+\frac{1}{u^2 \sin^2 v}\frac{\partial \rho}{\partial w}\,\mathbf{e}_3$$

Let $\mathbf{a}$ be a vector which has the representation

$$\mathbf{a}=a^1\,\mathbf{e}_1+a^2\,\mathbf{e}_2+a^3\,\mathbf{e}_3$$

relative to the unitary vectors of the spherical coordinate system. Then, from Eq. (3-21),

$$\text{div } \mathbf{a}=\frac{1}{\sqrt{g}}\frac{\partial}{\partial u^i}\,(\sqrt{g}\,a^i)$$
$$=\frac{1}{u^2 \sin v}\left[\frac{\partial}{\partial u}\,(a^1\,u^2 \sin v)+\frac{\partial}{\partial v}\,(a^2\,u^2 \sin v)+\frac{\partial}{\partial w}\,(a^3\,u^2 \sin v)\right]$$
$$=\frac{2}{u}\,a^1+\frac{\partial a^1}{\partial u}+\cot v\;a^2+\frac{\partial a^2}{\partial v}+\frac{\partial a^3}{\partial w}$$

Similarly, from Eq. (3-24), we have

$$\text{curl } \mathbf{a}=\frac{1}{\sqrt{g}}\left\{\left[\frac{\partial}{\partial v}\,(g_{33}\,a^3)-\frac{\partial}{\partial w}\,(g_{22}\,a^2)\right]\mathbf{e}_1+\left[\frac{\partial}{\partial w}\,(g_{11}\,a^1)-\frac{\partial}{\partial u}\,(g_{33}\,a^3)\right]\mathbf{e}_2+\left[\frac{\partial}{\partial u}\,(g_{22}\,a^2)-\frac{\partial}{\partial v}\,(g_{11}\,a^1)\right]\mathbf{e}_3\right\}$$
$$=\frac{1}{u^2 \sin v}\left\{\left[\frac{\partial}{\partial v}\,(u^2 \sin^2 v\;a^3)-\frac{\partial}{\partial w}\,(u^2\,a^2)\right]\mathbf{e}_1+\left[\frac{\partial a^1}{\partial w}-\frac{\partial}{\partial u}\,(u^2 \sin^2 v\;a^3)\right]\mathbf{e}_2+\left[\frac{\partial}{\partial u}\,(u^2\,a^2)-\frac{\partial a^1}{\partial v}\right]\mathbf{e}_3\right\}$$

$$\text{curl } \mathbf{a}=\left[2a^3 \cos v+\sin v\,\frac{\partial a^3}{\partial v}-\frac{1}{\sin v}\frac{\partial a^2}{\partial w}\right]\mathbf{e}_1+\left[\frac{\partial a^1}{\partial w}-\frac{2 \sin v}{u}\,a^3-\sin v\,\frac{\partial a^3}{\partial u}\right]\mathbf{e}_2+$$

$$\left[\frac{2a^2}{u \sin v}+\frac{1}{\sin v}\frac{\partial a^2}{\partial u}-\frac{\partial a^1}{\partial v}\right]\mathbf{e}_3$$

Finally, if ρ is a scalar function, it follows from Eq. (3-25) that

$$\nabla^2\rho=\frac{1}{u^2 \sin v}\left[\frac{\partial}{\partial u}\left(u^2 \sin v\, \frac{\partial\rho}{\partial u}\right)+\frac{\partial}{\partial v}\left(u^4 \sin v\, \frac{\partial\rho}{\partial v}\right)+\frac{\partial}{\partial w}\left(\frac{1}{\sin v}\frac{\partial\rho}{\partial w}\right)\right]$$

3-4. Orthogonal Coordinate Systems. Although we have so far imposed no restrictions on the generalized coordinate systems for which our results are valid, there are only a limited number of so-called orthogonal systems which are of practical interest. This results from the fact that most of the physical problems for which vector analysis offers any simplification ultimately result in partial differential equations, and the resulting equations are solvable only if the vectors are expressed in a limited number of coordinate systems.

In this section, we shall specialize the results of the previous sections to the case of orthogonal coordinates. In Appendix 1, we shall list in detail some of the properties of the most frequently encountered orthogonal systems. The restriction to orthogonal systems provides substantial simplification in the theory.

An orthogonal system is defined by the requirement that the unitary vectors ($\mathbf{e}_1$, $\mathbf{e}_2$, $\mathbf{e}_3$) are everywhere mutually orthogonal,

$$\mathbf{e}_i\cdot\mathbf{e}_j=g_{ij}\,\delta_{ij} \tag{3-26}$$

Under these conditions, the reciprocal unitary vector $\mathbf{e}^i$ is parallel to $\mathbf{e}_i$, and is given by

$$\mathbf{e}^i=\frac{1}{\mathbf{e}_i\cdot\mathbf{e}_i}\,\mathbf{e}_i=\frac{1}{g_{ii}}\,\mathbf{e}_i$$

It follows from Eq. (3-26) that the metric coefficients g_{ij} vanish whenever $i\neq j$. For an orthogonal curvilinear coordinate system, it is customary to define a set of scale factors

$$h_i=\sqrt{g_{ii}} \tag{3-27}$$

In the case of an orthogonal system, the set of second metric coefficients are given by

$$g^{ij}=\frac{1}{g_{ij}}=\frac{1}{h_i^2}\,\delta_{ij}$$

From the form of the first metric coefficients, Eq. (3-17), it follows that the scale factors are calculated by

$$h_i^2=\left(\frac{\partial x^1}{\partial u^i}\right)^2+\left(\frac{\partial x^2}{\partial u^i}\right)^2+\left(\frac{\partial x^3}{\partial u^i}\right)^2 \tag{3-28}$$

The elementary cell which is bounded by the coordinate surfaces is the rectangular region bounded by the edges of length

$$ds_1=h_1du^1,\ ds_2=h_2du^2,\ ds_3=h_3du^3$$

The volume of this elementary cell is

$$dv=h_1h_2h_3\ du^1\ du^2\ du^3$$

It is easy to see that this is a special case of the previous result for the elementary volume in a generalized coordinate system, $dv=\sqrt{g}\ du^1\ du^2\ du^3$. In the case of an orthogonal system, all of the off diagonal terms in det (g_{ij}) are zero, and consequently $g=h_1^2 h_2^2 h_3^2$.

In a fixed coordinate system the advantages of using either covariant or contravariant components of a given vector do not appear. It is then convenient to express a given vector **a**, in terms of its physical components relative to the orthogonal unit base $(\mathbf{i}_1, \mathbf{i}_2, \mathbf{i}_3)$, where

$$\mathbf{e}_i=h_i\ \mathbf{i}_i,\ \mathbf{e}^i=\frac{1}{h_i}\ \mathbf{i}_i$$

For an orthogonal system, we have the relatively simple relations between the covariant, contravariant, and physical components of the vector **a**,

$$a^i=\frac{1}{h_i}\ A_i,\ a_i=h_i\ A_i$$

The differential operators relative to an orthogonal curvilinear coordinate system can be immediately deduced as special cases of the general results of Sec. 3-3. It follows from Eq. (3-19) that in an orthogonal system, the gradient of a scalar field has the representation

$$\text{grad}\ \phi=\sum_{j=1}^{3}\frac{1}{h_j}\frac{\partial \rho}{\partial u^j}\ \mathbf{i}_j \tag{3-29}$$

The divergence of a vector field in terms of an orthogonal curvilinear system can be obtained as a special case of Eq. (3-21),

$$\text{div}\ \mathbf{a}=\frac{1}{h_1h_2h_3}\left\{\frac{\partial}{\partial u^1}\ (h_2h_3A_1)+\frac{\partial}{\partial u^2}\ (h_3h_1A_2)+\frac{\partial}{\partial u^3}\ (h_1h_2A_3)\right\} \tag{3-30}$$

Similarly, it follows from (3-24) that in an orthogonal curvilinear system, the curl of a vector field has the representation

$$\text{curl}\ \mathbf{a}=\frac{1}{h_2h_3}\left[\frac{\partial}{\partial u^2}(h_3A_3)-\frac{\partial}{\partial u^3}\ (h_2A_2)\right]\mathbf{i}_1+$$

$$\frac{1}{h_3h_1}\left[\frac{\partial}{\partial u^3}(h_1A_1)-\frac{\partial}{\partial u^1}(h_3A_3)\right]\mathbf{i}_2+$$
$$\frac{1}{h_1h_2}\left[\frac{\partial}{\partial u^1}(h_2A_2)-\frac{\partial}{\partial u^2}(h_1A_1)\right]\mathbf{i}_3 \qquad (3\text{-}31)$$

In Eqs. (3-30) and (3-31), the quantities A_i are the physical components of the vector **a** relative to the coordinate system (u^1, u^2, u^3). We note that Eq. (3-31) is the formal expansion of the determinant

$$\text{curl } \mathbf{a}=\frac{1}{h_1h_2h_3}\begin{vmatrix} h_1\mathbf{i}_1 & h_2\mathbf{i}_2 & h_3\mathbf{i}_3 \\ \dfrac{\partial}{\partial u^1} & \dfrac{\partial}{\partial u^2} & \dfrac{\partial}{\partial u^3} \\ h_1A_1 & h_2A_2 & h_3A_3 \end{vmatrix}$$

Finally, with respect to an orthogonal curvilinear system, the Laplacian of a scalar field has the representation

$$\nabla^2\rho=\frac{1}{h_1h_2h_3}\left\{\frac{\partial}{\partial u^1}\left(\frac{h_2h_3}{h_1}\frac{\partial\rho}{\partial u^1}\right)+\frac{\partial}{\partial u^2}\left(\frac{h_1h_3}{h_2}\frac{\partial\rho}{\partial u^2}\right)+\frac{\partial}{\partial u^3}\left(\frac{h_1h_2}{h_3}\frac{\partial\rho}{\partial u^3}\right)\right\} \qquad (3\text{-}32)$$

EXAMPLE 6. A particle of mass m is constrained to move on the surface of a right circular cylinder of radius r_0. In order to analyze the motion, it is convenient to use the circular cylinder coordinates (r_0, ϕ, z) which are defined by

$$x=r_0\cos\phi$$
$$y=r_0\sin\phi$$
$$z=z$$

where ϕ and z are assumed to be functions of the time t. Since Newton's second law is a vector relation, we need to calculate the components of the acceleration in terms of the cylindrical coordinates. From Appendix 2, the unit vectors for cylindrical coordinates are

$$\mathbf{i}_1=\cos\phi\,\mathbf{i}+\sin\phi\,\mathbf{j}$$
$$\mathbf{i}_2=-\sin\phi\,\mathbf{i}+\cos\phi\,\mathbf{j}$$
$$\mathbf{i}_3=\mathbf{k}$$

Solving the first pair of these equations for the cartesian unit vectors, **i** and **j**,

$$\mathbf{i}=\cos\phi\,\mathbf{i}_1-\sin\phi\,\mathbf{i}_2$$
$$\mathbf{j}=\sin\phi\,\mathbf{i}_1+\cos\phi\,\mathbf{i}_2$$

Now, the displacement vector **r** is given by

$$\mathbf{r}=x\,\mathbf{i}+y\,\mathbf{j}+z\,\mathbf{k}$$
$$=r_0\cos\phi\,(\cos\phi\,\mathbf{i}_1-\sin\phi\,\mathbf{i}_2)+r_0\cos\phi\,(\sin\phi\,\mathbf{i}_1+\cos\phi\,\mathbf{i}_2)+z\,\mathbf{i}_3$$

$$\mathbf{r}=r_0\,\mathbf{i}_1+z\,\mathbf{i}_3$$

Then, by definition,

$$\mathbf{v}=\frac{d\mathbf{r}}{dt}=r_0\frac{d\mathbf{i}_1}{dt}+\frac{dz}{dt}\,\mathbf{i}_3$$

However,

$$\frac{d\mathbf{i}_1}{dt}=\frac{d}{dt}(\cos\phi\,\mathbf{i}+\sin\phi\,\mathbf{j})=(-\sin\phi\,\mathbf{i}+\cos\phi\,\mathbf{j})\frac{d\phi}{dt}$$

$$=\frac{d\phi}{dt}\,\mathbf{i}_2$$

and hence,

$$\mathbf{v}=r_0\frac{d\phi}{dt}\,\mathbf{i}_2+\frac{dz}{dt}\,\mathbf{i}_3$$

The acceleration is defined to be

$$\mathbf{a}=\frac{d\mathbf{v}}{dt}=r_0\frac{d^2\phi}{dt^2}\,\mathbf{i}_2+r_0\frac{d\phi}{dt}\frac{d\mathbf{i}_2}{dt}+\frac{d^2z}{dt^2}\,\mathbf{i}_3$$

However,

$$\frac{d\mathbf{i}_2}{dt}=\frac{d}{dt}(-\sin\phi\,\mathbf{i}+\cos\phi\,\mathbf{j})=\frac{d\phi}{dt}(-\cos\phi\,\mathbf{i}-\sin\phi\,\mathbf{j})$$

$$=-\frac{d\phi}{dt}\,\mathbf{i}_1$$

Substituting this result in the expression for the acceleration, we obtain

$$\mathbf{a}=-r_0\left(\frac{d\phi}{dt}\right)^2\mathbf{i}_1+r_0\frac{d^2\phi}{dt^2}\,\mathbf{i}_2+\frac{d^2z}{dt^2}\,\mathbf{i}_3$$

The physical components of the acceleration are then identified as

$$a_1=r_0\,(d\phi/dt)^2$$
$$a_2=d^2\phi/dt^2$$
$$a_3=d^2z/dt^2$$

EXAMPLE 7. The basic equation governing the behaviour of a quantum mechanical particle is the time independent Schrödinger equation

$$\nabla^2\psi+\frac{2m}{\hbar^2}(E-V)\,\psi=0$$

where m is the mass of the particle, E its total energy, V its potential energy, and $\hbar$ is Planck's constant divided by 2π. The dependent variable ψ is known as the wave function for the particle, and is related to the probability of finding the particle at a particular point in space.

One model of the hydrogen atom consists of a massive central nucleus with a positive electric charge e, and a much less massive electron with electric charge $-e$ in equilibrium motion under the electrostatic force between the two particles. The potential energy V of the electron, measured in MKS units, is the electrostatic potential V between two equal and opposite charges separated by a distance r

$$V=-\frac{e^2}{4\pi\varepsilon_0 r}$$

Since the mass of the nucleus is much greater than that of the electron ($m_n=1836\ m_e$), we can, in the first approximation, regard the nucleus as being fixed at the origin. Then, the potential energy of the electron depends only on the distance of the electron from the origin. This suggests that some simplification will result from expressing the wave function as a function of the spherical coordinates (r, θ, φ) and solving the Schrödinger equation in terms of these coordinates. We have previously calculated an expression for the scalar Laplacian operator in spherical coordinates (Example 5 of Sec. 3-3). Hence, the Schrödinger equation for the wave function of the electron in the hydrogen atom is

$$\frac{1}{r^2}\frac{\partial}{\partial r}\left(r^2\frac{\partial\psi}{\partial r}\right)+\sin\theta\frac{\partial}{\partial\theta}\left(\sin\theta\frac{\partial\psi}{\partial\theta}\right)+\frac{1}{r^2\sin^2\theta}\frac{\partial^2\psi}{\partial\phi^2}+\frac{2m}{\hbar^2}\left(\frac{e^2}{4\pi\varepsilon_0 r}+E\right)\psi=0$$

EXAMPLE 8. One of the important operators in quantum mechanics is the operator corresponding to the angular momentum **L**, which is defined to be

$$\hat{\mathbf{L}}=-i\,(\mathbf{r}\times\text{grad})$$

where **r** is the coordinate operator with cartesian components (x, y, z).

In cartesian coordinates, the angular momentum operator $\hat{\mathbf{L}}$ has components

$$\hat{L}_x=-i\left(y\frac{\partial}{\partial z}-z\frac{\partial}{\partial y}\right)$$

$$\hat{L}_y=-i\left(z\frac{\partial}{\partial x}-x\frac{\partial}{\partial z}\right)$$

$$\hat{L}_z=-i\left(x\frac{\partial}{\partial y}-y\frac{\partial}{\partial x}\right)$$

Of particular interest are the operator $\hat{L}_z$, and the two linear combinations

$$\hat{L}_{+}=\hat{L}_x+i\,\hat{L}_y$$
$$\hat{L}_{-}=\hat{L}_x-i\,\hat{L}_y$$

These two linear combinations are known as the raising and lowering operators respectively.

We frequently require expressions for the raising and lowering operators and the operator $\hat{L}_z$ in spherical coordinates. The operator $\hat{L}_z$ corresponds, of course, to the component of angular momentum along the polar axis. In spherical coordinate (see Appendix 5)

$$\mathbf{r}=r\,\mathbf{i}_1$$

$$\text{grad}=\mathbf{i}_1\frac{\partial}{\partial r}+\mathbf{i}_2\frac{1}{r}\,\frac{\partial}{\partial\theta}+\mathbf{i}_3\,\frac{1}{r\sin\theta}\,\frac{\partial}{\partial\varphi}$$

and hence,

$$\hat{\mathbf{L}}=-i\,(\mathbf{r}\times\text{grad})=i\left(\mathbf{i}_2\,\frac{1}{\sin\theta}\,\frac{\partial}{\partial\varphi}-\mathbf{i}_3\,\frac{\partial}{\partial\theta}\right)$$

However,

$$\mathbf{i}_2=\cos\theta\,\cos\varphi\,\mathbf{i}+\cos\theta\,\sin\varphi\,\mathbf{j}-\sin\theta\,\mathbf{k}$$
$$\mathbf{i}_3=-\sin\varphi\,\mathbf{i}+\cos\varphi\,\mathbf{j}$$

where (**i**,**j**,**k**) are the cartesian unit vectors. It then follows that

$$\hat{\mathbf{L}}=-i\Bigg[\left(-\cot\theta\cos\varphi\,\frac{\partial}{\partial\varphi}-\sin\varphi\,\frac{\partial}{\partial\theta}\right)\mathbf{i}+\left(-\cot\theta\sin\varphi\,\frac{\partial}{\partial\varphi}+\cos\varphi\,\frac{\partial}{\partial\theta}\right)\mathbf{j}+\frac{\partial}{\partial\varphi}\,\mathbf{k}\Bigg]$$

Then,

$$\hat{L}_{+}=\hat{L}_x+i\,\hat{L}_y=(\cos\varphi+i\sin\varphi)\left(i\cot\theta\,\frac{\partial}{\partial\varphi}+\frac{\partial}{\partial\theta}\right)$$
$$=e^{i\varphi}\left(\frac{\partial}{\partial\theta}+i\cot\theta\,\frac{\partial}{\partial\varphi}\right)$$

$$\hat{L}_{-}=\hat{L}_x-i\,\hat{L}_y=-(\cos\varphi-i\sin\varphi)\left(-i\cot\theta\,\frac{\partial}{\partial\varphi}+\frac{\partial}{\partial\varphi}\right)$$
$$=-e^{-i\varphi}\left(\frac{\partial}{\partial\theta}+i\cot\theta\,\frac{\partial}{\partial\varphi}\right)$$

$$\hat{L}_z=-i\,\frac{\partial}{\partial\varphi}$$

EXAMPLE 9. For the class of physical problems involving

two force centres, the use of prolate spheroidal coordinates simplifies the analysis. As a concrete example of this type of problem, we consider the problem of the hydrogen molecule ion. This ion is a system consisting of two protons which are separated by a fixed distance $2a$ and an electron moving in the electrostatic field of the two protons. If we label the protons 1 and 2, we can denote the distance of the electron from proton 1 by r_1; its distance from proton 2 by r_2; and the interproton distance by $r_{12}=2a$. Introducing a rectangular cartesian system with origin at the midpoint between the two protons and with the axis oriented so that the two protons are along the z-axis (see Fig. 3-6), we have

$$r_1=[x^2+y^2+(a-z)^2]^{1/2}$$
$$r_2=[x^2+y^2+(a+z)^2]^{1/2}$$

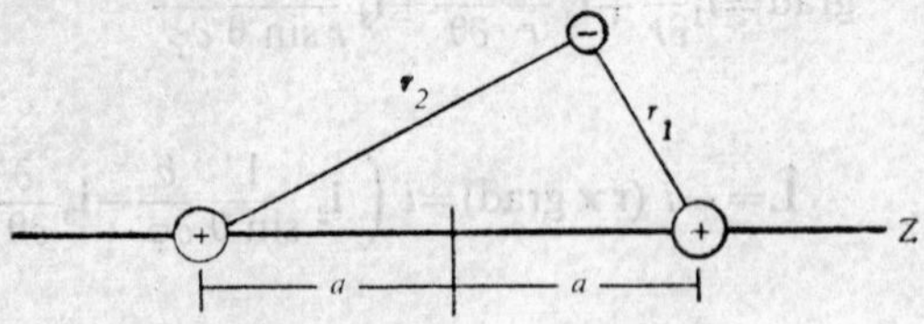

Figure 3-6.

This Schrödinger equation for the system is

$$\nabla^2\psi+\frac{2m}{\hbar^2}\left[E+\frac{e^2}{4\pi\varepsilon_0 r_1}+\frac{e^2}{4\pi\varepsilon_0 r_2}-\frac{e^2}{4\pi\varepsilon_0 r_{12}}\right]\psi=0$$

We introduce prolate spheroidal coordinates by the transformation equation (see Appendix 8)

$$x=a\,[(u^2-1)\,(1-v^2)]^{1/2}\cos w$$
$$y=a\,[(u^2-1)\,(1-v^2)]^{1/2}\sin w$$
$$z=auv$$

Substituting these transformation equations in the definitions of r_1 and r_2, we obtain

$$\begin{aligned}r_1&=a\,[(u^2-1)\,(1-v^2)\,(\cos^2 w+\sin^2 w)+1-2uv+u^2v^2]^{1/2}\\&=a\,(u-v)\end{aligned}$$
$$\begin{aligned}r_2&=a\,[(u^2-1)\,(1-v^2)\,(\cos^2 w+\sin^2 w)+1+2uv+u^2v^2]^{1/2}\\&=a\,(u+v)\end{aligned}$$

Then,

$$\frac{e^2}{4\pi\varepsilon_0}\left(\frac{1}{r_1}+\frac{1}{r_2}\right)=\frac{e^2}{4\pi\varepsilon_0 a}\left(\frac{1}{u-v}+\frac{1}{u+v}\right)$$

$$=\frac{2e^2}{4\pi\varepsilon_0 a}\frac{u}{u^2-v^2}$$

Using the explicit form for the scalar Laplacian operator in prolate spheroidal coordinates, the Schrödinger equation can be written

$$\frac{1}{u^2-v^2}\left\{\frac{\partial}{\partial u}\left[(u^2-1)\frac{\partial\psi}{\partial u}\right]+\frac{\partial}{\partial v}\left[(1-v^2)\frac{\partial\psi}{\partial v}\right]\right\}+\frac{1}{(u^2-1)(1-v^2)}\frac{\partial^2\psi}{\partial w^2}+$$
$$\frac{2ma}{\hbar^2}\left[\frac{2e^2}{4\pi\varepsilon_0}\frac{u}{u^2-v^2}+E'\right]\psi=0$$

where

$$E'=a\mathrm{E}-\frac{e^2}{8\pi\varepsilon_0}$$

This form of the Schrödinger equation can be solved by the method of separation of variables.

EXAMPLE 10. When a hydrogen atom is placed in an electric field the electron experiences an additional potential due to its interaction with the electric field. This is experimentally observed as a small shift in the wavelengths of the characteristic radiation from hydrogen and is known as the Stark effect.

If we choose the origin of the coordinate system at the proton, which we assume to be fixed, and orient the coordinate axes so that the z-axis is along the constant electric field E_0, the additional potential energy of the electron is

$$V'=-eE_0 z$$

In this case the Schrödinger equation for the electrons is

$$\nabla^2\psi+\frac{2m}{\hbar^2}\left[E+\frac{e^2}{4\pi\varepsilon_0 r}+eE_0 z\right]\psi=0$$

The presence of the z-dependent term destroys the spherical symmetry and limits the utility of spherical coordinates in the analysis of this problem. If we introduce parabolic coordinates (Appendix 7) by the transformation equations

$$x=(uv)^{1/2}\cos w$$
$$y=(uv)^{1/2}\sin w$$
$$z=\frac{u-v}{2}$$

we have

$$r=(x^2+y^2+z^2)^{1/2}=\left[uv\cos^2 w+uv\sin^2 w+\frac{u^2-2uv+v^2}{4}\right]^{1/2}$$

$$=\frac{u+v}{2}$$

In terms of this coordinate system, the Schrödinger equation has the explicit form

$$\frac{4}{u+v}\left[\frac{\partial}{\partial u}\left(u\frac{\partial\psi}{\partial u}\right)+\frac{\partial}{\partial v}\left(v\frac{\partial\psi}{\partial v}\right)\right]+\frac{1}{uv}\frac{\partial^2\psi}{\partial w^2}+$$
$$\frac{2m}{\hbar^2}\left[E+\frac{2e^2}{4\pi\varepsilon_0}\frac{1}{u+v}+\frac{eE_0}{2}(u-v)\right]\psi=0$$

The Schödinger equation in this form can be easily solved by the method of separation of variables.

PROBLEMS

3-1. Determine (a) the constant coordinate surfaces, (b) the unitary vectors, and (c) the reciprocal unitary vectors for the coordinate systems defined by the following transformations:

(i) Bispherical coordinates:

$$x=\frac{aw\sqrt{1-v^2}}{u-v},\quad 1\leqslant u<\infty$$

$$y=\frac{a\sqrt{(1-v^2)(1-w^2)}}{w-v},\quad -1\leqslant v\leqslant 1$$

$$z=\frac{a\sqrt{u^2-1}}{u-v},\quad -1\leqslant w\leqslant 1$$

(ii) Exponential coordinates:

$$u=\ln(x^2+y^2)-z$$
$$v=\tfrac{1}{2}(x^2+y^2)+z,\quad x^2+y^2\neq 0$$
$$w=\tan^{-1}(y/x)$$

3-2. Determine the metric coefficients for the coordinate systems given in Problem 3-1.

3-3. Find the elements of arclength, surface area, and volume in the coordinate system defined by

$$u=\ln x$$
$$v=\ln y$$
$$w=z$$

3-4. Show that the parabolic cylinder coordinates

$$x=\frac{u-v}{2}$$
$$y=\sqrt{uv}$$
$$z=w$$

are an orthogonal system.

3-5. Calculate grad ρ, div **a**, and curl **a** in terms of the elliptic cylinder coordinates

$$x=a\sqrt{(u^2-1)(1-v^2)}$$
$$y=auv$$
$$z=w$$

3-6. Use the identity

$$\text{curl curl } \mathbf{a}=\text{grad div } \mathbf{a}=\nabla^2 \mathbf{a}$$

to calculate ∇^2 **a** in terms of the circular cylinder coordinates

$$x=\rho \cos \phi$$
$$y=\rho \sin \phi$$
$$z=w$$

Show by direct calculation that the z-component of this result is equal to $\nabla^2 A_z$ where the scalar Laplacian operator is appropriate to the given coordinate system. Discuss the other two components.

3-7. Calculate the components of acceleration with respect to the oblate spheroidal coordinates

$$x=a\,uv \cos w$$
$$y=a\,uv \sin w$$
$$z=a\sqrt{(u^2-1)(1-v^2)}$$

3-8. Show that the transformation

$$r'=r$$
$$\theta'=\pi-\theta$$
$$\varphi'=\pi+\varphi$$

reflects the point whose spherical coordinates are (r, θ, φ) through the origin.

3-9. Express the operator $\hat{\mathbf{L}}=-i\,(\mathbf{r} \times \text{grad})$ in spherical coordinates and show that

(i) $\hat{\mathbf{L}}\times\hat{\mathbf{L}}=i\hat{\mathbf{L}}$

(ii) $\text{grad}=\dfrac{\mathbf{r}}{|\mathbf{r}|}\dfrac{\partial}{\partial r}-i\,\dfrac{\mathbf{r}\times\hat{\mathbf{L}}}{r^2}$

(iii) $r\nabla^2-\text{grad}\left(1+r\,\dfrac{\partial}{\partial r}\right)=i \text{ curl } \hat{\mathbf{L}}$

3-10. A perfectly conducting wire along the z-axis carries a steady electric current I. This results in a vector potential

$$\mathbf{A}=\mathbf{k}\frac{\mu I}{2} \ln (1/\rho)$$

where ρ is the radial distance from the wire. Show that this gives rise to the magnetic induction field

$$\mathbf{B}=\text{curl } \mathbf{A}=\mathbf{i}_\varphi \frac{\mu I}{2\pi\rho}$$

where $\mathbf{i}_\varphi$ is the unit vector along the φ-coordinate in cylindrical coordinates.

3-11. Use prolate spheroidal coordinates to show that

$$I=\frac{1}{\pi a_0^3}\int \exp\,[(r_1+r_2)/a_0]\; dv$$

$$=\left(1+\frac{2a}{a_0}+\frac{4a^2}{3a_0^2}\right)e^{-2a/a_0}$$

where the integral is over all space, and

$$r_1=\sqrt{x^2+y^2+(a-z)^2}$$

$$r_2=\sqrt{x^2+y^2+(a+z)^2}$$

3-12. In physical problems involving two cylindrical axes, there is some advantage in introducing the bipolar coordinates

$$x=\frac{a\sinh v}{\cosh v-\cos u}$$

$$y=\frac{a\sin u}{\cosh v-\cos u}$$

$$z=w$$

(i) Describe the constant coordinate surfaces.

(ii) Calculate the scale factors.

(iii) Consider three points $(a, 0)$, $(-a, 0)$, (x, y) and let $\mathbf{r}_1$ be the vector from $(a, 0)$ to (x, y) which makes an angle θ_1 with respect to the positive x-axis; $\mathbf{r}_2$ be the vector from $(-a, 0)$ to (x, y) which makes the angle θ_2 with respect to the positive x-axis. Show that

$$u=\theta_1-\theta_2$$

$$v=\ln(|\mathbf{r}_2|/|\mathbf{r}_1|)$$

CHAPTER 4

Linear Transformations

In the previous chapter, we have considered the description of vectors, and the operations of differentiation of vectors in terms of generalized coordinate systems. We have not, however, considered how the components of a given vector change as we go from one coordinate system to another. This problem is one of the fundamental problems of vector analysis. We shall not treat the general problem of determining the behaviour of the components of a given vector as we transform from one general coordinate system to another until we consider the tensor formalism. In this chapter, we shall consider the less general problem of the behaviour of vector under linear transformations from one cartesian system to another.

4-1. Orthogonal Transformations. The most general linear transformation from the set of orthogonal cartesian coordinates (x^1, x^2, x^3) to the set of cartesian coordinates $(\bar{x}^1, \bar{x}^2, \bar{x}^3)$, not necessarily orthogonal, is defined by the set of equations

$$\bar{x}^i = a^i_j\, x^j + b^i \tag{4-1}$$

where the a^i_j and the b^i are real constants. The terms (b^1, b^2, b^3) represent a translation of the origin through the vector $\mathbf{b} = (b^1, b^2, b^3)$. The particular transformation (4-1) is called inhomogeneous, because of the translation terms, and if these terms are absent, the transformation is called homogeneous. We have previously seen, Sec. 1-2, that a vector is invariant under a translation of the origin. We need, therefore, consider only the homogeneous transformation

$$\bar{x}^i = a^i_j\, x^j \tag{4-2}$$

One of the simplest linear transformations, and one of the most useful in the consideration of physical problems, is the orthogonal transformation. This transformation carries one system of orthogonal cartesian coordinates into another system of orthogonal cartesian axes without distortion or change of scale. As we shall see later, the general orthogonal transformation corresponds to a rotation about some axis followed by a reflection of the coordinate axes in the origin. In order to specify an orthogonal transformation, we require the coefficients in Eq. (4-2) to satisfy the auxiliary conditions

$$\sum_{i=1}^{3} a_j^i \, a_k^i = \delta_{jk}; \quad j, k = 1, 2, 3 \tag{4-3}$$

We now wish to investigate the effect of an orthogonal transformation of coordinates on the components of a vector. Let the origin of the original coordinate system be the initial point of the vector **f**, and let the terminal point of **f** be at the point P, whose cartesian coordinates are (x^1, x^2, x^3). It is clear that the cartesian components of **f** relative to this set of coordinates are

$$f^1 = x^1, \quad f^2 = x^2, \quad f^3 = x^3$$

If we now make an orthogonal transformation of coordinates, so that the new coordinates of the point P are $(\bar{x}^1, \bar{x}^2, \bar{x}^3)$, the components of **f** relative to the barred coordinate system are,

$$\bar{f}^1 = \bar{x}^1, \quad \bar{f}^2 = \bar{x}^2, \quad \bar{f}^3 = \bar{x}^3$$

Thus, under an orthogonal transformation of coordinates, the components of a vector transform in exactly the same way as the coordinates. We shall define a cartesian vector as a set of three quantities which, under an orthogonal transformation of coordinates, transform in the same way as the coordinates.

We now show that a cartesian vector is invariant, i.e. has the same magnitude and direction in space, under an orthogonal transformation. Consider the vector **f**, whose components relative to the orthogonal cartesian coordinates (x^1, x^2, x^3) are (f^1, f^2, f^3). The square of the magnitude of **f** is, by definition

$$|\mathbf{f}|^2 = (f^1)^2 + (f^2)^2 + (f^3)^2$$

If we make an orthogonal transformation to the set of orthogonal cartesian coordinates $(\bar{x}^1, \bar{x}^2, \bar{x}^3)$, the components of **f** relative to the barred coordinates are related to the components of **f** in the original system by

$$\bar{f}^i = a_j^i \, f^j \tag{4-4}$$

In terms of the barred coordinates, the magnitude of **f** is given by

$$|\bar{\mathbf{f}}|^2 = (\bar{f}^1)^2 + (\bar{f}^2)^2 + (\bar{f}^3)^2$$

Squaring and adding the expressions (4-4), we obtain

$$|\bar{\mathbf{f}}|^2 = \sum_{i=1}^{3} a^i_j a^i_k f^j f^k$$
$$= (f^1)^2 + (f^2)^2 + (f^3)^2$$

as a consequence of the orthogonality relations (4-3). Hence, the magnitude of the vector $\bar{\mathbf{f}}$ is the same when $\mathbf{f}$ is referred to different orthogonal cartesian systems which are related by an orthogonal transformation. In order to specify the direction of a given vector $\mathbf{f}$ in space, it is customary to specify the direction cosines of the directed line segment representing $\mathbf{f}$ relative to some set of orthogonal cartesian axes. Let the direction angles of $\mathbf{f}$ relative to the set of cartesian coordinates (x^1, x^2, x^3) be α, β, and γ respectively. Then, if $\mathbf{i}$, $\mathbf{j}$, and $\mathbf{k}$ are the set of unit vectors along the three coordinate axes,

$$\cos\alpha = \frac{\mathbf{f}\cdot\mathbf{i}}{|\mathbf{f}|}, \quad \cos\beta = \frac{\mathbf{f}\cdot\mathbf{j}}{|\mathbf{f}|}, \quad \cos\gamma = \frac{\mathbf{f}\cdot\mathbf{k}}{|\mathbf{f}|}$$

or, in terms of the components of $\mathbf{f}$ relative to the coordinate set (x^1, x^2, x^3)

$$\cos\alpha = \frac{f^1}{|f|}, \quad \cos\beta = \frac{f^2}{|f|}, \quad \cos\gamma = \frac{f^3}{|f|}$$

If we make an orthogonal transformation to the coordinate sytem

$$\bar{x}^i = a^i_j x^j$$

the direction cosines of $\mathbf{f}$ relative to the barred coordinates are

$$\cos\bar{\alpha} = \frac{\bar{f}^1}{|\bar{f}|}, \quad \cos\bar{\beta} = \frac{\bar{f}^2}{|\bar{f}|}, \quad \cos\gamma = \frac{\bar{f}^2}{\bar{f}|}$$

It will be left as an exercise to show that the angles $\bar{\alpha}$, $\bar{\beta}$, and $\bar{\gamma}$ describe the same orientation in space as the angles α, β, and γ. The geometrical relations are shown in Fig. 4-1.

An orthogonal transformation of coordinates is completely specified by the set of constant coefficients a^i_j, (which are the direction cosines of the new axes relative to the old), and the orthogonality conditions (4-3). In order to study the mathematical properties of the transformation, we need only consider the array of coefficients

$$\begin{pmatrix} a^1_1 & a^1_2 & a^1_3 \\ a^2_1 & a^2_2 & a^2_3 \\ a^3_1 & a^3_2 & a^3_3 \end{pmatrix}$$

and the orthogonality conditions. A rectangular array of quantities with m rows and n columns is known as an $m \times n$ matrix. We shall consider only arrays with the same number of rows and columns, i.e. square matrices. The individual quantities in the array are called the matrix elements. In particular, we shall refer to the array which represents a linear transformation as the transformation matrix, which will be denoted by (A) or just A. The transformation matrix corresponding to an orthogonal transformation is called an orthogonal matrix.

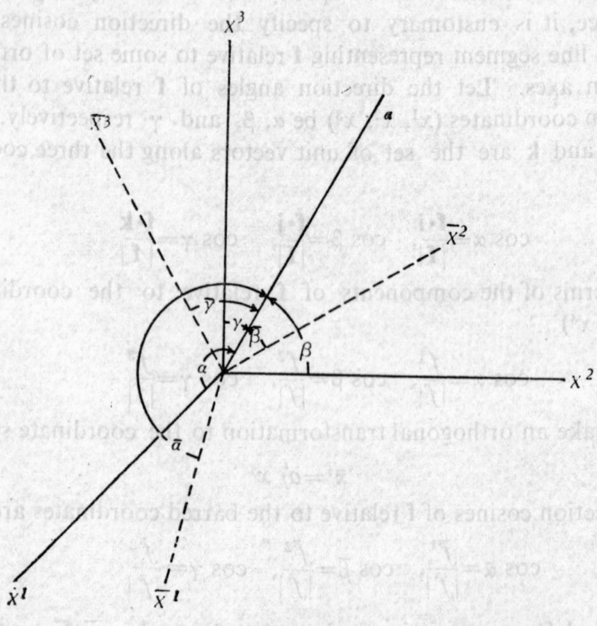

Figure 4-1.

Let us now consider some of the formal properties of the transformation matrix. We shall denote the matrix corresponding to the orthogonal transformation

$$\bar{x}^k = b^k_j\, x^j \tag{4-5}$$

by B. Similarly, we shall denote the second orthogonal transformation

$$y^i = a^i_k\ \bar{x}^k \tag{4-6}$$

by A. A direct relation between the coordinate set (x^1, x^2, x^3) and the set (y^1, y^2, y^3) is obtained by combining Eqs. (4-5) and (4-6)

$$\begin{aligned} y^i &= a^i_k\ b^k_j\ x^j \\ &= c^i_j\ x^j \end{aligned}$$

where

$$c^i_j = a^i_k \ b^k_j \tag{4-7}$$

The matrix C whose elements are given by Eq. (4-7) is called the matrix product of the two matrices A and B, written

$$C = AB$$

The definition of matrix multiplication as expressed by Eq. (4-7) is general, and does not depend on the orthogonality condition. It can be shown that matrix multiplication is associative and distributive, but is not, in general, commutative. Let A and B be matrices which represent orthogonal transformations. Then

$$\begin{aligned}\sum_{i=1}^{3} c^i_j \ c^i_k &= \sum_{i=1}^{3} (a^i_m \ b^m_j)(a^i_n \ b^n_k)\\ &= b^m_j \ b^n_k \sum_{i=1}^{3} a^i_m \ a^i_n\\ &= b^m_j \ b^n_k \ \delta_{mn}\\ &= \sum_{m=1}^{3} b^m_j \ b^m_k = \delta_{jk}\end{aligned}$$

and hence, the matrix C represents an orthogonal transformation. Thus, for orthogonal transformations, we have:

PROPERTY 1. *The successive application of two orthogonal transformations results in an orthogonal transformation. Technically, this property is known as closure.*

Consider the orthogonal transformation

$$\bar{x}^i = a^i_j \, x^j, \ \sum_{i=1}^{3} a^i_j \ a^i_k = \delta_{jk} \tag{4-8}$$

Let A^{-1} be the matrix which represents the inverse transformation to (4-8), and let the elements of A^{-1} be a'^i_j. Then,

$$x^i = a'^i_j \, \bar{x}^j \tag{4-9}$$

Combining Eqs. (4-8) and (4-9), we obtain

$$\begin{aligned}x^i &= a'^i_j \ (a^j_k \, x^k)\\ &= (a'^i_j \ a^j_k) \ x^k\end{aligned}$$

Since the x^i are a set of independent coordinates, the sum on j must vanish for $i \neq k$, and must be equal to unity for $i = k$. Hence

$$a'^i_j \ a^j_k = \delta^i_k \tag{4-10}$$

The δ^i_k are the elements of a matrix known as the identity matrix, I,

$$I=\begin{pmatrix} 1 & 0 & 0 \\ 0 & 1 & 0 \\ 0 & 0 & 1 \end{pmatrix}$$

Hence, we may write

$$A^{-1}\,A=I \tag{4-11}$$

In obtaining Eq. (4-11), we have not made use of the orthogonality conditions, so that the relations defines the inverse of any square matrix if this inverse exists. Now, consider the double sum

$$\sum_{i=1}^{3} a^i_k \; a^i_j \; a'^{\,j}_{\;m}$$

Summing first on the index i,

$$\sum_{i=1}^{3} a^i_k \; a^i_j \; a'^{\,j}_{\;m} = a'^{\,j}_{\;m} \sum_{i=1}^{3} a^i_k \; a^i_j = a'^{\,j}_{\;m} \; \delta^k_j = a'^{\,k}_{\;m}$$

On the other hand, if we sum first with respect to the index j,

$$\sum_{i=1}^{3} a^i_k \, a^i_j \; a'^{\,j}_{\;m} = \sum_{i=1}^{3} a^i_k \, (a^i_j \; a'^{\,j}_{\;m}) = \sum_{i=1}^{3} a^i_k \, \delta^i_m = a^m_k$$

Thus, the elements of the orthogonal matrix A and its inverse A^{-1} are related in a simple way,

$$a'^{\,k}_{\;m} = a^m_k \tag{4-12}$$

Corresponding to every square matrix A there is a number called the determinant of A, which is defined in the usual way

$$\det A = |\,A\,| = \begin{vmatrix} a^1_1 & a^1_2 & a^1_3 \\ a^2_1 & a^2_2 & a^2_3 \\ a^3_1 & a^3_2 & a^3_3 \end{vmatrix}$$

It is easy to show that the determinant of the product of two matrices is equal to the product of the determinants (Prob. 4-3). It then follows from Eq. (4-11) that

$$|\,A^{-1}\,|\,|\,A\,|=1$$

If A is orthogonal, then the elements of A^{-1} are obtained from the elements of A by interchanging rows and columns. However, the value of a determinant is invariant under such an interchange, and it follows that

$$|A\,|^2=1$$

or

$$|\,A\,|=\pm 1, \; A \text{ orthogonal} \tag{4-13}$$

Equation (4-13) may also be used as the defining relation for an orthogonal matrix. If the matrix A represents an orthogonal transformation, and if $|A|=+1$, the transformation is known as a proper transformation or pure rotation. On the other hand, if $|A|=-1$, the transformation is called an improper rotation. The proper rotations correspond to rotations about some axis through the origin, whereas an improper rotation can be decomposed into a proper rotation followed by an inversion of the coordinates through the origin. The two types of transformations are illustrated in Fig. 4-2.

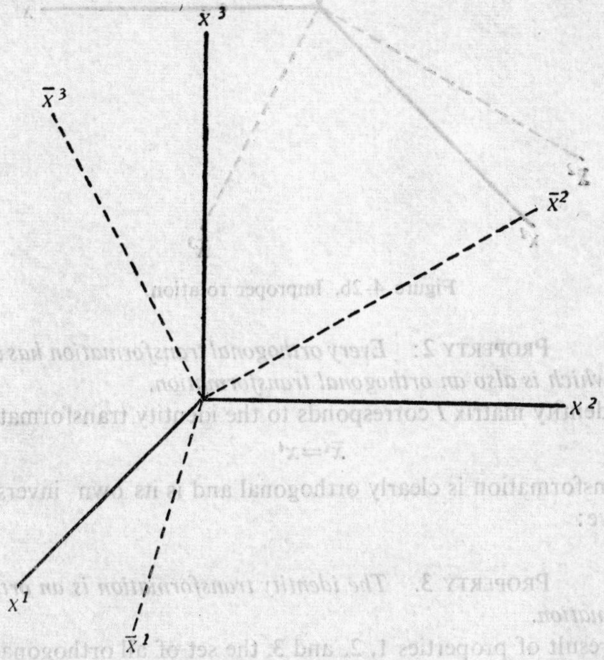

Figure 4-2a. Proper rotation.

The matrix obtained from a given matrix A by interchanging the rows and columns of A is known as the transpose of A, written A^T. Thus, for orthogonal matrices, we have the simple relation

$$A^{-1}=A^T$$

Since the value of a determinant is invariant under an interchange of its rows and columns, it follows that if A represents an orthogonal transformation, then $A^{-1}=A^T$ must also represent an orthogonal transformation. Hence, for orthogonal transformations, we have:

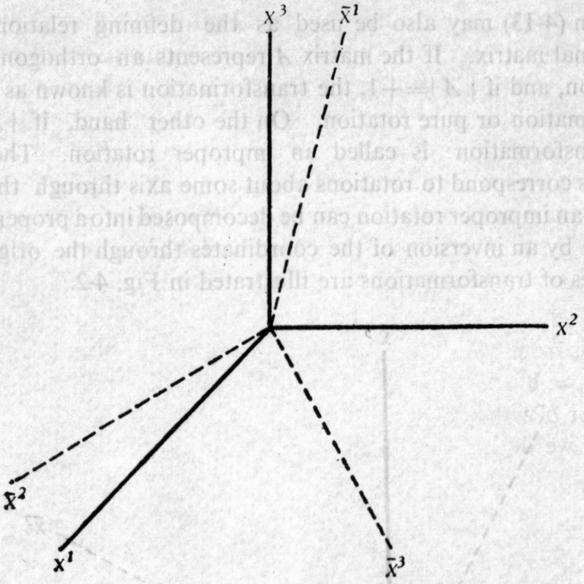

Figure 4–2b. Improper rotation.

PROPERTY 2: *Every orthogonal transformation has a unique inverse which is also an orthogonal transformation.*

The identity matrix I corresponds to the identity transformation

$$\bar{x}^i = x^i$$

This transformation is clearly orthogonal and is its own inverse. We thus have:

PROPERTY 3. *The identity transformation is an orthogonal transformation.*

As a result of properties 1, 2, and 3, the set of all orthogonal transformations forms what is known as a group, and the individual transformations are called the elements of the group. In the particular case we are considering, the group of transformations is known as the three-dimensional orthogonal group. Since there are infinitely many orthogonal transformations, the orthogonal group has infinitely many elements and is said to be an infinite-dimensional group. It is left as an exercise (Prob. 4-5) to show that the set of proper orthogonal transformations is a group, known as the three-dimensional rotation group, but that the set of improper orthogonal transformations is not a group. Since there is a one-to-one correspondence between the three-dimensional orthogonal transformations and the 3×3 orthogonal matrices, the set of 3×3 orthogonal matrices is said to be a representation of the three-dimensional orthogonal group.

There is an alternate interpretation of the orthogonal transformations which is useful in the applications of vector analysis. Suppose that the coordinate system (x^1, x^2, x^3) is rigidly fixed in space. Then, the application of an orthogonal transformation with coefficients a^i_j to the vector **b** with components b^i results in a new vector **b′** with components b'^i, where the two sets of components are related by

$$b'^i = a^i_j \, b^j \tag{4-14}$$

It follows from the orthogonality conditions on the transformation, that $|\mathbf{b}'| = |\mathbf{b}|$. Hence, the effect of the transformation is to rotate the vector b with respect to the fixed set of coordinate axes (see Fig. 4-3). If we denote the vectors **b** and **b′** by the column matrices

$$\mathbf{b} = \begin{pmatrix} b^1 \\ b^2 \\ b^3 \end{pmatrix}, \qquad \mathbf{b}' = \begin{pmatrix} b'^1 \\ b'^2 \\ b'^3 \end{pmatrix} \tag{4-15}$$

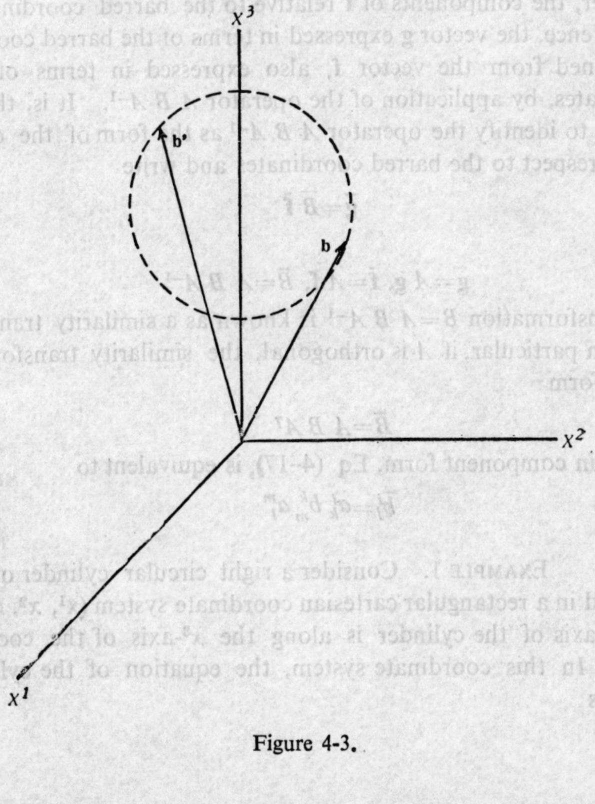

Figure 4-3.

the transformation (4-14) can be written in the form

$$\mathbf{b}' = A\,\mathbf{b}$$

where A is the matrix of the transformation, and multiplication is defined in the usual way. The column matrices defined in Eq. (4-15) are known as column vectors. We now show that the matrix A is a linear operator with respect to column matrices. Let $\mathbf{f}$ and $\mathbf{g}$ be two vectors, and let α and β be two scalar constants. Then, it can be shown by direct calculation that

$$A\,(\alpha\mathbf{f}+\beta\mathbf{g}) = \alpha\,A\,\mathbf{f} + \beta\,A\,\mathbf{g}$$

Now, let B be any 3×3 matrix with real elements and let $\mathbf{f}$ and $\mathbf{g}$ be two vectors such that

$$\mathbf{g} = B\,\mathbf{f}$$

with respect to some coordinate system (x^1, x^2, x^3). If we transform to a coordinate system $(\bar{x}^1, \bar{x}^2, \bar{x}^3)$ by means of the orthogonal transformation A, the components of $\mathbf{g}$ relative to the barred coordinates are given by

$$A\,\mathbf{g} = A\,B\,\mathbf{f} = A\,B\,A^{-1}\,(A\,\mathbf{f})$$

However, the components of $\mathbf{f}$ relative to the barred coordinates are $A\,\mathbf{f}$. Hence, the vector $\mathbf{g}$ expressed in terms of the barred coordinates is obtained from the vector $\mathbf{f}$, also expressed in terms of barred coordinates, by application of the operator $A\,B\,A^{-1}$. It is, therefore, possible to identify the operator $A\,B\,A^{-1}$ as the form of the operator B with respect to the barred coordinates and write

$$\bar{\mathbf{g}} = \bar{B}\,\bar{\mathbf{f}} \tag{4-16}$$

where

$$\bar{\mathbf{g}} = A\,\mathbf{g},\quad \bar{\mathbf{f}} = A\,\mathbf{f},\quad \bar{B} = A\,B\,A^{-1}$$

The transformation $B = A\,B\,A^{-1}$ is known as a similarity transformation. In particular, if A is orthogonal, the similarity transformation has the form

$$\bar{B} = A\,B\,A^{T} \tag{4-17}$$

Written in component form, Eq. (4–17), is equivalent to

$$\bar{b}^i_j = a^i_k\,b^k_m\,a^m_j \tag{4-18}$$

EXAMPLE 1. Consider a right circular cylinder of radius ρ, located in a rectangular cartesian coordinate system (x^1, x^2, x^3) such that the axis of the cylinder is along the x^3-axis of the coordinate system. In this coordinate system, the equation of the cylindrical surface is

$$\phi(x^1, x^2, x^3) = (x^1)^2 + (x^2)^2 - \rho^2 = 0$$

Then, the vector grad ϕ, which has the cartesian components

$$(2x^1, 2x^2, 0)$$

is everywhere normal to the surface of the cylinder.

Now, consider a coordinate system $(\bar{x}^1, \bar{x}^2, \bar{x}^3)$ which is related to the unbarred system by the orthogonal transformation whose matrix is

$$\begin{pmatrix} 1 & 0 & 0 \\ 0 & \sqrt{3}/2 & 1/2 \\ 0 & -1/2 & \sqrt{3}/2 \end{pmatrix}$$

This transformation corresponds to a rotation of the coordinate system through an angle of $\pi/6$ about the x^1-axis of the unbarred system. With respect to barred coordinates, the equation of the surface of the cylinder is

$$\begin{aligned} \bar{\phi}(\bar{x}^1, \bar{x}^2, \bar{x}^3) &= \phi(x^1, x^2, x^3) \\ &= (\bar{x}^1)^2 + \frac{3(\bar{x}^2)^2 + 2\sqrt{3}\bar{x}^2\bar{x}^3 + (\bar{x}^3)^2}{4} - \rho^2 = 0 \end{aligned}$$

In terms of the barred coordinates, the gradient of ϕ has the cartesian components

$$\left(2\bar{x}^1, \frac{3\bar{x}^2 + \sqrt{3}\,\bar{x}^3}{4}, \frac{\sqrt{3}\bar{x}^2 + \bar{x}^3}{4}\right)$$

Since grad ϕ is known to be a cartesian vector, the components relative to the barred coordinates can be obtained directly by applying Eq. (4-4). We immediately obtain

$$\overline{(\text{grad }\phi)}^1 = (\text{grad }\phi)^1 = 2x^1 = 2\bar{x}^1$$

$$\overline{(\text{grad }\phi)}^2 = \frac{\sqrt{3}}{2}(\text{grad }\phi)^2 = \sqrt{3}\,x^2 = \frac{3\bar{x}^2 + \sqrt{3}\,\bar{x}^3}{4}$$

$$\overline{(\text{grad }\phi)}^3 = -\frac{1}{2}(\text{grad }\phi)^2 = -x^2 = \frac{\sqrt{3}\,\bar{x}^2 + \bar{x}^3}{4}$$

as expected.

EXAMPLE 2. Equation (4-4) can be used to determine the vectorial character of a given triplet of functions. As an example, let us consider the triplet of functions

$$f^1 = x^1 - x^2$$

$$f^2 = x^1 + x^2$$

$$f^3 = 0$$

defined in a given cartesian system (x^1, x^2, x^3), and the orthogonal transformation

$$A=\begin{pmatrix} -1/4 & \sqrt{3}/4 & \sqrt{3}/2 \\ -\sqrt{3}/2 & -1/2 & 0 \\ \sqrt{3}/4 & -3/4 & 1/2 \end{pmatrix}$$

If (f^1, f^2, f^3) are the cartesian components of a cartesian vector, then with respect to the barred coordinate system,

$$\bar{f}^1=-\frac{1}{4}f^1+\frac{\sqrt{3}}{4}f^2+\frac{\sqrt{3}}{2}f^3$$

$$=-\frac{1}{4}(x^1-x^2)+\frac{\sqrt{3}}{4}(x^1+x^2)$$

$$\bar{f}^2=-\frac{\sqrt{3}}{2}f^1-\frac{1}{2}f^2$$

$$=-\frac{\sqrt{3}}{2}(x^1-x^2)-\frac{1}{2}(x^1+x^2)$$

$$\bar{f}^3=\frac{\sqrt{3}}{4}f^1-\frac{3}{4}f^2+\frac{1}{2}f^3$$

$$=\frac{\sqrt{3}}{4}(x^1-x^2)-\frac{3}{4}(x^1+x^2)$$

Since A is orthogonal,

$$A^{-1}=\begin{pmatrix} -1/4 & -\sqrt{3}/2 & \sqrt{3}/4 \\ \sqrt{3}/4 & -1/2 & -3/4 \\ \sqrt{3}/2 & 0 & 1/2 \end{pmatrix}$$

and

$$(x^1-x^2)=-\frac{1+\sqrt{3}}{4}\bar{x}^1+\frac{1-\sqrt{3}}{4}\bar{x}^2+\frac{3+\sqrt{3}}{4}\bar{x}^3$$

$$(x^1+x^2)=-\frac{1-\sqrt{3}}{4}\bar{x}^1-\frac{1+\sqrt{3}}{4}\bar{x}^2-\frac{3-\sqrt{3}}{4}\bar{x}^3$$

Hence,

$$\bar{f}^1=\frac{1}{4}\bar{x}^1-\frac{1}{2}\bar{x}^2-\frac{\sqrt{3}}{4}\bar{x}^3$$

$$\bar{f}^2=\frac{1}{2}\bar{x}^1+\bar{x}^2-\frac{\sqrt{3}}{4}\bar{x}^3$$

$$\bar{f}^3=-\frac{\sqrt{3}}{4}\bar{x}^1+\frac{\sqrt{3}}{2}\bar{x}^2+\frac{3}{4}\bar{x}^3$$

Now consider the point whose coordinates with respect to the unbarred system are (1, 1, 1). At this point,

$$(f^1)^2+(f^2)^2+(f^3)^2=4$$

In the barred coordinate system, the coordinates of the given point are

$$\left(\frac{3\sqrt{3}-1}{4}, -\frac{\sqrt{3}+1}{4}, \frac{\sqrt{3}-1}{4}\right)$$

Thus,

$$(\bar{f}^1)^2+(\bar{f}^2)^2+(\bar{f}^3)^2=7-3\sqrt{3}\neq 4$$

Since the sum of the squares of the elements of the triplet is not invariant when these elements are transformed as the components of a cartesian vector, we conclude that the triplet of functions are not the components of a cartesian vector. We have considered only a particular transformation. In the general case, the general orthogonal transformation must be used to test the vectorial properties of a given triplet of functions.

4-2. Euler's Theorem. In the preceeding section, we have indicated that the general homogeneous orthogonal transforma-consists of a rotation about some axis through the origin accompanied by an inversion in the origin. The proof of this proposition was originally given by Euler in connection with the kinematics of rigid body motion. We shall prove the theorem in a slightly different form. If the matrix A, corresponding to a homogeneous orthogonal transformation represents a rotation about some axis, then any vector $\mathbf{R}$ along this axis is invariant under the transformation, i.e.

$$\bar{\mathbf{R}}=A\,\mathbf{R}=\mathbf{R} \tag{4-19}$$

Equation (4-19) represents a special case of what is known as the eigenvalue problem. For any operator A which represents an orthogonal transformation, there are certain vectors $\mathbf{E}$, known as the eigenvectors of A, which are changed only in magnitude by the operator A. For such vectors,

$$A\,\mathbf{E}=a\,\mathbf{E} \tag{4-20}$$

where a is a scalar, either real or complex. The values of the scalar a for which Eq. (4-20) is satisfied are called the eigenvalues of the operator A. Although we shall not consider the eigenvalue problem in general, it is convenient to state Euler's theorem in terms of such a problem. In order to prove our assertion concerning the orthogonal transformation A, it is sufficient to show that any matrix representing a proper orthogonal transformation has one and only one eigenvalue equal to $+1$.. The proof depends on three lemmas.

LEMMA 1: *The eigenvalues of A are all of unit magnitude.*

Although the elements of A are all real, this does not exclude the possibility of complex eigenvalues. The eigenvector $\mathbf{E}$ corresponding to a complex eigenvalue must have components which are complex qualities. In this case, the norm of $\mathbf{E}$ is defined to be

$$|\mathbf{E}|^2 = \mathbf{E}^* \cdot \mathbf{E}$$

where $\mathbf{E}^*$ is the vector whose components are the complex conjugates of the components of E. Since the elements of A are all real, if $\mathbf{E}$ is an eigenvector associated with the complex eigenvalue a such that

$$A\,\mathbf{E} = a\,\mathbf{E}$$

then $\mathbf{E}^*$ must satisfy

$$A\,\mathbf{E}^* = a^*\,\mathbf{E}^*$$

i.e. $\mathbf{E}^*$ is an eigenvector of A with eigenvalue a^*. One of the important properties of orthogonal transformations, which we discussed in Sec. 4-1, is that the norm of a vector is invariant under the orthogonal transformations. Hence,

$$\mathbf{E}^* \cdot \mathbf{E} = (A\,\mathbf{E}^*) \cdot (A\,\mathbf{E}) = (a^*\,\mathbf{E}^*) \cdot (a\,\mathbf{E}) = a^* a\,\mathbf{E}^* \cdot \mathbf{E}$$

and it follows that

$$a^* a = |a|^2 = 1$$

LEMMA 2: *The operator A has at least one real eigenvalue.*

In order to prove this lemma, we must consider the method whereby the eigenvalues of A are calculated. If $\mathbf{E}$ is an eigenvector of A with eigenvalues a, then Eq. (4-20) can be written in the form

$$(A - aI)\mathbf{E} = 0 \tag{4-21}$$

Denote the components of $\mathbf{E}$ by (E^1, E^2, E^3). Then Eq. (4-21) is expressed in component form by

$$\begin{aligned} (a_1^1 - a)\,E^1 + a_2^1\,E^2 + a_3^1\,E^3 &= 0 \\ a_1^2\,E^1 + (a_2^2 - a)\,E^2 + a_3^2\,E^3 &= 0 \\ a_1^3\,E^1 + a_2^3\,E^2 + (a_3^3 - a)\,E^3 &= 0 \end{aligned}$$

This system of homogeneous linear equations has a non-trivial solution for the E^i only if the determinant of the coefficients vanishes

$$|A - aI| = \begin{vmatrix} (a_1^1 - a) & a_2^1 & a_3^1 \\ a_1^2 & (a_2^2 - a) & a_3^2 \\ a_1^3 & a_2^3 & (a_3^3 - a) \end{vmatrix} = 0 \tag{4-22}$$

Equation (4-22) is known as the secular equation for the operator A, and the values of a which satisfy this equation are the desired eigenvalues. The left-hand side of (4-22) is a cubic polynomial in a with all real coefficients, and can be written in the form

$$P_3(a)=a^3+b_1\,a^2+b_2\,a+b_3=0$$

where the b_i are all real. If a is sufficiently large and negative, the polynomial is negative; if a is sufficiently large and positive, the polynomial is positive. Hence, for at least one real value of a, the polynomial is zero, and this value of a satisfies the secular equation. It further follows from Lemma 1, that the real eigenvalue of A must be ± 1. If there is only one real solution of the secular equation, say a_1, then the other two solutions, (a_2, a_3) are complex conjugates.

LEMMA 3: *The product of the eigenvalues of the operator A is $+1$.*

Let us label the roots of the secular equation, i.e. the eigenvalues of the operator A, as (a_1, a_2, a_3), and denote the corresponding eigenvectors by $(\mathbf{E}_1, \mathbf{E}_2, \mathbf{E}_3)$. Then a typical component equation of Eq. (4-21) can be

$$a^i_j\; E^j_k = a_k\; E^i_k = E^i_k\; \delta^j_k\, a_k, \text{ (no sum on } k) \tag{4-23}$$

If we interpret the E^i_j as the elements of a matrix E, then we may write Eq. (4-23) in the form

$$AE=EH$$

where H is the diagonal matrix with elements $\delta^i_k\; a_k$. Multiplying from the left by E^{-1}, and rearranging, we find that H is obtained from A by the similarity transformation

$$H=E^{-1}AE$$

where the k-th row of E is formed by the components of the k-th eigenvector of A. Now, the determinant of the transformation A is $+1$, since A represents a proper orthogonal transformation. Hence,

$$|H|=|E^{-1}AE|=|E^{-1}||A||E|=|E^{-1}||E|=1$$

but,

$$|H|=a_1a_2a_3$$

which establishes the lemma.

With these lemmas, Euler's theorem is easily established. All of the eigenvalues of A cannot be real and distinct, since by Lemma 1, $|a|^2=1$. If the three roots of the secular equation are all real, it follows from Lemma 3 that the distinct eigenvalue must be $+1$. From Lemma 2, there cannot be more than two complex roots of the secular equation, which are complex conjugates, or else all three roots are real. Further, the complex eigenvalues must satisfy the relation $a^*a=1$, and hence, the real eigenvalue must be $+1$.

If A represents an improper orthogonal transformation, we may write the matrix A in the form

$$A=SA'$$

where

$$S=\begin{pmatrix} -1 & 0 & 0 \\ 0 & -1 & 0 \\ 0 & 0 & -1 \end{pmatrix}$$

and $|A'|=+1$. It is clear that S represents an inversion of the coordinates, and it follows from Euler's theorem that A' represents a rotation about some axis through the origin. Hence, the improper, homogeneous, orthogonal transformation consists of a rotation about some axis, accompanied by a coordinate inversion.

If we consider the proper orthogonal transformation described by the matrix A, it is not difficult to determine either the orientation of the axis of rotation or the angle of rotation in terms of the elements of the matrix A. The axis of rotation is obtained by setting $a=+1$ in Eq. (4-21). The system of component equations does not have a unique solution, since only two of them are independent. This means that only the ratios of the components of **E** can be uniquely determined. These ratios do specify the orientation of a vector along the axis of rotation. It is usually easier to obtain the components of the vector **E** by requiring that **E** be a unit vector. We then obtain the additional equation

$$(E^1)^2+(E^2)^2+(E^3)^2=1$$

The solution of this equation and any pair from (4-21) yields the components of a unit vector **E** which is along the axis of rotation relative to the fixed set of cartesian coordinates (x^1, x^2, x^3).

In order to find the angle of rotation about the axis specified by the unit vector **E**, we first note that there exists some orthogonal cartesian coordinate system $(\bar{x}^1, \bar{x}^2, \bar{x}^3)$, such that the unit vector **E** is directed along the $\bar{x}^3$-axis. If $\bar{A}$ denotes the form of the transformation matrix A relative to the barred coordinate system, then $\bar{A}$ is related to A by a similarity transformation. In terms of the barred coordinate system, the matrix $\bar{A}$ represents a rotation through the angle θ about the $\bar{x}^3$-axis, and hence is of the form

$$\bar{A}=\begin{pmatrix} \cos\theta & \sin\theta & 0 \\ -\sin\theta & \cos\theta & 0 \\ 0 & 0 & 1 \end{pmatrix}$$

The trace of any matrix is defined to be the sum of its diagonal elements, and hence the trace of $\bar{A}$ is

$$\text{Tr}\ \bar{A}=1+2\cos\theta$$

It can be shown that the trace of a matrix is invariant under a similarity transformation, and hence,

$$1+2\cos\theta=a_i^i \tag{4-24}$$

Equation (4-24) expresses the angle of rotation in terms of the diagonal elements of the transformation matrix A.

EXAMPLE 3. As an example of Euler's theorem, consider the orthogonal transformation specified by the matrix

$$A=\begin{pmatrix} 0.1996 & 0.8542 & 0.4800 \\ -0.9328 & 0.0157 & 0.3600 \\ 0.3000 & -0.5196 & 0.8000 \end{pmatrix}$$

The angle of rotation is immediately determined, since

$$\cos\theta=\tfrac{1}{2}(\text{Tr}\ A-1)=\tfrac{1}{2}(1.0153-1)=0.0077$$

and

$$\theta=1.5630 \text{ rad.}$$

The calculation of the orientation of the axis of rotation is equally straightforward, although the calculations are somewhat more difficult. It follows from (4-21), that the components, with respect to the original cartesian coordinates, of the unit vector along the axis of rotation satisfy the system of equations

$$-0.8004\ E^1+0.8542\ E^2+0.4800\ E^3=0$$
$$-0.9328\ E^1-0.9843\ E^2+0.3600\ E^3=0$$
$$0.3000\ E^1-0.5196\ E^2-0.2000\ E^3=0$$

and the normalization condition

$$(E^1)^2+(E^2)^2+(E^3)^2=1$$

Multiply the third of the eigenvector equations by 2.400 and add to the first to eliminate E^3,

$$-0.0804\ E^1-0.3938\ E^2=0$$

Then,

$$E^2=0.2042\ E^1$$

It then follows from the third equation that

$$E^3=2.0305\ E^1$$

The same result can be obtained by eliminating E^3 between the second and third equations. It follows from the normalization condition that

$$E^1[1+(.2042)^2+(2.0305)^2]^{1/2}=1$$

from which

$$E^1=0.4299$$

Then, the cartesian components, with respect to the original coordinate system, of the unit vector along the axis of rotation are

$$(0.4299, -0.0899, 0.8934)$$

4-3. Representations of the Rotation Group. We have previously seen that the set of 3×3 orthogonal matrices constitute a representation of the three-dimensional rotation group. In this section, we shall obtain two explicit representations of this group. The first representation is particularly useful in describing the dynamics of rigid body motion. The second representation is fundamental to the study of angular momentum in quantum mechanics, and in the theory of the spin of fundamental particles.

A 3×3 matrix is completely defined by its nine elements. In the case of an orthogonal matrix, the nine elements are related by the six orthogonality conditions

$$\sum_{i=1}^{3} a_j^i\, a_k^i=\delta_{jk}, \quad j, k=1, 2, 3$$

so that only three independent parameters are required to completely specify the matrix. For the representation of the rotation group, a useful set of parameters are the three Euler angles. In order to define the Euler angles, we decompose the total rotation into three simple rotations. We first rotate the original set of coordinates (x^1, x^2, x^3) through the angle φ in a counter clockwise sense about the x^3-axis, and denote the resulting set of cartesian coordinates by (ξ^1, ξ^2, ξ^3). The set of axes (ξ^1, ξ^2, ξ^3) are then rotated in a counter clockwise sense through an angle θ about the ξ^1-axis to obtain a new set of cartesian axes (η^1, η^2, η^3). The final transformation rotates the (η^1, η^2, η^3) axes in a counter clockwise sense about the η^3-axis through the angle ψ into the final set of cartesian coordinates $(\bar{x}^1, \bar{x}^2, \bar{x}^3)$. The successive rotations are illustrated in Fig. 4-4. Since the three independent angles (φ, θ, ψ) specify the orientation of the $(\bar{x}^1, \bar{x}^2, \bar{x}^3)$ axes relative to the (x^1, x^2, x^3) axes, they completely determine the orthogonal transformation

$$\bar{x}^1=a_j^i\, x^j$$

The first rotation

$$\xi^i=a_j^i\, x^j$$

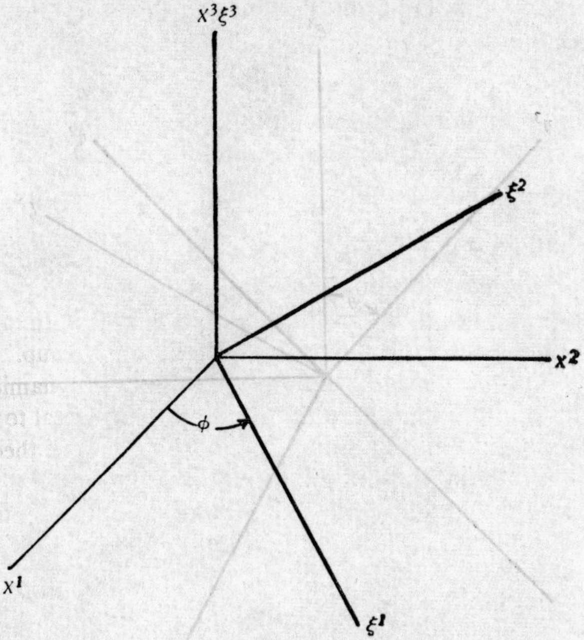

Figure 4-4a.

is represented by the matrix

$$A^{(1)}=\begin{pmatrix} \cos\varphi & \sin\varphi & 0 \\ -\sin\varphi & \cos\varphi & 0 \\ 0 & 0 & 1 \end{pmatrix} \tag{4-25}$$

Similarly, the transformation

$$\eta^i = a^i_j \, \xi^i$$

is represented by the matrix

$$A^{(2)}=\begin{pmatrix} 1 & 0 & 0 \\ 0 & \cos\theta & \sin\theta \\ 0 & -\sin\theta & \cos\theta \end{pmatrix} \tag{4-26}$$

Finally, the transformation

$$\bar{x}^i = a^i_j \, \eta^j$$

is represented by the matrix

$$A^{(3)}=\begin{pmatrix} \cos\psi & \sin\psi & 0 \\ -\sin\psi & \cos\psi & 0 \\ 0 & 0 & 1 \end{pmatrix} \tag{4-27}$$

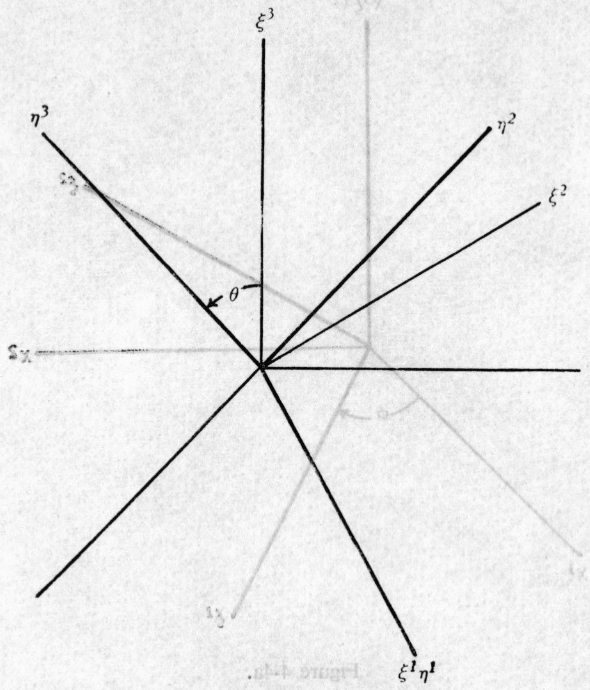

Figure 4-4b.

The composite transformation from (x^1, x^2, x^3) to $(\bar{x}^1, \bar{x}^2, \bar{x}^3)$ is represented by the product of (4-25), (4-26), and (4-27) taken in the proper sequence. Explicitly, the transformation has representation

$$A = A^{(3)} A^{(2)} A^{(1)} =$$

$$= \begin{pmatrix} \cos\psi\cos\varphi - \sin\psi\cos\theta\sin\varphi, & \cos\psi\sin\varphi + \sin\psi\cos\theta\cos\varphi, & \sin\psi\sin\theta \\ -\sin\psi\cos\varphi - \cos\psi\cos\theta\sin\varphi, & -\sin\psi\sin\varphi + \cos\psi\cos\theta\cos\varphi, & \cos\psi\sin\theta \\ \sin\theta\sin\varphi, & -\sin\theta\cos\varphi, & \cos\theta \end{pmatrix} \tag{4-28}$$

The inverse transformation

$$x^i = a'^i_j \bar{x}^j$$

is represented by the transpose of (4-28)

$$A^{-1} = A^T = A^{(1)T} A^{(2)T} A^{(3)T}$$

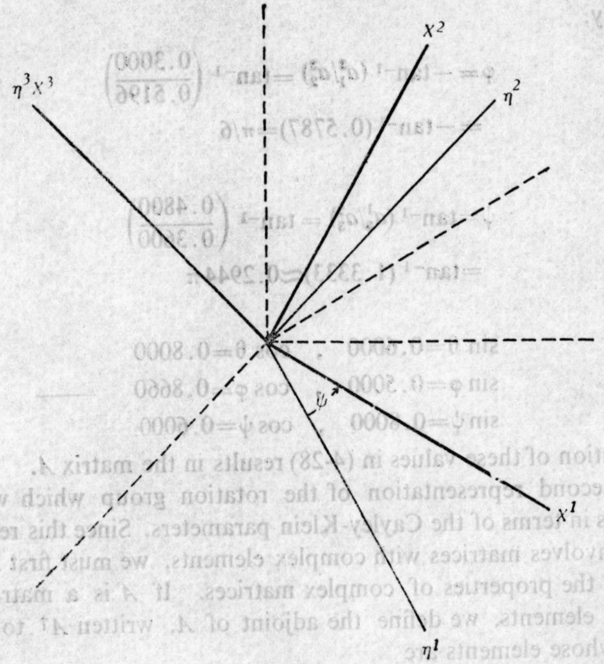

Figure 4-4c.

$$
=\begin{pmatrix}
\cos\psi\cos\varphi-\sin\psi\cos\theta\sin\varphi, & -\sin\psi\cos\varphi-\cos\psi\cos\theta\sin\varphi, & \sin\psi\sin\varphi \\
\cos\psi\sin\varphi+\sin\psi\cos\theta\cos\varphi, & -\sin\psi\sin\varphi+\sin\psi\cos\theta\cos\varphi, & -\sin\varphi\cos\varphi \\
\sin\psi\sin\theta, & \cos\psi\sin\theta, & \cos\theta
\end{pmatrix} \qquad (4\text{-}29)
$$

EXAMPLE 4. To illustrate the calculation of the Euler angles corresponding to a given orthogonal transformation, we again consider the transformation of Example 3,

$$
A=\begin{pmatrix} 0.1996 & 0.8542 & 0.4800 \\ -0.9328 & 0.0157 & 0.3600 \\ 0.3000 & -0.5196 & 0.8000 \end{pmatrix}
$$

We see from (4-28) that

$$a_3^3=\cos\theta$$

and hence

$$\theta=\cos^{-1} a_3^3=\cos^{-1} 0.8000\approx 0.206\,\pi$$

Similarly,

$$\varphi = -\tan^{-1}(a_1^3/a_2^3) = \tan^{-1}\left(\frac{0.3000}{0.5196}\right)$$
$$= -\tan^{-1}(0.5787) = \pi/6$$

and

$$\psi = \tan^{-1}(a_3^1/a_3^2) = \tan^{-1}\left(\frac{0.4800}{0.3600}\right)$$
$$= \tan^{-1}(1.3333) \approx 0.2944\,\pi$$

Then,

$$\sin\theta = 0.6000 \quad , \quad \cos\theta = 0.8000$$
$$\sin\varphi = 0.5000 \quad , \quad \cos\varphi = 0.8660$$
$$\sin\psi = 0.8000 \quad , \quad \cos\psi = 0.6000$$

Substitution of these values in (4-28) results in the matrix A.

The second representation of the rotation group which we shall obtain is in terms of the Cayley-Klein parameters. Since this representation involves matrices with complex elements, we must first look at some of the properties of complex matrices. If A is a matrix with complex elements, we define the adjoint of A, written $A^\dagger$ to be the matrix whose elements are

$$(A^\dagger)^i_j = a^j_i{}^*$$

where the asterisk indicates complex conjugation. In matrix notation, this is usually written

$$A^\dagger = (A^T)^* \tag{4-30}$$

The complex analog of an orthogonal matrix is known as a unitary matrix, which is defined by the condition

$$A^\dagger A = I \tag{4-31}$$

The similarity transformation of an arbitrary complex matrix B by a unitary matrix A is of the form

$$\bar{B} = ABA^\dagger$$

If A is a complex matrix such that

$$A = A^\dagger \tag{4-32}$$

then A is said to be self adjoint or hermitian. It can be shown that the hermitian property is preserved under a similarity transformation by a unitary matrix. It can also be shown that the trace and determinant of an arbitrary complex matrix are invariant under a similarity transformation by a unitary matrix.

Now, consider a two-dimensional space with complex coordinate axes (u, v). The general linear transformation of such a space is of the form

$$\bar{u}=au+bv$$
$$\bar{v}=cu+dv \tag{4-33}$$

where a, b, c, and d are complex constants. The transformation (4-33) is represented by the complex matrix

$$U=\begin{pmatrix} a & b \\ c & d \end{pmatrix}$$

The properties of transformation are greatly simplified if we restrict U to be a unitary matrix with determinant $+1$. The unitarity condition requires

$$\begin{pmatrix} a & b \\ c & d \end{pmatrix} \begin{pmatrix} a^* & c^* \\ b^* & d^* \end{pmatrix} = \begin{pmatrix} aa^*+bb^*, & ac^*+bd^* \\ a^*c+b^*d, & cc^*+dd^* \end{pmatrix} = \begin{pmatrix} 1 & 0 \\ 0 & 1 \end{pmatrix}$$

which is equivalent to the three conditions

$$|a|^2+|b|^2=1$$
$$|c|^2+|d|^2=1 \tag{4-34}$$
$$ac^*+bd^*=0$$

The left-hand side of the last of (4-34) is a complex quantity, so that the real and imaginary parts must separately be equated to zero, giving two real relations. The condition we have imposed on the determinant of U implies that

$$ad-bc=1 \tag{4-35}$$

The general complex transformation (4-33) involves eight parameters, since each of the four constants a, b, c, and d is complex. These eight parameters are, however, connected by the five independent relations (4-34) and (4-35). Hence, the two-dimensional unitary matrix with determinant $+1$ involves only three independent real parameters, as is the case of the three-dimensional orthogonal matrix with determinant $+1$. This offers the possibility of obtaining a relation between the 3×3 orthogonal matrices and the 2×2 unitary matrices, and hence describing the three-dimensional proper rotations in terms of the two-dimensional unitary transformations.

In order to obtain the correspondence between the 3×3 orthogonal matrices, and the 2×2 unitary matrices, let (x^1, x^2, x^3) be the coordinates of a point in the real three-dimensional space referenced to some orthogonal cartesian coordinate system. In the two-dimensional complex space, define the 2×2 hermitian matrix

$$H=\begin{pmatrix} x^3, & x^1-ix^2 \\ x^1+ix^2, & -x^3 \end{pmatrix} \tag{4-36}$$

The hermitian matrix (4-36) has zero trace, and determinant

$$-((x^1)^2+(x^2)^2+(x^3)^2)$$

We now perform a similarity transformation of H by the unitary matrix U to obtain

$$\overline{H}=UHU^\dagger$$

Since H must also be a hermitian matrix with zero trace, it can be written in the form

$$\overline{H}=\begin{pmatrix} \bar{x}^3 & \bar{x}^1-i\bar{x}^2 \\ \bar{x}^1+i\bar{x}^2, & -\bar{x}^3 \end{pmatrix}$$

where $\bar{x}^1$, $\bar{x}^2$, and $\bar{x}^3$ are all real quantities. The determinant of H must be invariant under the similarity transformation, so that

$$|\overline{H}|^2=-((\bar{x}^1)^2+(\bar{x}^2)^2+(\bar{x}^3)^2)$$
$$=-((x^1)^2+(x^2)^2+(x^3)^2)=|H|^2$$

If we interpret the real quantities x^1, x^2, x^3 as being the components of a vector relative to the original orthogonal cartesian coordinates in the real three-dimensional space, and the real quantities $\bar{x}^1$, $\bar{x}^2$, and $\bar{x}^3$ as the components of the same vector relative to some barred coordinate system in the real three-dimensional space, we see that the magnitude of the vector is left invariant under the transformation from unbarred to barred coordinates. This however, can occur only if the transformation of the real three-dimensional cartesian coordinates is an orthogonal transformation. Hence, to every unitry transformation in the complex two-dimensional space, there corresponds an orthogonal transformation in the real three-dimensional space.

Let A represent an orthogonal transformation from the orthogonal cartesian coordinates (x^1, x^2, x^3) to the orthogonal cartesian coordinates $(\bar{x}^1, \bar{x}^2, \bar{x}^3)$. Corrosponding to A is the 2×2 unitary matrix U_1, which transforms the hermitian matrix (4-36) according to

$$\overline{H}=U_1HU_1{}^\dagger$$

If B represents an orthogonal transformation from $(\bar{x}^1, \bar{x}^2, \bar{x}^3)$ to (y^1, y^2, y^3), then there is a corresponding 2×2 unitary matrix U_2, which transforms $\overline{H}$ according to

$$G=U_2\overline{H}U_2{}^\dagger$$

There is, of course, the single orthogonal transformation from (x^1, x^2, x^3) to (y^1, y^1, y^3), namely

$$C=BA$$

Similarly,

$$G=U_2U_1H\ U_1{}^\dagger\ U_2{}^\dagger=(U_2U_1)\,H(U_2U_1)^\dagger=U_3H\ U_3{}^\dagger \qquad (4\text{-}37)$$

where $U_3=U_2U_1$.

Thus, for every operation among the 3×3 orthogonal matrices with determinant $+1$, there is a corresponding operation among the 2×2 unitary matrices with determinant $+1$. Further, it can be shown that the 2×2 unitary matrices with determinant $+1$ form an infinite-dimensional three parameter group. Whenever the elements of two groups are related in such a way that for every element of one group there is at least one corresponding element of the other group, and any relation among the elements of one group is satisfied by the corresponding elements of the other, the two groups are said to be homomorphic. The homomorphism between the 3×3 orthogonal matrices and the 2×2 unitary matrices does not imply that we can obtain a rotation of a three-dimensional orthogonal coordinate system by the application of a 2×2 unitary matrix. All that the homomorphism means is that there is a definite correspondence between the two groups of matrices. The 3×3 orthogonal matrices operate on column vectors with three components to produce column vectors with three components, whereas the 2×2 unitary matrices operate on 2×2 hermitian matrices to produce new 2×2 hermitian matrices. The apparent correspondence between the column vectors and the hermitian matrices arises because we have constructed the hermitian matrices from the elements of the column vectors in a particular way. There is, however, no particular relation between the real three-dimensional space of the 3×3 orthogonal matrices and the complex two-dimensional space of the 2×2 unitary matrices.

In order to complete the representation of the three-dimensional rotation group in terms of the Cayley-Klein parameters a, b, c, and d, we shall express the elements of the unitary matrix U in terms of the three Euler angles. We first note that the unitary conditions, Eq. (4-34) are satisfied by the two conditions

$$a=d^*, \quad b=-c^*, \quad aa^*+bb^*=1$$

Hence, we need only to determine a and b in terms of the Euler angles. Define

$$x_+=x^1+ix^2, \quad x_-=x^1-ix^2$$

so that the hermitian matrix (4-36) can be written in the form

$$H=\begin{pmatrix} \bar{x}^3 & \bar{x}_- \\ \bar{x}_+ & -\bar{x}^3 \end{pmatrix}=\begin{pmatrix} a & b \\ c & d \end{pmatrix}\begin{pmatrix} x^3 & x_- \\ x_+ & -x^3 \end{pmatrix}\begin{pmatrix} a^* & c^* \\ b^* & d^* \end{pmatrix}$$

$$=\begin{pmatrix} a & b \\ c & d \end{pmatrix}\begin{pmatrix} x^3 & x_- \\ x_+ & -x^3 \end{pmatrix}\begin{pmatrix} d & -b \\ -c & a \end{pmatrix}$$

$$\begin{pmatrix} (ad+bc)\,x^3-ac\,x_-+bd\,x_+, & -2ab\,x^3+a^2\,x_--b^2\,x_+ \\ 2cd\,x^3-c^2\,x_-+d^2\,x_+, & -(ad+bc)\,x^3+ac\,x_--bd\,x_+ \end{pmatrix}$$

Hence,

$$\begin{aligned}\bar{x}_- &= 2cd\, x^3 - c^2\, x_- + d^2\, x_+ \\ \bar{x}_+ &= -2ab\, x^3 + a^2\, x_- - b^2\, x_+ \\ \bar{x}^3 &= (ad+bc)\, x^3 - ac\, x_- + bd\, x_+\end{aligned} \tag{4-38}$$

The quantities $\bar{x}_-$, $\bar{x}_+$, and $\bar{x}^3$ can also be obtained by the proper combinations of the coordinates $(\bar{x}^1, \bar{x}^2, \bar{x}^3)$ which are the result of applying the real orthogonal transformation A to the set of coordinates (x^1, x^2, x^3). Let us now consider the three successive transformations which define the Euler angles. The angle φ is defined by a rotation about the x^3-axis,

$$\begin{aligned}\bar{x}^1 &= x^1 \cos\varphi + x^2 \sin\varphi \\ \bar{x}^2 &= -x^1 \sin\varphi + x^2 \cos\varphi \\ \bar{x}^3 &= x^3\end{aligned}$$

The transformation equations can be written in complex form as

$$\begin{aligned}\bar{x}_+ &= x_+\, e^{-i\varphi} \\ \bar{x}_- &= x_-\, e^{i\varphi} \\ \bar{x}^3 &= x^3\end{aligned} \tag{4-39}$$

In order that Eqs. (4-38) and (4-39) correspond to the same transformation of the coordinate axes, we must set

$$c = b = 0, \quad a = e^{i\varphi/2}, \quad d = e^{-i\varphi/2}$$

Hence, the unitary matrix which corresponds to the orthogonal matrix $A^{(1)}$ is

$$U^{(1)} = \begin{pmatrix} e^{i\varphi/2} & 0 \\ 0 & e^{-i\varphi/2} \end{pmatrix}$$

Similarly, the unitary matrices which correspond to $A^{(2)}$ and $A^{(3)}$ are

$$U^{(2)} = \begin{pmatrix} \cos(\theta/2) & i\sin(\theta/2) \\ i\sin(\theta/2) & \cos(\theta/2) \end{pmatrix}$$

and

$$U^{(3)} = \begin{pmatrix} e^{i\psi/2} & 0 \\ 0 & e^{-i\psi/2} \end{pmatrix}$$

respectively. We have already seen that the general three-dimensional rotation is represented in terms of the Euler angles by the matrix

$$A = A^{(3)}\, A^{(2)}\, A^{(1)}$$

Then, since the 2×2 unitary matrices with determinant $+1$ are homomorphic to the 3×3 orthogonal matrices with determinant $+1$, it follows that

$$U = U^{(3)}\, U^{(2)}\, U^{(1)}$$

$$U=\begin{pmatrix} e^{i(\varphi+\psi)/2}\cos(\theta/2) & , & i\, e^{i(\varphi-\psi)/2}\sin(\theta/2) \\ +i\, e^{-i(\varphi-\psi)/2}\sin(\theta/2), & & e^{-i(\varphi+\psi)/2}\cos(\theta/2) \end{pmatrix} \tag{4-40}$$

is a representation of the general three-dimensional rotation. The Cayley-Klein parameters can be immediately identified from Eq. (4-40).

EXAMPLE 5. Once the Euler angles corresponding to a given orthogonal transformation are known, it is a relatively simple calculation to obtain the Cayley-Klein parameters. Expanding the complex exponentials in Eq. (4-40).

$$a=\cos\frac{\varphi}{2}\cos\frac{\psi}{2}\cos\frac{\theta}{2}-\sin\frac{\varphi}{2}\sin\frac{\psi}{2}\cos\frac{\theta}{2}+ i\left(\sin\frac{\varphi}{2}\cos\frac{\psi}{2}\cos\frac{\theta}{2}+\cos\frac{\varphi}{2}\sin\frac{\psi}{2}\cos\frac{\theta}{2}\right)$$

$$b=\cos\frac{\varphi}{2}\sin\frac{\psi}{2}\sin\frac{\theta}{2}-\sin\frac{\varphi}{2}\cos\frac{\psi}{2}\sin\frac{\theta}{2}+ i\left(\cos\frac{\varphi}{2}\cos\frac{\psi}{2}\sin\frac{\theta}{2}+\sin\frac{\varphi}{2}\sin\frac{\psi}{2}\sin\frac{\theta}{2}\right)$$

$$c=-b^*$$

$$d=a^*$$

Once again, consider the orthogonal transformation specified by

$$A=\begin{pmatrix} 0.1996 & 0.8542 & 0.4800 \\ -0.9328 & 0.0157 & 0.3600 \\ 0.3000 & -0.5196 & 0.8000 \end{pmatrix}$$

In Example 4, we determined the Euler angles corresponding to this transformation, and found that

$$\sin\theta=0.6000, \quad \cos\theta=0.8000$$
$$\sin\varphi=0.5000, \quad \cos\varphi=0.8660$$
$$\sin\psi=0.8000, \quad \cos\psi=0.6000$$

Then, using the half-angle formulas

$$\sin\frac{\alpha}{2}\sqrt{\frac{1-\cos\alpha}{2}}, \quad \cos\frac{\alpha}{2}=\sqrt{\frac{1+\cos\alpha}{2}}$$

we find

$$a=0.7047+i\,0.6294 \quad , \quad b=0.0634+i\,0.3098$$
$$c=-0.0634+i\,0.3098, \quad d=0.7097-i\,0.6294$$

The half angles which appear in the representation of the three-dimensional rotation group by the 2×2 unitary matrices lead to some rather interesting results. For example, if we consider a counter clockwise rotation through the angle 2π about the x^3-axis, the cartesian set (x^1, x^2, x^3) is left invariant. The 3×3 orthogonal matrix which corresponds to this rotation is

$$A^{(1)}(2\pi)=\begin{pmatrix} \cos 2\pi & \sin 2\pi & 0 \\ -\sin 2\pi & \cos 2\pi & 0 \\ 0 & 0 & 1 \end{pmatrix}=\begin{pmatrix} 1 & 0 & 0 \\ 0 & 1 & 0 \\ 0 & 0 & 1 \end{pmatrix}=I$$

On the other hand

$$U^{(1)}(2\pi)=\begin{pmatrix} e^{i\pi} & 0 \\ 0 & e^{-i\pi} \end{pmatrix}=\begin{pmatrix} -1 & 0 \\ 0 & -1 \end{pmatrix}=-I$$

However, since $U^{(1)}(2\pi)$ is homomorphic to $A^{(1)}(2\pi)$, then $U^{(1)}(2\pi)$ must also be equal to the identity. In general, if U corresponds to a given rotation, then $-U$ must correspond to the same rotation. The representation of the three-dimensional rotation group by the 2×2 unitary matrices with determinant $+1$ is, therefore, double valued. If U defines a proper linear transformation of the complex two-dimensional (u, v) space, then $-U$ represents the same proper rotation followed by a coordinate inversion. Thus, corresponding to every proper rotation in the real three-dimensional space (x^1, x^2, x^3), there is a proper rotation in the two-dimensional space (u, v); however, there is also this same rotation in the complex space followed by a coordinate inversion which also corresponds to the proper rotation in the three-dimensional space. We shall investigate this no further.

One other consequence of the homomorphism between the three-dimensional rotation group and the two-dimensional proper unitary group is the definition of a type of quantities which are known as spinors. Corresponding to every vector in the three-dimensional space (x^1, x^2, x^3), whose components transform as the coordinates under a proper rotation, is a vector like quantity with complex components defined in the two-dimensional complex space (u, v) whose components transform as the complex coordinates under the corresponding proper unitary transformation. These vectors like quantities with complex components are known as spinors of the first order and first kind.

It is sometimes convenient to introduce a set of four independent 2×2 hermitian matrices, known as the Pauli spin matrices. These are defined to be

$$\sigma_1=\begin{pmatrix} 0 & 1 \\ 1 & 0 \end{pmatrix},\quad \sigma_2=\begin{pmatrix} 0 & -i \\ i & 0 \end{pmatrix}$$
$$\sigma_3=\begin{pmatrix} i & 0 \\ 0 & -i \end{pmatrix},\quad \sigma_4=\begin{pmatrix} 1 & 0 \\ 0 & 1 \end{pmatrix}=I \qquad (4\text{-}41)$$

Note that the first three of these matrices all have zero trace. Hence, any 2×2 hermitian matrix with zero trace can be expressed as a linear combination of σ_1, σ_2. and σ_3. For example, the hermitian matrix (4-36) can be written in the form

$$H=x^1\,\sigma_1+x^2\,\sigma_2+x^3\,\sigma_3$$

The set of Pauli matrices satisfy the multiplication table

$$\begin{aligned}\sigma_1\sigma_2&=i\,\sigma_3, \quad \sigma_2\sigma_1=-i\,\sigma_3\\ \sigma_2\sigma_3&=i\,\sigma_1, \quad \sigma_3\sigma_2=-i\,\sigma_1\\ \sigma_3\sigma_1&=i\,\sigma_2, \quad \sigma_1\sigma_3=-i\,\sigma_2\end{aligned} \tag{4-42}$$

Since the matrices are independent, any 2×2 matrix with four parameters can be expressed as a linear combination of the four Pauli matrices. The three unitary matrices which correspond to the three simple Euler angle rotations are

$$\begin{aligned}U^{(1)}&=\sigma_4\cos(\varphi/2)+i\,\sigma_3\sin(\varphi/2)\\ U^{(2)}&=\sigma_4\cos(\theta/2)+i\,\sigma_1\sin(\theta/2)\\ U^{(3)}&=\sigma_4\cos(\psi/2)+i\,\sigma_3\sin(\psi/2)\end{aligned} \tag{4-43}$$

The product of the three expressions (4-43) taking in the proper sequence gives a representation of the three-dimensional rotation group in terms of the Pauli spin matrices.

4-4. The General Linear Transformation. We now consider the general, homogeneous linear transformation

$$\bar{x}^i=a^i_j\,x^j \tag{4-44}$$

where the only restriction we impose is that the determinant of the coefficients of the transformation is different from zero,

$$\det(a^i_j)\neq 0 \tag{4-45}$$

The condition (4-45) insures that the linear transformation has, as we shall see, an inverse transformation. Linear transformations with non-vanishing determinant are known as non-singular transformations. It is clear that the product AB specifies a non-singular transformation, since

$$\det(AB)=\det(A)\det(B)\neq 0$$

if neither $\det(A)$ nor $\det(B)$ vanishes. The system of Eq. (4-44) can be solved for the unbarred variables to obtain

$$x^i=\frac{A^i_k\,\bar{x}^k}{|A|} \tag{4-46}$$

where A^i_k is the cofactor of a^i_k in $\det(A)$, i.e. A^i_k is equal to $(-1)^{i+k}$

times the determinant obtained by omitting the i-th row and k-th column of $\det(A)$. Since (4-46) is the inverse transformation to (4-44), the matrix whose elements are

$$\frac{A^i_k}{|A|}$$

is A^{-1}. In general, for any square matrix A, such that $\det(A) \neq 0$,

$$(A^{-1})^i_k = \frac{A^k_i}{|A|} \tag{4-47}$$

Now,

$$A\,A^{-1} = A^{-1}\,A = I$$

by definition, and hence $\det(A\,A^{-1}) = \det(A)\det(A^{-1}) = 1$. Therefore, $\det(A^{-1}) \neq 0$, and A^{-1} represents a non-singular linear transformation. Finally, the identity transformation

$$\bar{x}^i = x^i$$

is a non-singular transformation. Thus, the set of all linear transformations whose matrices have a non-vanishing determinant forms a group, which is generally known as the affine or general linear group. The orthogonal transformations form a subgroup of the affine group.

One of the simplest, non-orthogonal, linear transformations is an extension or contraction of the coordinate axes. Such a transformation is characterized by three positive real numbers k_1, k_2, and k_3,

$$\bar{x}^1 = k_1 x^1, \quad \bar{x}^2 = k_2 x^2, \quad \bar{x}^3 = k_2 x^3 \tag{4-48}$$

The matrix corresponding to this transformation is the diagonal matrix

$$A = \begin{pmatrix} k_1 & 0 & 0 \\ 0 & k_2 & 0 \\ 0 & 0 & k_3 \end{pmatrix}$$

with determinant $\det(A) = k_1 k_2 k_3$. Since $\det(A) \neq 0$, the inverse transformation exists, and it follows from (4–47) that

$$A = \begin{pmatrix} 1/k_1 & 0 & 0 \\ 0 & 1/k_2 & 0 \\ 0 & 0 & 1/k_3 \end{pmatrix}$$

If (b^1, b^2, b^3) are the components of the vector $\mathbf{b}$ relative to the original orthogonal cartesian axes, then, under the transformation (4-48), these components are extended or contracted according to

$$\bar{b}^1 = k_1 b^1, \quad \bar{b}^2 = k_2 b^2, \quad \bar{b}^3 = k_3 b^3$$

If we interpret $(\bar{b}^1, \bar{b}^2, \bar{b}^3)$ as the components of a new vector $\bar{\mathbf{b}}$ relative to the original orthogonal cartesian coordinates, we see that the vector produced by the linear transformation has neither the direction nor magnitude of the original vector. There is a special case of the transformation which produces a vector with the same direction as the original vector. This is the transformation

$$\bar{x}^1 = k\,x^1, \quad \bar{x}^2 = k\,x^2, \quad \bar{x}^3 = k\,x^3$$

which is represented by the matrix

$$A = \begin{pmatrix} k & 0 & 0 \\ 0 & k & 0 \\ 0 & 0 & k \end{pmatrix} = k \begin{pmatrix} 1 & 0 & 0 \\ 0 & 1 & 0 \\ 0 & 0 & 1 \end{pmatrix} = kI$$

The inverse of this matrix is

$$A^{-1} = (1/k)\,I$$

We have previously seen that any three non-zero, non-coplanar vectors provide a basis for the description of a vector in the sense that any vector can be expressed as a linear combination of the basic set. In particular, in the discussion of curvilinear coordinates, the basic set of vectors at a point was chosen to be either the set of unitary vectors or reciprocal unitary vectors. The unitary vectors were defined to be the projections of unit distances along the coordinate lines on the tangents to the coordinate lines through the point of interest (see Fig. 4-5). If we start with a rectangular cartesian system (x^1, x^2, x^3), then the coordinate lines for any coordinate system derived from (x^1, x^2, x^3) by a linear transformation are necessarily straight lines, and the set of unitary vectors $(\mathbf{e}_1, \mathbf{e}_2, \mathbf{e}_3)$ measure unit distances along the coordinate axes $(\bar{x}^1, \bar{x}^2, \bar{x}^3)$. It is not difficult to obtain representations of the unitary vectors $(\mathbf{e}_1, \mathbf{e}_2, \mathbf{e}_3)$ in terms of the set of cartesian unit vectors $(\mathbf{i}, \mathbf{j}, \mathbf{k})$. Suppose that the coordinate systems (x^1, x^2, x^3) and $(\bar{x}^1, \bar{x}^2, \bar{x}^3)$ are related by the non-singular linear transformation

$$\bar{x}^i = a^i_j\, x^j$$

Then, the inverse transformation is given by

$$x^i = a'^i_j\, \bar{x}^j \tag{4-49}$$

where $a'^i_j = A^j_i/\det(A)$. If the initial points of the unitary vectors are at the origin of the barred coordinates, then their terminal points are a unit distance along the $\bar{x}^1$, $\bar{x}^2$, and $\bar{x}^3$ axes respectively. In terms of the barred coordinates, the terminal point of $\mathbf{e}_1$ has the coordinates $(1, 0, 0)$. The coordinates of the terminal point of $\mathbf{e}_1$ referenced to the original cartesian system are obtained from the inverse transformation (4-49)

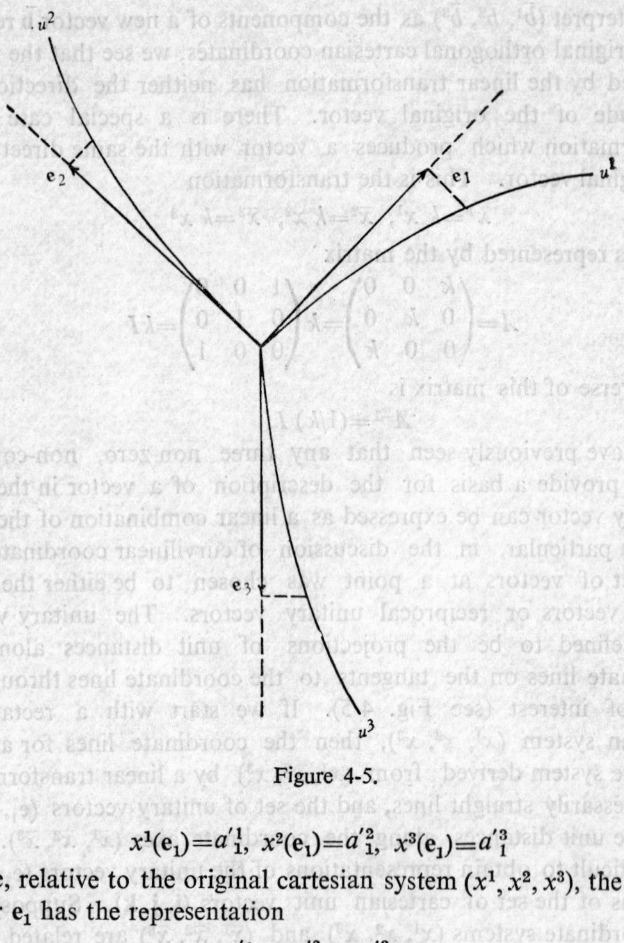

Figure 4-5.

$$x^1(\mathbf{e}_1)=a'^1_1,\ x^2(\mathbf{e}_1)=a'^2_1,\ x^3(\mathbf{e}_1)=a'^3_1$$

Hence, relative to the original cartesian system (x^1, x^2, x^3), the unitary vector $\mathbf{e}_1$ has the representation

$$\mathbf{e}_1=a'^1_1\,\mathbf{i}+a'^2_1\,\mathbf{j}+a'^3_1\,\mathbf{k}$$

Similarly,

$$\mathbf{e}_2=a'^1_2\,\mathbf{i}+a'^2_2\,\mathbf{j}+a'^3_2\,\mathbf{k}$$
$$\mathbf{e}_3=a'^1_3\,\mathbf{i}+a'^2_3\,\mathbf{j}+a'^3_3\,\mathbf{k}$$

The description of a vector in terms of a non-cartesian system is also conveniently made in terms of the set of reciprocal unitary vectors $(\mathbf{e}^1, \mathbf{e}^2, \mathbf{e}^3)$, which are defined by the condition

$$\mathbf{e}^i\cdot\mathbf{e}_j=\delta^i_j$$

In order to calculate the representations of the reciprocal unitary vectors in terms of the original cartesian coordinates, it is somewhat easier to use the explicit relations

$$\mathbf{e}^1 = \frac{\mathbf{e}_2 \times \mathbf{e}_3}{\mathbf{e}_1 \cdot \mathbf{e}_2 \times \mathbf{e}_3}, \quad \mathbf{e}^2 = \frac{\mathbf{e}_3 \times \mathbf{e}_1}{\mathbf{e}_1 \cdot \mathbf{e}_2 \times \mathbf{e}_3} \tag{4-50}$$

$$\mathbf{e}^3 = \frac{\mathbf{e}_1 \times \mathbf{e}_2}{\mathbf{e}_1 \cdot \mathbf{e}_2 \times \mathbf{e}_3}$$

If we substitute the representations of the unitary vectors in the first of Eq. (4-50), we obtain

$$\mathbf{e}^1 = \frac{A'^1_1}{\det(A')}\,\mathbf{i} + \frac{A'^2_1}{\det(A')}\,\mathbf{j} + \frac{A'^3_1}{\det(A')}\,\mathbf{k}$$

where A'^i_j is the cofactor of a'^i_j in $\det(A') = \det(A^{-1})$. We have already seen that $A'^j_i/\det(A^{-1})$ is the ij-th element of $(A^{-1})^{-1}$, i.e. a^i_j. Hence

$$\mathbf{e}^1 = a^1_1\,\mathbf{i} + a^1_2\,\mathbf{j} + a^1_3\,\mathbf{k}$$

Similarly,

$$\mathbf{e}^2 = a^2_1\,\mathbf{i} + a^2_2\,\mathbf{j} + a^2_3\,\mathbf{k}$$

$$\mathbf{e}^3 = a^3_1\,\mathbf{i} + a^3_2\,\mathbf{j} + a^3_3\,\mathbf{k}$$

Now, let $\mathbf{b}$ be a vector from the origin which has the representation

$$\mathbf{b} = b_1\,\mathbf{i} + b_2\,\mathbf{j} + b_3\,\mathbf{k}$$

relative to the orthogonal cartesian coordinate system (x^1, x^2, x^3). We seek a representation of $\mathbf{b}$ relative to the set of coordinates

$$\bar{x}^i = a^i_j\, x^j, \; \det(a^i_j) \neq 0$$

If we choose the set of unitary vectors as our basis, a representation of $\mathbf{b}$ relative to the barred coordinates is

$$\mathbf{b} = \bar{b}^i\, \mathbf{e}_i$$

where the $\bar{b}^i$ are the contravariant components of $\mathbf{b}$ relative to the coordinate system $(\bar{x}^1, \bar{x}^2, \bar{x}^3)$. The contravariant components are given by

$$\bar{b}^i = \mathbf{b} \cdot \mathbf{e}^i \tag{4-51}$$

Substituting the representations of $\mathbf{b}$ and the reciprocal unitary vectors in (4-51), we find that

$$\bar{b}^i = a^i_j\, b^j \tag{4-52}$$

where we have written the components of $\mathbf{b}$ relative to (x^1, x^2, x^3) in contravariant form. We may do this since in an orthogonal cartesian system, there is no distinction between the contravariant, covariant and physical components of a given vector. We see from (4-52) that the contravariant components of $\mathbf{b}$ transform in the same way as the coordinates under an affine transformation. Let (x'^1, x'^2, x'^3) and

(y^1, y^2, y^3) be two sets of coordinates, neither of which is necessarily orthogonal, which are related by an affine transformation. Let (b^1, b^2, b^3) and $(\bar{b}^1, \bar{b}^2, \bar{b}^3)$ two sets of quantities defined in the X' and Y coordinate systems respectively. If the $\bar{b}^i$ and b^i are related by

$$\bar{b}^i = a^i_j \, b^j \tag{4-53}$$

where the a^i_j are the coefficients of the affine transformation, we define the $\bar{b}^i$ (or the b^i) to be the contravariant affine components of a vector **b**. The transformation law (4-53) is known as transformation by cogradience.

Again, let (x^1, x^2, x^3) be an orthogonal cartesian system and let $(\bar{x}^1, \bar{x}^2, \bar{x}^3)$ be an arbitrary coordinate system related to the orthogonal system by the affine transformation

$$\bar{x}^i = a^i_j \, x^j, \ \det (a^i_j) \neq 0$$

Let **b** be a vector with the representation

$$\mathbf{b} = b_1 \, \mathbf{i} + b_2 \, \mathbf{j} + b_3 \, \mathbf{k}$$

relative to the orthogonal cartesian system. In the barred coordinate system, a representation of **b** is

$$\bar{\mathbf{b}} = b_i \, \mathbf{e}^i \tag{4-54}$$

where the

$$\bar{b}_i = \mathbf{b} \cdot \mathbf{e}_i \tag{4-55}$$

are known as the covariant components of **b** relative to the barred coordinate system. If we substitute the representations of **b** and the unitary vectors relative to the orthogonal cartesian set in (4-55), we find that the covariant components of **b** relative to the barred coordinates and the cartesian components of **b** relative to the orthogonal cartesian set are related by equations of the form

$$\bar{b}_i = a'^i_j \, b_j \tag{4-56}$$

Here, we have written the cartesian component of **b** in covariant form, which is permissible since they are referred to an orthogonal cartesian system. In the transformation (4-56), the a'^j_i are the elements of $(A^{-1})^T = (A^T)^{-1}$. If A represents an orthogonal transformation of coordinates, $A^T = A^{-1}$, so that $(A^T)^{-1} = A$, and $\bar{b}^i = \bar{b}_i$. In general, any set of three quantities (b_1, b_2, b_3) which transform according to (4-56) under the affine group are defined to be the affine covariant components of a vector **b**. The transformation law (4-56) is known as transformation by contragradience, and the matrix $(A^T)^{-1}$ is known as the contragradient of the matrix A.

In order to complete our discussion of the behaviour of the components of a vector under the affine group, we need to obtain an explicit relation between the contravariant and the covariant components of a given vector relative to the same coordinate system. (x^1, x^2, x^3) define an orthogonal cartesian system, and let the coordinate system $(\bar{x}^1, \bar{x}^2, \bar{x}^3)$ be derived from the orthogonal system by an affine transformation with coefficients a^i_j. Let **b** be an arbitrary vector with components (b^1, b^2, b^3) or (b_1, b_2, b_3) relative to the unbarred coordinates. In the barred coordinate system, the vector **b** has the two equivalent representations

$$\mathbf{b}=\bar{b}^i\,\mathbf{e}_i=\bar{b}_i\,\mathbf{e}^i \tag{4-57}$$

If we form the inner product of (4-57) with $\mathbf{e}_j$, we obtain

$$\bar{b}_j=\bar{b}^i\,(\mathbf{e}_i\cdot\mathbf{e}_j) \tag{4-58}$$

If we do not substitute the representations of $\mathbf{e}_i$ and $\mathbf{e}_j$ relative to the orthogonal cartesian system in (4-58), we have

$$\bar{b}_j=\bar{b}^i\sum_{k=1}^{3} a'^k_i\, a'^k_j=\bar{b}^i\, g_{ij} \tag{4-59}$$

where the g_{ij} are the first metric coefficients for the barred coordinate system, which we defined in Sec. 3-1. It is clear from Eq. (4-59) that the first metric coefficients for the coordinate system $(\bar{x}^1, \bar{x}^2, \bar{x}^3)$ are

$$g_{ij}=\sum_{k=1}^{3} a'^k_i\, a'^k_j \tag{4-60}$$

Similarly,

$$\bar{b}^i=g^{ij}\bar{b}_j$$

where

$$g^{ij}=\sum_{k=1}^{3} a^i_k\, a^j_k \tag{4-61}$$

are the second metric coefficients for the coordinate system $(\bar{x}^1, \bar{x}^2, \bar{x}^3)$. The result (4-60) could, of course, have been obtained directly by applying Eq. (3-17) to the set of transformation equations.

In our discussion of curvilinear coordinates, we defined the inner product of two vectors **a** and **b** as the sum

$$\mathbf{a}\cdot\mathbf{b}=a^i\, b_i=a_i\, b^i$$

We shall now show that the magnitude of a given vector **b** is invariant under the affine group. Let **b** have components (b_1, b_2, b_3) relative to an orthogonal cartesian system (x^1, x^2, x^3), and let $(\bar{x}^1, \bar{x}^2, \bar{x}^3)$ be a set of coordinates derived from the orthogonal cartesian set by an affine transformation with coefficients a^i_j. The covariant and contravariant components of **b** relative to the barred coordinate system are given by

$$\bar{b}_i = a'^j_i \, b_j, \qquad \bar{b}^i = a^i_j \, b^j$$

respectively. Then,

$$|\bar{\mathbf{b}}|^2 = \bar{\mathbf{b}} \cdot \bar{\mathbf{b}} = \bar{\mathbf{b}}_i \bar{\mathbf{b}}^i = a'^j_i \, a^i_k \, b_j \, b^k = \delta^j_k \, b_j \, b^k = b_j \, b^j = |\mathbf{b}|^2 \tag{4-62}$$

The argument which leads to Eq. (4-62) is in fact a tautology, since we have assumed the invariance of the vector **b** in deriving the transformation laws for its covariant and contravariant components. However, it is possible to define the covariant and contravariant affine components of a vector by the transformation laws, and it is then necessary to show that this definition leads to an invariant magnitude for the vector under the affine group.

Now, let O be a linear operator defined in an orthogonal cartesian coordinate system (x^1, x^2, x^3), such that for some two vectors **b** and **c** defined in the same coordinate system

$$\mathbf{c} = O \, \mathbf{b} \tag{4-63}$$

We now seek a representation of O relative to the coordinate system $(\bar{x}^1, \bar{x}^2, \bar{x}^3)$ which is derived from (x^1, x^2, x^3) by the affine transformation with coefficients a^i_j. The exact form of O relative to the barred coordinates will depend on the choice of representation used for the vectors **b** and **c**. If we use the contravariant representations of **b** and **c**, in the barred coordinate system, the contravariant components of **c** are given by

$$A \, \mathbf{c} = A \, (O \, \mathbf{b}) = A \, O \, A^{-1} \, (A\mathbf{b}) \tag{4-64}$$

and we identify the representation of O in the barred coordinate system as AOA^{-1}. On the other hand, the covariant components of **c** relative to the barred coordinates are given by

$$(A^T)^{-1} \, \mathbf{c} = (A^T)^{-1} \, (O\mathbf{b}) = (A^T)^{-1} O A^T \, (A^T)^{-1}\mathbf{b} \tag{4-65}$$

so that the matrix VOV^{-1}, where $V = (A^T)^{-1}$, is also a representation of the operator O relative to the barred coordinates. Since the covariant and contravariant components of a given vector in a fixed coordinate system are related in a definite way, it should be possible to relate the two different representations of the operator O. We may write (4-63) in terms of contravariant components as

$$c^i = O^i_j \, b^j \tag{4-66}$$

where the O^i_j are the elements of the matrix which represents the operator O, the upper index being the row index and lower index the column index. Since there is no distinction between covariant and contravariant components of a vector in an orthogonal cartesian system, we may also write (4-63) in the form

$$c_i = O_i^j b_j \tag{4-67}$$

Hence, if O is the matrix (with elements O_j^i) which represents the operator, then O^T also represents the operator, and

$$O = O^T$$

If $\bar{O} = AOA^T$ is a representations of the operator in the barred coordinate system, then $\bar{O}^T$ must also represent the same operator relative to the barred coordinate system. Now,

$$\bar{O}^T = (AOA^{-1})^T = (A^{-1})^T O^T A^T = VOV^{-1}$$

which is in agreement with (4-65).

In general, any set of nine quantities with transform according to

$$\bar{t}_j^i = a'^n_j \, t_n^m \, a_m^i \tag{4-68}$$

under the affine group, are defined to be the mixed affine components of a tensor of order two. We now seek other affine representations of tensors of order two. Consider the nine quantities t_{ij} defined in some coordinate system (x^1, x^2, x^3), and form the sum

$$t_{ij} \, b^i \, c^j \tag{4-69}$$

where b^i and c^j are the affine contravariant components of two arbitrary vectors relative to the given coordinate system. If we derive a new coordinate system $(\bar{x}^1, \bar{x}^2, \bar{x}^3)$ by the affine transformation with coefficients a_j^i, the sum (4-69) has the form

$$\bar{t}_{ij} \, \bar{b}^i \, \bar{c}^j$$

relative to the barred coordinates. If the t_{ij} are such that (4-69) is invariant under the affine group, then the t_{ij} are defined to be the covariant affine components of a tensor of order two. Since b^i and c^j are affine contravariants of two vectors, we have

$$b^i = a'^i_k \, \bar{b}^k, \quad c^j = a'^j_m \, \bar{c}^m$$

The invariance of the sum (4-69) then requires that

$$\bar{t}_{ij} \, \bar{b}^i \, \bar{c}^j = t_{km} \, a'^k_i \; a'^m_j \, \bar{b}^i \, \bar{c}^j$$

so that the t_{ij} satisfy the transformation law

$$\bar{t}_{ij} = t_{km} \, a'^k_i \, a'^m_j \tag{4-70}$$

under the affine group of transformations.

Now, let b_i and c_j be covariant affine components of two arbitrary vectors **b** and **c** relative to some coordinate system (x^1, x^2, x^3), and consider the sum

$$t^{ij}\, b_i\, c_j \tag{4-71}$$

where the t^{ij} are a set of nine specified quantities in the coordinate system (x^1, x^2, x^3). If we make the affine transformation

$$\bar{x}^i = a^i_j\, x^j, \quad \det(a^i) \neq 0$$

the sum (4-71) has the representation

$$\bar{t}^{ij}\, \bar{b}_i\, \bar{c}_j$$

relative to the barred coordinates. If we require the sum (4-71) to be invariant under the affine group, the set of nine quantities t^{ij} are defined to be the contravariant affine components of a tensor of order two. Under the affine transformation,

$$b_i = a^k_i\, \bar{b}_k, \quad c_j = a^m_j\, \bar{c}_m$$

so that the invariance of (4-71) implies that the t^{ij} satisfy the transformation law

$$\bar{t}^{ij} = t^{km}\, a^i\, a^j \tag{4-72}$$

under the affine group.

In a given coordinate system, there are definite relations between the mixed affine components, the covariant affine components, and the contravariant affine components of a given tensor of order two. The relations between the various components can be expressed in terms of the two sets of metric coefficients for the given coordinate system. The calculation of these relations is left as an exercise.

PROBLEMS

4-1. Let α, β, and γ be the direction angles of a vetcor **b** relative to an orthogonal cartesian coordinate system (x^1, x^2, x^3), and let $\bar{\alpha}$, $\bar{\beta}$, and $\bar{\gamma}$ be the direction angles of the same vector relative to the orthogonal cartesian system $(\bar{x}^1, \bar{x}^2, \bar{x}^3)$, where the barred and unbarred coordinates are related through an orthogonal transformation. Show that α, β, and γ describe the same direction in space as the angles $\bar{\alpha}$, $\bar{\beta}$, and $\bar{\gamma}$.

4-2. If $\phi(x, y, z)$ is an invariant, i.e. if $\bar{\phi}(\bar{x}, \bar{y}, \bar{z}) = \phi(x, y, z)$ under all orthogonal transformations, show that grad ϕ is a cartesian vector.

4-3. If **A** is a cartesian vector, show that div **A** is an invariant.

4-4. Determine whether or not the triplet of functions $(x+y,\ y+z,\ z+x)$ are the cartesian components of a cartesian vector.

4-5. A matrix C is said to be the sum of the two matrices A and B if

$c^i_j = a^i_j + b^i_j$. Show that matrix addition is

(i) Commutative

(ii) Associative

(iii) Distributive under matrix multiplication

4-6. Show that the determinant of the product of two matrices is equal to the product of the determinants.

4-7. If A and B are matrices, show that

(i) $(AB)^T = B^T A^T$

(ii) $(AB)^{-1} = B^{-1} A^{-1}$

4-8. Show that the set of all proper orthogonal transformations is a group, but that the set of all improper orthogonal transformations is *not* a group.

4-9. Let M be any 3×3 matrix, and let A be a 3×3 orthogonal matrix. If $\overline{M} = AMA^T$, show that

(i) $\det(\overline{M}) = \det(M)$

(ii) $\operatorname{Tr} \overline{M} = \operatorname{Tr} M$

4-10. Consider the proper rotation

$$\bar{x}^1 = -0.250\, x^1 + 0.433\, x^2 + 0.866\, x^3$$

$$\bar{x}^2 = -0.866\, x^1 - 0.500\, x^2$$

$$\bar{x}^3 = 0.433\, x^1 - 0.750\, x^2 + 0.500\, x^3$$

Determine the axis of rotation and the angle of rotation corresponding to the given orthogonal transformation.

4-11. Find the Euler angles corresponding to the orthogonal transformation

$$\bar{x}^1 = -0.080\, x^1 + 0.786\, x^2 + 0.612\, x^3$$

$$\bar{x}^2 = -0.862\, x^1 - 0.362\, x^2 + 0.354\, x^3$$

$$\bar{x}^3 = 0.500\, x^1 + 0.500\, x^2 + 0.707\, x^3$$

4-12. Determine the Cayley-Klein parameters for the orthogonal transformation

$$\bar{x}^1 = 0.096\, x^1 + 0.872\, x^2 + 0.480\, x^3$$

$$\bar{x}^2 = -0.896\, x^1 - 0.096\, x^2 + 0.360\, x^3$$

$$\bar{x}^3 = 0.360\, x^1 + 0.048\, x^2 + 0.880\, x^3$$

4-13. Let U be a 2×2 unitary matrix, and let H be a 2×2 hermitian matrix. If $\overline{H} = UHU^\dagger$, show that

(i) $\overline{H} = \overline{H}^\dagger$

(ii) $\operatorname{Tr} \overline{H} = \operatorname{Tr} H$

(iii) $\det(\overline{H}) = \det(H)$

4-14. Use Eq. (4-43) and the multiplication table (4-42) to obtain a representation of the three-dimensional rotation group in terms of the Pauli matrices.

4-15. Show that the transformation law for the mixed affine components of a tensor of order two can be obtained by requiring the bilinear form

$$t^i_j\, b_i\, c^j$$

where b_i and c^j are the components of two arbitrary vectors, to be invariant under the affine group.

4-16. Obtain relations between the covariant affine components, the contravariant affine components, and the mixed affine components of a given order two tensor by requiring the bilinear forms

$$t^i_j\, b_i\, c^j, \quad t^{ij}\, b_i\, c_j, \quad t_{ij}\, b^i\, c^j$$

to be equal in any given coordinate system.

Chapter 5

Tensor Formalism

In Chapter 4, we discussed the behaviour of the components of a vector under transformations from one cartesian coordinate system to another. At that time, we introduced the covariant, contravariant, and mixed affine components of an order two tensor, by requiring the invariance of certain bilinear forms under the affine group. In this chapter, we shall generalize the tensor concept, and study the behaviour of the components of vectors and tensors under more general transformations.

5-1. The Admissible Transformations. In the discussion of curvilinear coordinates, we assumed that the coordinate systems under consideration could be compared to an orthogonal cartesian system. This would seem to imply that in three dimensions there is a natural preference for orthogonal cartesian coordinates. In a sense this is true for the type of geometrical spaces we considered there. The reason for this preference will be made clear somewhat later in this chapter. Three-dimensional orthogonal cartesian coordinates and those coordinate systems derived from them exist in what is known as a Euclidean three space. This is the space of our ordinary experience. However, there are spaces of both mathematical and physical interest which are neither three-dimensional nor Euclidean. In the discussion of linear transformations, we considered only those coordinate systems which were related to an orthogonal cartesian system by means of linear or affine transformations. In general, the sets of all coordinates which are related to one another by an affine transformation define what we shall call an affine space. This space may or may not be Euclidean.

In order to develop and use the full power of vector and tensor analysis, it is necessary to consider spaces which are more general than either Euclidean or affine spaces. The coordinate systems used to describe the points in these spaces cannot, in general, be referred to an orthogonal cartesian coordinate system. We shall first consider such spaces in three dimensions, and then extend the discussion to spaces of an arbitrary, but finite, dimensionality. Consider the set of three independent variables (u^1, u^2, u^3). Each set of values of these variables define a point in a general three-dimensional space. A new set of three variables u'^i may be introduced by the functional transformations

$$u'^i = u'^i(u^j) \; ; \; i = 1, 2, 3 \tag{5-1}$$

where it is assumed that the u'^i have continuous derivatives to some order for some range of the variables u^j. It will also be assumed that the set of functional transformations (5-1) is reversible for some range of the variables u^j. The set of all points defined by the variables u^j and all other sets of three variables u'^i derived from the u^j in the manner described above define what we shall call an X_3 manifold. The sets of coordinates which are related to one another by transformations of the type (5-1) are called the admissible coordinates in X_3.

Let us now examine some of the analytic properties of the functional transformations between admissible coordinate systems in an X_3 manifold. First of all, it is required that the quantities

$$\frac{\partial^n u'^i}{(\partial u^j)^n} \; ; \; i, j = 1, 2, 3$$

be continuous for some n and some range of the variables u^j. Then, in some sufficiently small region about the point (u_0^1, u_0^2, u_0^3), we can expand the functions u'^i as the first n terms of a Taylor series about u_0^j to obtain

$$u'^i(u^j) \approx u'^i(u_0^j) + \sum_{j=1}^{3} \left[\left(\frac{\partial u'^i}{\partial u^j} \right)_{u_0^j} (u^j - u_0^j) + \dots + \left(\frac{\partial^n u'^i}{\partial (u^j)^n} \right)_{u_0^j} \frac{(u^j - u_0^j)^n}{n!} \right]$$

If we translate the origin to u_0^j and consider only those points which are sufficiently near the origin that we may ignore all but the linear term in the Taylor series for u'^i, we have

$$u'^i(u^j) \approx u'^i(0) + \left(\frac{\partial u'^i}{\partial u^j} \right)_0 u^j \tag{5-2}$$

Hence, in any sufficiently small region about any point (u_0^j), the transformation (5-1) is linear with coefficients $(\partial u'^i / \partial u^j) u_0^j$. Such a transformation is said to be locally linear.

Since the transformation is locally linear, a necessary and sufficient condition for the existence of the inverse transformation in a sufficiently small region about the point (u_0^j) is that

$$\begin{vmatrix} (\partial u'^1/\partial u^1)_{u_0} & (\partial u'^2/\partial u^1)_{u_0} & (\partial u'^3/\partial u^1)_{u_0} \\ (\partial u'^1/\partial u^2)_{u_0} & (\partial u'^2/\partial u^2)_{u_0} & (\partial u'^3/\partial u^2)_{u_0} \\ (\partial u'^1/\partial u^3)_{u_0} & (\partial u'^2/\partial u^3)_{u_0} & (\partial u'^3/\partial u^3)_{u_0} \end{vmatrix}$$

Hence, the transformation (5-1) is reversible in a neighbourhood of any point (u^j) such that

$$\begin{vmatrix} \partial u'^1/\partial u^1 & \partial u'^2/\partial u^1 & \partial u'^3/\partial u^1 \\ \partial u'^1/\partial u^2 & \partial u'^2/\partial u^2 & \partial u'^3/\partial u^2 \\ \partial u'^1/\partial u^3 & \partial u'^2/\partial u^3 & \partial u'^3/\partial u^3 \end{vmatrix} \neq 0 \tag{5-3}$$

The functional determinant which appears in Eq. (5-3) is known as the Jacobian of the transformation $u'^i = u'^i(u^j)$, and is usually written

$$J(u', u) = J(\partial u'^i/\partial u^j) = \det(\partial u'^i/\partial u^j) \tag{5-4}$$

In order for the Jacobian to be a continuous function of the u^j, it is clearly necessary that the u'^i have continuous first derivatives with respect to the u^j. Thus, a necessary and sufficient condition for the coordinate system (u'^i) to be an admissible coordinate system at the point (u^j) is that $J(u', u)$ be continuous and non-zero at the point (u^j).

The transformation which connects two admissible coordinate systems is known as an admissible transformation. It is not difficult to show that the set of all admissible transformations in a given manifold X_3 is a group, which we shall call the admissible group for the given manifold. Those points, if there are any, at which a given transformation is not admissible are called singular points of the transformation.

The extension of these remarks from three-dimensional manifolds to manifolds of dimensionality n is obvious. In a given n-dimensional manifold X_n, points are defined by the ordered n-tuples of quantities $(u^1, \ldots, u^n)$. Admissible coordinate systems in X_n are defined by the functional transformations

$$u'^i = u'^i(u^j) \quad ; \quad i, j = 1, \ldots, n \tag{5-5}$$

where we again assume continuous derivatives of u'^i to some order, and require the transformation to be reversible for some range of the variables u^j. The necessary and sufficient condition for the admissibility of the transformation (5-5) in X_n is the continuity and non-vanishing of the Jacobian of the transformation. In this case, the Jacobian is the $n \times n$ determinant with elements $\partial u'^i/\partial u^j$. The set of all admissible transformations in X_n is called the admissible group for the given manifold.

When we are dealing with an n-dimensional manifold, we shall modify the summation convention to the extent that any summed index is to be summed from one to the dimensionality of the manifold. If the summed index is over any other range, we shall either indicate the range of the summed index, or write out the summation explicitly.

EXAMPLE 1. In three-dimensional Euclidean space, the coordinate transformation defined by

$$r=\sqrt{x^2+y^2+z^2}$$

$$\theta=\tan^{-1}(\sqrt{x^2+y^2}/z)$$

$$\varphi=\tan^{-1}(y/x)$$

is an admissible transformation for all values of (x, y, z) such that

$$J=\begin{vmatrix} \partial r/\partial x & \partial r/\partial y & \partial r/\partial z \\ \partial\theta/\partial x & \partial\theta/\partial y & \partial\theta/\partial z \\ \partial\varphi/\partial x & \partial\varphi/\partial y & \partial\varphi/\partial z \end{vmatrix}$$

$$J=\begin{vmatrix} \dfrac{x}{r} & \dfrac{y}{r} & \dfrac{z}{r} \\ \dfrac{xz}{r\sqrt{x^2+y^2}} & \dfrac{yz}{r\sqrt{x^2+y^2}} & \dfrac{-\sqrt{x^2+y^2}}{r} \\ \dfrac{-y}{x^2+y^2} & \dfrac{x}{x^2+y^2} & 0 \end{vmatrix}$$

$$=\frac{1}{\sqrt{x^2+y^2}}$$

is finite and non-zero, i.e. for all (x, y, z) with x and y not both zero.

On the other hand, if we wish to describe the points on the surface of a sphere, we may use the spherical coordinates (θ, φ) which are known as the co-latitude and longitude respectively. However, in this case, it is not possible to refer θ and φ to a rectangular cartesian system. Hence, the coordinates (θ, φ) define a non-Euclidean manifold X_2. We shall later obtain criteria for a given manifold to be Euclidean without the necessity of determining whether or not the manifold can be described by cartesian coordinates. The particular manifold X_2 defined here is known as a two-dimensional Riemannian manifold.

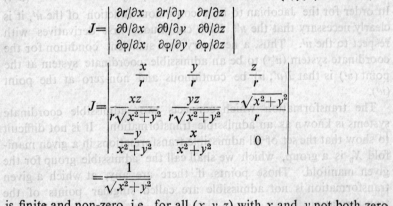

5-2. Transformation Laws. Consider an n-dimensional manifold X_n, and let $F(P)$ be a continuous function of the points P in some region of the manifold. If the points of the manifold are described in terms of some convenient coordinate system U, $P=P(u^1,\ldots, u^n)$, then in terms of this set of coordinates, the point function $F(P)$ has the form

$$F(P)=f_1(u^1,\ldots, u^n)$$

We may introduce another admissible coordinate system in X_n by the transformation

$$u'^i=u'^i\,(u^j), \quad i, j=1,\ldots, n, \quad J\,(u', u)\neq 0 \tag{5-6}$$

Since the coordinate system U' is admissible, the transformation (5-6) is reversible, and we can wirte

$$u^i=u^i\,(u'^j)$$

In terms of the coordinate system U', the function $F(P)$ has the representation

$$f_1[u^1(u'^j),\ldots, u^n(u'^j)]=f_2(u'^1,\ldots, u'^n)$$

If the functional values of f_2 are the same as the functional values of f_1 at each point in the manifold, then $F(P)$ is said to be invariant under the transformation (5-6). If $F(P)$ is invariant under the admissible group in X_n, the function $F(P)$ is defined to be a scalar invariant in X_n.

It is convenient to think of $f_1(u^i)$ as the component of the scalar $F(P)$ relative to the coordinate system U, and to regard $f_2(u'^i)$ as the component of the invariant $F(P)$ relative to the coordinate system U'. Thus, under an admissible transformation in X_n, the component of a scalar invariant transforms according to

$$T_0: f_1(u'^i)=f_2[u^j(u'^i)] \tag{5-7}$$

The transformation law (5-7) is known as transformation by invariance. If the representation of a scalar $F(P)$ is known in one admissible coordinate system of X_n, the representation of the same scalar in any other admissible coordinate system of X_n is obtained by the transformation (5-7). A scalar in X_n is defined to be the equivalence class of all its components in every admissible coordinate system of X_n, where each component is related to the others through a transformation by invariance.

Now, let $F(P)$ be a continuous scalar invariant defined in some region of the manifold X_n, and suppose that $F(P)$ has the component $f(u^1,\ldots, u^n)$ relative to an admissible coordinate system U of X_n. If $f(u)$ has continuous first derivatives in the given region, we form the set of n functions

$$g_i(u^j)=\frac{\partial f}{\partial u^i}$$

and look for a transformation law for the functions $g_i(u^j)$ under the transformation

$$u'^i = u'^i(u^j); \quad J(u', u) \neq 0$$

Since we shall later interpret the functions $g_i(u^j)$ as the components of the gradient of $F(P)$ referred to the U-coordinate system, we shall require that the transformed quantities $g_i'(u'^j)$, referred to the U'-system, be given by

$$g'(u'^j) = \frac{\partial f}{\partial u'^i} = \frac{\partial f}{\partial u^k}\frac{\partial u^k}{\partial u'^i}$$

Any set of n quantities $a_i(u^j)$ defined in a region of X_n, which transform according to

$$T_1: \; a_i'(u'^j) = \frac{\partial u^k}{\partial u'^i} a_k(u'^j) \tag{5-8}$$

under the admissible group in X_n, is defined to be the set of covariant components of the vector **a** in X_n. In the transformation (5-8), $a_k(u'^j)$ is obtained from $a_k(u^j)$ by the invariance transformation T_0. The transformation law T_1 is called transformation by covariance. The equivalence class of sets of covariant components in all admissible coordinate systems of X_n, where each set of covariant components is related to the others by the transformation law T_1, constitutes the covariant description of a vector in X_n.

Let $(u^1, \ldots, u^n)$ be an admissible coordinate system in X_n, and consider the n differentials du^i. We can interpret this set of differentials as the components of the differential $d\mathbf{r}$ which is the infinitesimal displacement of two points in X_n. If we make a transformation to the set of admissible coordinates $(u'^1, \ldots, u'^n)$ in X_n, we shall require the quantities in the primed system which correspond to the du^i to be the set of differentials du'^i. Now,

$$du'^i = \frac{\partial u'^i}{\partial u^j} du^j \tag{5-9}$$

Any set of n quantities $a^i(u^j)$, defined in a region of X_n, which transform according to

$$T_2: \; a'^i(u'^j) = \frac{\partial u'^i}{\partial u^k} a^k(u'^j)$$

under the admissible group in X_n, is defined to be the set of contravariant components of the vector **a** defined in the given region of X_n. The quantities $a^k(u'^j)$ are obtained by an invariance transformation from the quantities $a^k(u^j)$. The contravariant description of a vector in X_n is defined to be the equivalence class of sets of contravariant components in all admissible coordinate systems of X_n, where each set of components is related to the others by the transformation law T_2. The transformation law T_2 is called transformation by contravariance.

A rigorous and precise definition of a vector in X_n can be based on the covariant and contravariant descriptions. A vector in X_n is defined to be the equivalence class of its covariant and contravariant descriptions in X_n.

EXAMPLE 2. In this example, we shall apply the general covariant and contravariant transformation laws T_1 and T_2 to vectors defined in the affine space A_3. The admissible group in A_3 is the affine group

$$x'^i = a^i_j\, x^j, \quad \det(a^i) \neq 0$$

where each affine transformation has the inverse

$$x^i = a'^i_j\, x'^j$$

The affine covariant components of a vector **b** defined in A_3 transform according to

$$b'_i = \frac{\partial x^k}{\partial x'^i}\, b_k = a'^k_i\, b_k$$

which is the same as the transformation law (4-52). Similarly, the affine contravariant components of the same vector **b** transform according to

$$b'^i = \frac{\partial x'^i}{\partial x^k}\, b^k = a^t_k\, b^k$$

which is the transformation law (4-49).

The extension of the transformation laws to other geometrical entities defined in X_n is not difficult. Consider the set of n^m quantities $T^{i_1,\ldots,i_m}(u^1,\ldots,u^n)$; $i_1, i_2,\ldots, i_m = 1,\ldots, n$, defined in some admissible coordinate system $(u^1,\ldots,u^n)$ of X_n. If this set of quantities transforms according to

$$T'^{i_1,\ldots,i_m}(u'^1,\ldots,u'^n) = \left|\frac{\partial u'^i}{\partial u^j}\right|^W \frac{\partial u'^{i_1}}{\partial u^{j_1}} \cdots \frac{\partial u'^{i_m}}{\partial u^{j_m}}\, T^{j_1,\ldots,j_m}(u'; T_0; u) \tag{5-10}$$

under the admissible group in X_n, then the quantities $T^{i_1,\ldots,i_m}$ (or $T'^{i_1,\ldots,i_m}$) are defined to be the contravariant components of a tensor of weight W and order m defined in X_n. In Eq. (5-10), the notation $T^{j_1,\ldots,j_m}(u'; T_0; u)$ indicates that the $T^{j_1,\ldots,j_m}(u')$ are obtained from the quantities $T^{j_1,\ldots,j_m}(u)$ by the invariance transformation T_0. The equivalence class of all sets of contravariant components in all admissible coordinate systems of X_n, where each set of components is related to the others by transformations of the form (5-10), constitutes the contravariant description of a tensor of weight W and order m in X_n.

Next, let the set of n^m quantities $T_{i_1,\dots,i_m}(u^1,\dots,u^n)$ be defined in some admissible coordinate system U of X_n. The quantities $T_{i_1,\dots,i_m}$ are defined to be the covariant components of a tensor of weight W and order m relative to the coordinate system U, if they transform according to

$$T'_{i_1,\dots,i_m}(u') = \left|\frac{\partial u^i}{\partial u'^j}\right|^W \frac{\partial u^{j_1}}{\partial u'^{i_1}} \cdots \frac{\partial u^{j_m}}{\partial u'^{i_m}} T_{j_1,\dots,j_m}(u'; T_0; u) \qquad \text{(5-11)}$$

under the admissible group in X_n. The covariant description of a tensor in X_n, is the equivalence class of all sets of covariant components in all admissible coordinate systems of X_n, where each set of covariant components is related to the others by a transformation of the form (5-11).

Finally, consider the set of n^m quantities $T^{i_1,\dots,i_r}_{j_1,\dots,j_s}(u,\dots,u)$; $r+s=m$, which are defined in an admissible coordinate system of X_n. These quantities are the mixed components (covariant of order s and contravariant of order $r=m-s$) of a tensor of order m and weight W, if they transform according to

$$T'^{i_1,\dots,i_r}_{j_1,\dots,j_s}(u') = \left|\frac{\partial u^i}{\partial u'^j}\right|^W \frac{\partial u'^{i_1}}{\partial u^{p_1}} \cdots \frac{\partial u'^{i_r}}{\partial u^{p_s}} \frac{\partial u^{q_1}}{\partial u'^{j_1}} \cdots \frac{\partial u^{q_s}}{\partial u'^{j_s}} T^{p_1,\dots,p_r}_{q_1,\dots,q_s}(u'; T_0; u) \qquad \text{(5-12)}$$

The mixed description, covariant of order s and contravariant of order $r=m-s$, of a tensor of order m and weight W in X_n is the equivalence class of sets of mixed components (covariant of order s and contravariant of order r) in all admissible coordinate systems of X_n, each set of components being related to the others by transformations of the form (5-12). It is clear that a given tensor of order m has $(m-1)$ mixed descriptions. For example, in a fixed coordinate system of X_n, a given tensor of order four is described by any of the three sets of mixed components

$$T^{i_1i_2i_3}_{i_4}(u), \quad T^{i_1i_2}_{i_3i_4}(u), \quad T^{i_1}_{i_2i_3i_4}(u)$$

Finally, a tensor in X_n is the equivalence class of its covariant, contravariant and mixed descriptions in X_n. It is convenient to refer to the various descriptions as if they are tensors themselves, i.e. the covariant descriptions of order m will be called a covariant tensor of order m; the contravariant description the contravariant tensor; and the mixed description, covariant of order s and contravariant of order $m-s$ will be called a mixed tensor covariant of order s and contravariant of order $m-s$. This is merely a linguistic convenience, and does not imply that the different descriptions are different entities, but are only different descriptions of the same entity in X_n. We shall later develop relations between the covariant, contravariant, and mixed descriptions of a given tensor in a fixed coordinate system of X_n. Since a tensor is completely described by its set of components in a fixed coordinate system, we shall indicate the tensor by giving its general component relative to the given coordinate system. For example, we shall call

$$A^{ijk}(u),\ B_{ijk}(u),\ C^{i}_{jk}(u)$$

a contravariant tensor of order three; a covariant tensor of order three; and a mixed tensor, covariant of order two and contravariant of order one, respectively.

Tensors of order zero are scalars, and tensors of order one are vectors. Tensors of weight zero are called absolute tensors, and tensors of weight $W\neq 0$ are called relative tensors. Unless otherwise indicated, all of the tensors we shall consider are absolute. Absolute scalars are simply called scalars, while scalars of weight $W=\pm 1$ are known as pseudo-scalars. Absolute vectors are polar vectors, and vectors of weight $W=\pm 1$ are axial vectors.

EXAMPLE 3. As an example of a set of quantities which form a tensor defined in E_3 consider the set

$$\{dx^1/dt,\ dx^2/dt,\ dx^3/dt\}$$

defined in a cartesian coordinate system (x^1, x^2, x^3). These derivatives are defined to be the components of the velocity vector relative to (x^1, x^2, x^3). Let (u^1, u^2, u^3) be any admissible coordinate system in E_3 defined by the transformation

$$u^i = u^i(x^1, x^2, x^3)$$

Now, from the chain rule,

$$\frac{du^i}{dt} = \frac{\partial u^i}{\partial x^k}\frac{dx^k}{dt}$$

which is the transformation law for a tensor contravariant of order one. Hence, we can interpret the quantities du^i/dt as the components of velocity relative to the coordinates (u^1, u^2, u^3).

EXAMPLE 4. As an example of a set of quantities which

are a tensor under a restricted group of admissible transformations in E_3, consider the cartesian coordinates (x^1, x^2, x^3) of a point in E_3. If we restrict the admissible group to the affine group, the admissible coordinates are defined by

$$x'^i = a^i_j\, x^j$$

where the a^i_j are constants. Since $\partial x'^i/\partial x^j = a^i_j$, it is clear that

$$x'^i = \frac{\partial x'^i}{\partial x^j}\, x^j$$

and the coordinates of a point transform as the components of a contravariant tensor of order one under the affine group.

However, under the general admissible group in E_3

$$u^i = u^i(x^1, x^2, x^3)$$

we have $\partial u^i/\partial x^j = f^i_j\,(x^1, x^2, x^3)$, so that

$$\frac{\partial u^i}{\partial x^j}\, x^j = f^i(x^1, x^2, x^3)\, x^j \neq u^i(x^1, x^2, x^3)$$

Thus, the generalized coordinates (u^1, u^2, u^3) of a point in E_3 do not form the components of a vector under the general admissible group in E_3. On the other hand, the set of differentials (dx^1, dx^2, dx^3) are the components of a contravariant tensor of order one, since

$$du^i = \frac{\partial u^i}{\partial x^j}\, dx^j$$

which is the transformation law T_2. The reason for the somewhat anomalous situation is that the admissible transformations are locally linear, i.e. are linear in a sufficiently small neighbourhood of a given point, although they are not, in general, linear throughout the space E_3.

A second and somewhat more difficult example of a set of quantities which are a tensor only under a restricted class of admissible transformations is the set of second time derivatives of the coordinates of a point, $(d^2x^1/dt^2,\ d^2x^2/dt^2,\ d^2x^3/dt^2)$. In a cartesian system, these derivatives are the cartesian components of the acceleration vector, (a^1, a^2, a^3). We have already seen that the quantities dx^i/dt are the components of a vector under the admissible group in E_3. We might then expect that

$$\frac{d}{dt}\left(\frac{du^i}{dt}\right) = \frac{d^2u^i}{dt^2}$$

are the components of the acceleration vector in the admissible coordinate system

$$u^i = u^i(x^1, x^2, x^3)$$

This, however, is not the case, since

$$a^i(x) = \frac{d}{dt}\left(\frac{\partial x^i}{\partial u^k}\frac{du^k}{dt}\right) = \frac{\partial x^i}{\partial u^k}\frac{d^2u^k}{dt^2} + \frac{\partial^2 x^i}{\partial u^j\,\partial u^k}\frac{du^j}{dt}\frac{du^k}{dt}$$

It is clear that the set of quantities (d^2u^k/dt^2) are not the components of an order one tensor relative to the admissible coordinate system (u^1, u^2, u^3). We shall later calculate the generalized components of acceleration, and show at that time that the simple form of the acceleration components in cartesian coordinates is a result of the simple structure of the coordinate system.

EXAMPLE 5. Consider, in E_2, the tensor A_{ij} whose covariant components relative to a cartesian coordinate system (x, y) are

$$A_{11} = x^2,\ A_{12} = A_{21} = xy,\ A_{22} = y^2$$

We wish to calculate the components of A_{ij} relative to the admissible coordinate system (r, θ), which is defined by the admissible transformation

$$x = r\cos\theta,\ y = r\sin\theta$$

From the transformation law T_1,

$$\begin{aligned}
A'_{11} &= \left(\frac{\partial x}{\partial r}\right)^2 A_{11}(u;T_0;x) + 2\frac{\partial x}{\partial r}\frac{\partial y}{\partial r}A_{12}(u;T_0;x) + \left(\frac{\partial y}{\partial r}\right)^2 A_{22}(u;T_0;x)\\
&= r^2\,(\cos^4\theta + 2\sin^2\theta\cos^2\theta + \sin^4\theta)\\
&= r^2\,(\cos^2\theta + \sin^2\theta)^2\\
&= r^2\\
A'_{12} &= \frac{\partial x}{\partial r}\frac{\partial x}{\partial \theta}A_{11} + \frac{\partial x}{\partial r}\frac{\partial y}{\partial \theta}A_{12} + \frac{\partial x}{\partial \theta}\frac{\partial y}{\partial r}A_{21} + \frac{\partial y}{\partial r}\frac{\partial y}{\partial \theta}A_{22}\\
&= 0\\
A'_{21} &= A'_{12} = 0\\
A'_{22} &= \left(\frac{\partial x}{\partial \theta}\right)^2 A_{11} + 2\frac{\partial x}{\partial \theta}\frac{\partial y}{\partial \theta}A_{12} + \left(\frac{\partial y}{\partial \theta}\right)^2 A_{22}\\
&= 0
\end{aligned}$$

5-3. The Algebra of Tensors. The algebraic operations defined on tensors in X are addition, multiplication and contraction. The operation of addition is defined only for tensors of the same kind. Two tensors are said to be of the same kind if they have the same weight, the same number of covariant indices, and the same number of contravariant indices. If $A^{i_1,\ldots,i_r}_{j_1,\ldots,j_s}$ and $B^{i_1,\ldots,i_r}_{j_1,\ldots,j_s}$ are two tensors of the same kind, their sum is defined to be the set of n^{r+s} quantities

$$S(i_1,\ldots,i_r,j_1,\ldots,j_s)=A^{i_1,\ldots,i_r}_{j_1,\ldots,j_s}+B^{i_1,\ldots,i_r}_{j_1,\ldots,j_s} \tag{5-13}$$

The notation on the right-hand side of (5-13) indicates that the quantities S depend on the $r+s$ indices, but it has not yet been determined whether or not they are the components of a tensor. The difference of the two tensors is defined to be the set of n^{r+s} quantities

$$D(i_1\ldots,i_r,j_1,\ldots,j_s)=A^{i_1,\ldots,i_r}_{j_1,\ldots,j_s}-B^{i_1,\ldots,i_r}_{j_1,\ldots,j_s} \tag{5-14}$$

It can be shown (Prob. 5-5) that both $S(i_1,\ldots,i_r,j_1,\ldots,j_s)$ and $D(i_1,\ldots,i_r,j_1,\ldots,j_s)$ are tensors of the same kind as the two given tensors.

Let $A^{i_1,\ldots,i_r}_{j_1,\ldots,j_s}$ and $B^{p_1,\ldots,p_t}_{q_1,\ldots,q_v}$ be two tensors of weight W_1 and W_2 respectively, defined in the same admissible coordinate system U of X_n. The set of $n^{(r+s+t+v)}$ quantities

$$P(i_1,\ldots,i_r,j_1,\ldots,j_s,p_1,\ldots,p_t,q_1,\ldots,q_v)=A^{i_1\ldots i_r}_{j_1\ldots j_s}\times B^{p_1\ldots p_t}_{q_1\ldots q_v} \tag{5-15}$$

are defined to be the outer product of the two given tensors referred to the U coordinate system. If we make an admissible transformation to the coordinates U', the direct product of the two tensors referred to the U' coordinates is

$$P'(i_1\ldots i_r j_1\ldots j_s p_1\ldots p_t q_1\ldots q_v)=$$

$$\left|\frac{\partial u^i}{\partial u'^j}\right|^{W_1}\frac{\partial u'^{i_1}}{\partial u^{k_1}}\cdots\frac{\partial u'^{i_r}}{\partial u^{k_r}}\cdots\frac{\partial u^{m_1}}{\partial u'^{j_1}}\cdots\frac{\partial u^{m_s}}{\partial u'^{j_s}}\times$$

$$A^{k_1\ldots k_r}_{m_1\ldots m_s}(u';T_o;u)\left|\frac{\partial u^i}{\partial u'^j}\right|^{W_2}\frac{\partial u'^{p_1}}{\partial u^{n_1}}\cdots\frac{\partial u'^{p_t}}{\partial u^{n_t}}\frac{\partial u^{o_1}}{\partial u'^{q_1}}\cdots\frac{\partial u^{o_v}}{\partial u'^{q_v}}B^{n_1\ldots n_t}_{o_1\ldots o_v}(u';T_o;u)$$

$$=\frac{\partial u^i}{\partial u'^j}^{W_1+W_2}\frac{\partial u'^{i_1}}{\partial u^{k_1}}\cdots\frac{\partial u'^{i_r}}{\partial u^{k_r}}\frac{\partial u^{m_1}}{\partial u'^{j_1}}\cdots\frac{\partial u^{m_s}}{\partial u'^{j_s}}\frac{\partial u'^{p_1}}{\partial u^{n_1}}\cdots\frac{\partial u'^{p_t}}{\partial u^{n_t}}\frac{\partial u^{o_1}}{\partial u'^{q_1}}\cdots\frac{\partial u^{o_v}}{\partial u'^{q_v}}\times$$

$$P(k_1\ldots k_r m_1\ldots m_s n_1\ldots n_t o_1\ldots o_v;u';T_o;u) \tag{5-16}$$

Equation (5-16) is the transformation law of a tensor covariant of order $s+v$, contravariant of order $r+t$, and of weight W_1+W_2. Hence, the direct product of two tensors is a tensor with: (i) a number of covariant indices equal to the sum of the number of covariant indices of the two factors; (ii) a number of contravariant indices equal to the sum of the number of contravariant indices of the two factors; and (iii) a weight equal to the sum of the weights of the two factors. If we extend the direct product to include three or more factors, it is not difficult to show that the direct product is associative. Further, the direct product is distributive with respect to addition.

The third algebraic operation we can define for tensors is the operation of contraction. Let $A^{i_1 \ldots i_r}_{j_1 \ldots j_s}$ be a mixed tensor of weight W, covariant of order s and contravariant of order r. If we equate a covariant and a contravariant index, the resulting set of n^{r+s-2} sums is called the contraction of $A^{i_1 \ldots i_r}_{j_1 \ldots j_s}$. In a coordinate system U, if we contract the tensor on the index i_m, we obtain sums of the form

$$C(i_1 \ldots i_{m-1}\, i_{m+1} \ldots i_r\, j_1 \ldots j_{m-1}\, j_{m+1} \ldots j_s;\, u) = A^{i_1 \ldots i_m \ldots i_r}_{j_1 \ldots j_m \ldots j_s}(u)$$

Referred to another admissible coordinate system U', the sums have the form

$$C'(i_1 \ldots i_{m-1}\, i_{m+1} \ldots i_r\, j_1 \ldots j_{m-1}\, j_{m+1} \ldots, j_s;\, u') = \left|\frac{\partial u^i}{\partial u'^j}\right|^W \frac{\partial u'^{\,i_1}}{\partial u^{p_1}} \ldots \times$$

$$\frac{\partial u'^{i_{m-1}}}{\partial u^{p_{m-1}}} \frac{\partial u'^{i_{m+1}}}{\partial u^{p_{m+1}}} \cdots \frac{\partial u'^{i_r}}{\partial u^{p_r}} \frac{\partial u^{q_1}}{\partial u'^{j_1}} \cdots \frac{\partial u^{q_{m-1}}}{\partial u'^{j_{m-1}}} \frac{\partial u^{q_{m+1}}}{\partial u'^{j_{m+1}}} \cdots \frac{\partial u^{q_s}}{\partial u'^{j_s}} \times$$

$$C(p_1 \ldots p_{m-1}\, p_{m+1} \ldots p_r\, q_1 \ldots q_{m-1}\, q_{m+1} \ldots q_s;\, u';\, T_0;\, u)$$

This is, however, the transformation law for a tensor of weight W, covariant of order $s-1$, and contravariant of order $r-1$. The operation of contraction is not a unique operation, since a given tensor of covariant order s and contravariant order r can be contracted in $r \times s$ different ways.

EXAMPLE 6. The tensor A^{ij}_{kmn} can be contracted in six different ways to obtain the tensors

$$C^i_{km} = A^{ij}_{kmj},\ C^i_{kn} = A^{ij}_{kjn},\ C^i_{mn} = A^{ij}_{jmn}\ ,$$
$$C^j_{km} = A^{ij}_{kmi},\ C^j_{kn} = A^{ij}_{kin},\ C^j_{mn} = A^{ij}_{imn}$$

A tensor which is covariant of order s and contravariant of order r can be contracted r times if $r \geqslant s$, and s times if $r < s$. The result in the first case is a covariant tensor of order $s-r$, and if $r=s$, the result is a scalar. In the second case, the result is a tensor of order $r-s$. We shall call a purely covariant tensor (or a purely contravariant tensor) a completely contracted tensor.

If we form the direct product of two tensors, and then contract with respect to one or more pairs of indices, the resulting tensor is called the inner product of the two given tensors with respect to the given pair of indices. For example, the tensor

$$C^{\,i}_{\ mn} = A^{ij} B_{jmn}$$

is the inner product of the two tensors A^{ij} and B_{kmn} with respect to the pair of indices (j, k). In writing the inner product, it is sometimes convenient to indicate the contracted index by a dot, i.e.

$$C^{\,i}_{\ mn} = A^{i\cdot} B_{\cdot mn}$$

Although the operation of division is undefined for tensors, there are a number of algorithms, known as the tensor quotient laws, which are similar in form to the usual division algorithm. The quotient laws are particularly useful in establishing the tensorial properties of sets of quantities without the necessity of examining transformations of the set under the admissible group. Suppose that we are given a set of n^{r+s} quantities $A(i_1,\ldots, i_r, j_1,\ldots, j_s; u)$ defined in an admissible coordinate system U of X_n, and further suppose that

$$A(k, i_1,\ldots, i_r, j_1,\ldots, j_s; u)\, b_k(u) \tag{5-17}$$

where $b_k(u)$ is an arbitrary covariant vector, is a tensor covariant of order s and contravariant of order $r-1$. If we transform to another admissible coordinate system U' of X_n, the quantities corresponding to the products (5–17) are

$$A'(k, i_2\ldots i_i\, j_1\ldots j_s; u')\, b'_k(u') = \frac{\partial u'^{i_2}}{\partial u^{p_2}}\cdots\frac{\partial u'^{i_r}}{\partial u^{p_r}}\frac{\partial u^{q_1}}{\partial u'^{j_1}}\cdots\frac{\partial u^{q_s}}{\partial u'^{j_s}}\times$$
$$A(m, p_2\ldots p_r\, q_1\ldots q_s; u'; T_0; u)\, b_m(u'; T_0; u)$$

Now

$$b_m(u') = \frac{\partial u'^k}{\partial u^m}\, b'_k(u')$$

so that

$$\left(A'(k, i_2\ldots i_r\, j_1\ldots j_s; u') - \frac{\partial u'^k}{\partial u^m}\frac{\partial u'^{i_2}}{\partial u^{p_2}}\cdots\frac{\partial u'^{i_r}}{\partial u^{p_r}}\frac{\partial u^{q_1}}{\partial u'^{j_1}}\cdots\frac{\partial u^{q_s}}{\partial u'^{j_s}}\times\right.$$
$$\left. A(m, p_2\ldots p_r q_1\ldots q_s; u'; T_0; u)\right) b'_k(u') = 0$$

Since $b'(u')$ is an arbitrary vector, the bracket must vanish and hence, $A(k, i_2\ldots i_r\, j_1\ldots j_s)$ transforms as a tensor covariant of order s and contravariant of order r. A similar result is obtained if the products of the n^{r+s} quantities $B(i_1\ldots i_r, k, j_2\ldots j_s; u)$ with an arbitrary contravariant vector c^k are the components of a tensor covariant of order $s-1$ and contravariant of order r. These results can be extended in a simple way to consider the products of sets of n^{r+s} quantities with an arbitrary tensor covariant of order $s-m$ and contravariant of order $r-k$. The similarity of this type of quotient law and the division algorithm is evident if we symbollically write

$$A^{k, i_2\ldots i_r}_{j_1\ldots j_s}(u) = \frac{A^{i_2\ldots i_r}_{j_1\ldots j_s}(u)}{b_k(u)}$$

There are many other forms of the quotient law which can be easily derived.

There are certain properties of a tensor which are invariant under the admissible group in X_n. One of these invariant properties is symmetry. A tensor, say $A_{i_1 \ldots i_m \ldots i_n \ldots i_r}(u)$ is said to be symmetric with respect to the pair of indices (i_m, i_n) if an interchange of these indices does not alter the values of the components of the tensor, i.e. if

$$A_{i_1 \ldots i_m \ldots i_n \ldots i_r}(u) = A_{i_1 \ldots i_n \ldots i_m \ldots i_r}(u) \tag{5-18}$$

In order to show that the symmetry is invariant under the admissible group, we write Eq. (5-18) in the form

$$A_{i_1 \ldots i_m \ldots i_n \ldots i_r}(u) - A_{i_1 \ldots i_n \ldots i_m \ldots i_r}(u) = 0 \tag{5-19}$$

Now, the difference of two tensors is a tensor, and a tensor which has all zero components in one admissible coordinate system of X_n has all zero components in every admissible coordinate system of X_n. Thus, Eq. (5-19) holds in every admissible coordinate system, and the symmetry property is invariant under the admissible group. If a tensor is symmetric with respect to every pair of indices, it is said to be completely symmetric.

A tensor, again say $A_{i_1 \ldots i_m \ldots i_n \ldots i_r}(u)$ is called anti-symmetric or skew-symmetric with respect to a given pair of indices (i_m, i_n) if an interchange of these indices changes the sign of every component of the tensor. This property is expressed by the condition

$$A_{i_1 \ldots i_n \ldots i_m \ldots i_r}(u) = -A_{i_1 \ldots i_n \ldots i_m \ldots i_r}(u) \tag{5-20}$$

The invariance of skew-symmetry under the admissible group of X_n follows in the same way as the invariance of the symmetry property. If a tensor is skew-symmetric with respect to every change of indices, it is said to be completely skew-symmetric.

If we have any symmetric covariant tensor of order two, say a_{ij}, and any tensor $A^{i_1 \ldots i_r}_{j_1 \ldots j_s}$, then the tensor formed by the inner product

$$B^{i_1 \ldots i_{m-1}\, i_{m+1} \ldots i_r}_{j_1 \ldots j_s k} = a_{k i_m} A^{i_1 \ldots i_m \ldots i_r}_{j_1 \ldots j_s} \tag{5-21}$$

is defined to be the tensor associated with the given tensor through the tensor a_{ij}. Similarly, if a^{ij} is a symmetric tensor contravariant of order two, the tensor

$$C^{k, i_1 \ldots i_r}_{j_1 \ldots j_{m-1} j_{m+1} \ldots j_s} = a^{kjm} A^{i_1 \ldots i_r}_{j_1 \ldots j_m \ldots j_s} \tag{5-22}$$

is said to be associated with the given tensor through the tensor a^{ij}.

EXAMPLE 7. It is not possible to define the cross product of two vectors **A** and **B** in manifolds other than E_3. However, it is possible to define a rank two tensor which is related to the cross product in E_3. Let a^i and b^j be two contravariant vectors defined in the manifold X_n. We then define

$$A^{ij} = a^i b^j - a^j b^i$$

It is clear that A^{ij} is a rank two tensor, since the direct products a^ib^j and a^jb^i are both tensors of rank two and the difference of two tensors of the same kind is again a tensor of the same kind. An interchange of the indices i and j immediately shows that the tensor A^{ij} is skew-symmetric.

EXAMPLE 8. One of the fundamental quantities in the theory of elasticity is the strain tensor u_j^i. This tensor can be derived in the following way. Under the action of applied forces, any solid body exhibits elastic deformations. Relative to some cartesian coordinate system (x^1, x^2, x^3), the position of a particular point in the solid body before deformation is described by the radius vector x^i. After the deformation, the same point in the body is described by the radius vector x'^i relative to the same coordinate system. The displacement of the given point is

$$y^i = x^i - x'^i$$

When a body is deformed, the distance between points in the body changes. Let us consider two points in the body which are infinitesimally near. This restriction is necessary, since as we have seen, the quantities y^i are components of a vector only in affine spaces, although their differentials dy^i are components of a vector under the entire admissible group in E_3. If the radius vector joining the two points before the deformation is dx^i, then after the deformation, the radius vector joining the same two points is

$$dx'^i = dx^i + dy^i$$

Before the deformation, the distance between the two points is

$$ds^2 = dx^i\, dx_i$$

where dx_i is the covariant component of the radius vector between the two points. After the deformation, the distance between the two points is

$$\begin{aligned} ds'^2 &= dx'^i\, dx'_i = (dx^i + dy^i)\,(dx_i + dy_i) \\ &= ds^2 + (dy^i\, dx_i + dy_i\, dx^i) + dy^i\, dy_i \end{aligned}$$

In general, the deformation is non-uniform, so that the displacement vector dy^i is a function of the undeformed point x^i. Similarly for the covariant component dy_i. Hence,

$$dy^i = \frac{\partial y^i}{\partial x^k}\, dx^k \quad , \quad dy_i = \frac{\partial y_i}{\partial x_k}\, dx_k$$

Then,

$$ds'^2 = ds^2 + \left(\frac{\partial y^i}{\partial x^k}\, dx^k\, dx_i + \frac{\partial y_i}{\partial x_k}\, dx^i\, dx_k \right) + \frac{\partial y^i}{\partial x^m}\, \frac{\partial y_i}{\partial x_k}\, dx^m\, dx_k$$

In the second term on the right, the sum is over both i and k, so that we can write

$$\frac{\partial y^i}{\partial x^k}\, dx^k\, dx_i = \frac{\partial y^k}{\partial x^i}\, dx^i\, dx_k$$

and in the third term, we can interchange i and m to obtain

$$\frac{\partial y^m}{\partial x^i}\frac{\partial y_m}{\partial x_k}\, dx^i\, dx_k$$

Then,

$$ds'^2 = ds^2 + \left(\frac{\partial y^k}{\partial x^i} + \frac{\partial y_i}{\partial x_k} + \frac{\partial y^m}{\partial x^i}\frac{\partial y_m}{\partial x_k}\right) dx^i\, dx_k$$

Now, $dx^i dx_k$ is a tensor covariant of order one and contravariant of order one. However, the product is an invariant, and it follows from the quotient law that

$$u_i^k = \tfrac{1}{2}\left(\frac{\partial y^k}{\partial x^i} + \frac{\partial y_i}{\partial x_k} + \frac{\partial y^m}{\partial x^i}\frac{\partial y_m}{\partial x_k}\right)$$

is a tensor covariant of order one and contravariant of order one. If we restrict the admissible group to the orthogonal group, there is no distinction between covariant indices, and the strain tensor has the somewhat more symmetric form

$$u_{ik} = \tfrac{1}{2}\left(\frac{\partial y_k}{\partial x^i} + \frac{\partial y_i}{\partial x^k} + \frac{\partial y^m}{\partial x^i}\frac{\partial y_m}{\partial x^k}\right)$$

It is immediately clear from either the cartesian form or the general form that the strain tensor is symmetric.

The change in the distance between two neighbouring points, i.e. the strain, under the deformation is

$$ds'^2 - ds^2 = 2\, u_{ik}\, dx^i\, dx^k$$

5-4. The Metric Tensor. In our discussion of curvilinear coordinates in three dimensions, we found that all the measurable properties of the coordinate system are expressed in terms of the set of nine quantities g_{ij}. In particular, the distance between two points which are infinitesimally separated is given by

$$ds^2 = g_{ij}\, du^i\, du^j;\ i, j = 1, 2, 3$$

There is no difficulty in extending this idea to the case of an n-dimensional space. For the curvilinear coordinate which we considered in Chapter 3, it is always possible to describe the points of the space by a set of orthogonal cartesian coordinates. In this situation, the curvilinear coordinates u^i can be written as explicit functions of the cartesian set x^i, and vice versa. By imposing the requirement that the distance function ds^2 be invariant under the given coordinate transformation, we found that g_{ij} are given by

$$g_{ij}=\frac{\partial x^1}{\partial u^i}\frac{\partial x^1}{\partial u^j}+\frac{\partial x^2}{\partial u^i}\frac{\partial x^2}{\partial u^j}+\frac{\partial x^3}{\partial u^i}\frac{\partial x^3}{\partial u^j}$$

The extension to an n-dimensional space in which it is possible to imbed an n-dimensional set of orthogonal cartesian axes is trivial. In this case, the metric coefficients are given by

$$g_{ij}=\sum_{k=1}^{3}\frac{\partial x^k}{\partial u^i}\frac{\partial x^k}{\partial u^j} \tag{5-23}$$

In an orthogonal cartesian system, the distance function is

$$ds^2=\sum_{i=1}^{3}dx^i\,dx^i$$

and hence the metric coefficients for such a system are

$$g_{ij}=\delta_{ij}$$

The distance function ds is commonly called the element of arc, and the equation

$$ds^2=g_{ij}\,du^i\,du^j\,,\quad i,j=1,\ldots,n$$

is known as the fundamental quadratic form.

In the manifold X_n, it is not generally possible to imbed an orthogonal cartesian system, so that Eq. (5-23) is not applicable. If, however, for every admissible coordinate system U of X_n, it is possible to define a unique distance between the points $P(u^1,\ldots,u^n)$ and $Q(u^1+du^1,\ldots u^n+du^n)$ by the fundamental quadratic form

$$ds^2=g_{ij}\,du^i\,du^j \tag{5-24}$$

and if ds^2 is invariant under the admissible group of X_n, then X_n is said to be a metric manifold, and the fundamental quadratic form (5-24) is said to metrize the manifold. Although it is possible to construct non-metric manifolds, they are not particularly useful in the applications of tensor analysis to physical problems. Since, in a metric manifold X_n, the element of arc is invariant, it follows from the quotient laws that the $g_{ij}(u)$ are the components of a tensor covariant of order two. This tensor is known as the fundamental or metric tensor. The fundamental quadratic form (5-24) must be invariant under an interchange of the indices i and j, and hence the metric tensor g_{ij} is symmetric with respect to its pair of indices.

We have already seen that if it is possible to imbed an orthogonal cartesian coordinate system in the manifold X_n, then with respect to this coordinate system, $g_{ij}\,(x)=\delta_{ij}$ and the fundamental quadratic form reduces to the sum of squares

$$ds^2 = \sum_{i=1}^{n} dx^i\, dx^i$$

Now, suppose that there is some admissible coordinate system of X_n, such that with respect to this coordinate system, say Y, the fundamental quadratic form is

$$ds^2 = h_{ii}\, dy^i\, dy^i \tag{5-25}$$

where the h_{ii} are constants. Then, there exists a linear transformation from the coordinate system Y to a coordinate system Y', such that with respect to the primed system, the fundamental quadratic form has the representation

$$ds^2 = \sum_{i=1}^{n} dy'^i\, dy'^i$$

Thus, the coordinate system Y' is an orthogonal cartesian system. Any manifold X_n for which it is possible to imbed an orthogonal cartesian coordinate system, is called a Euclidean manifold. Hence, a necessary and sufficient condition for X_n to be a Euclidean manifold is that there exists in X_n an admissible coordinate system Y, with respect to which, the fundamental quadratic form is given by Eq. (5-25), and is positive definite. If the fundamental quadratic form is given by Eq. (5-25), and is indefinite, the manifold is called pseudo-Euclidean. The conditions under which it is possible to find a fundamental quadratic form of the form (5-25) is one of the principle problems of differential geometry, and is an important problem in many areas of theoretical physics. A manifold which is metrized by a positive definite fundamental quadratic form is called a Riemannian manifold.

Since the g_{ij} are the components of a tensor of order two, it is always possible to represent the metric tensor as an $n \times n$ matrix which we shall denote by g. We shall assume that the g_{ij} are at least once differentiable, and that det $(g) \neq 0$. We now form the set of n^2 quantities

$$g^{ij} = G^{ij}/\det(g)$$

where G^{ij} is the cofactor of g_{ij} in det (g). It can be shown (Prob. 5-6) that the quantities g^{ij} are the components of a tensor contravariant of order two. It further follows from the definition of the minor of a determinant that

$$g_{ij}\, g^{ik} = \delta_j^k$$

The tensor g^{ij} is symmetric since the determinant obtained by omitting the i-th row and j-th column of the symmetric determinant det (g) is the same as the determinant obtained by omitting the j-th row and the i-th column. The elements of the tensor g^{ij} are the second metric coefficients which we introduced in the discussion of curvilinear coordinates in a three-dimensional Euclidean space.

Let $P(u^i)$ and $Q(u^i+du^i)$ be two points in the metric manifold X_n. The element of arc between P and Q is, by definition

$$ds^2=g_{ij}\ du^i du^j$$

If $d\mathbf{r}$ is the vector between P and Q, it has contravariant components du^i and covariant components du_i. In terms of the vector $d\mathbf{r}$, the element of arc between P and Q is

$$ds^2=d\mathbf{r}\cdot d\mathbf{r}=du^i du_j=g_{ij}\ du^i du^j$$

Then, since the du^i are independent quantities, it follows that the covariant and contravariant components of dr are related by

$$du_i=g_{ij}\ du^j \tag{5-26}$$

If we form the inner product of both sides of Eq. (5-26) with g^{ik}, we obtain the alternative relation

$$du^i=g^{ij}\ du_j \tag{5-27}$$

The relations (5-26) and (5-27) are extended to arbitrary vectors in an obvious way. The covariant and contravariant components of a given vector in a fixed coordinate system U of the metric manifold X_n are related by

$$f_i(u)=g_{ij}f^j(u), \quad f^i(u)=g^{ij}f_j(u) \tag{5-28}$$

where $g_{ij}(u)$ and $g^{ij}(u)$ are the covariant and contravariant components of the metric tensor relative to the fixed coordinate system U.

The covariant, contravariant and mixed components of an order two tensor in a fixed coordinate system U of X_n are related by

$$\begin{aligned} t^{ij}&=g^{ik}\ g^{jm}\ t_{km} \\ t_{ij}&=g_{ik}\ g_{jm}\ t^{km} \\ t^i_j&=g^{ik}\ t_{kj}=g_{jm}\ t^{im} \end{aligned} \tag{5-29}$$

The relations (5-29) are easily proved by requiring the bilinear forms

$$t^{ij}\ b_i c_j; \quad t_{ij}\ b^i c^j; \quad t^i_j\ b_i c^j$$

where **b** and **c** are arbitrary vectors, to be the same invariant in any fixed coordinate system. Relations among the representations of higher order tensors can be obtained by considering the appropriate multilinear forms.

EXAMPLE 9. As we have previously mentioned (Example 1

of this chapter), the points on the surface of a sphere of radius r_0 can be described in terms of the coordinates (θ, φ), where θ is the co-latitude and φ is the longitude. The set of all such points is a non-Euclidean manifold X_2. It is relatively easy, however, to find the metric tensor for this manifold, since it can be imbedded in a three-dimensional Euclidean manifold. In E_3, we have made frequent use of the spherical polar coordinates (r, θ, φ), defined by

$$x = r \sin\theta \cos\varphi$$
$$y = r \sin\theta \sin\varphi$$
$$z = r \cos\theta$$

The form of the metric coefficients relative to the spherical polar system is

$$g_{11}=1; \quad g_{22}=r^2; \quad g_{33}=r^2 \sin^2\theta; \quad g_{ij}=0, \; i\neq j$$

We have also noted that the coordinate surfaces r=constant are spheres of radius r. Hence, for the surface of the sphere of radius r_0, we have the fundamental quadratic form

$$ds^2 = r_0^2 \,(d\theta)^2 + r_0^2 \sin^2\theta \,(d\varphi)^2$$

Since θ and φ are real variables, ds^2 is positive definite and the manifold is Riemannian. We shall later show that the manifold is non-Euclidean.

EXAMPLE 10. In a Euclidean manifold E_2, consider the tensor whose covariant components relative to an orthogonal cartesian system are

$$A_{11}=x^2, \quad A_{12}=A_{21}=0, \quad A_{22}=y^2$$

If we introduce the two-dimensional parabolic coordinates (u, v),

$$x=\frac{u-v}{2}, \quad y=\sqrt{uv}$$

a simple calculation yields the metric coefficients

$$g_{11}=\frac{u+v}{4u}, \quad g_{12}=g_{21}=0, \quad g_{22}=\frac{u+v}{4v}$$
$$g^{11}=\frac{4u}{u+v}, \quad g^{12}=g^{21}=0, \quad g^{22}=\frac{4v}{u+v}$$

It is also not difficult to show that the covariant components of the tensor relative to the coordinate system (u, v) are

$$\bar{A}_{11}=\tfrac{1}{16}(u^2-2uv+5v^2); \quad \bar{A}_{12}=\bar{A}_{21}=-\tfrac{1}{4}(u^2-3uv+v^2)$$
$$\bar{A}_{22}=\tfrac{1}{16}(5u^2-2uv+v^2)$$

From the first of Eqs. (5-29), the contravariant components of the tensor relative to the (u, v) coordinates are

$$\bar{A}^{11}=g^{11}\,g^{11}\bar{A}_{11}=\frac{u^2(u^2-2uv+5v^2)}{(u+v)^2}$$

$$\bar{A}^{12}=\bar{A}^{21}=g^{11}\,g^{22}\bar{A}_{12}=-\frac{4uv(u^2-3uv+v^2)}{(u+v)^2}$$

$$\bar{A}^{22}=g^{22}\,g^{22}\bar{A}_{22}=\frac{v^2(5u^2-2uv+v^2)}{(u+v)^2}$$

5-5. The Christoffel Symbols. If $\Phi(P)$ is a continuously differentiable function in the metric manifold X_n, whose component relative to the admissible coordinate system U is $\varphi(u)$, then the n derivatives $\partial\varphi/\partial u^i$ are the covariant components of the vector grad Φ, relative to the coordinate system U. Let us now consider the set of n^2 second derivatives

$$\frac{\partial}{\partial u^j}\frac{\partial\varphi}{\partial u^i}$$

defined in the coordinate system U. If we transform to another admissible coordinate system U', the second derivatives of Φ are given by

$$\frac{\partial^2\varphi'}{\partial u'^j\,\partial u'^i}=\frac{\partial}{\partial u'^j}\left(\frac{\partial\varphi}{\partial u^k}\frac{\partial u^k}{\partial u'^i}\right)$$

$$=\frac{\partial^2\varphi'}{\partial u^m\,\partial u^k}\frac{\partial u^m}{\partial u'^j}\frac{\partial u^k}{\partial u'^i}+\frac{\partial\varphi}{\partial u^k}\frac{\partial^2 u^k}{\partial u'^j\,\partial u'^i} \tag{5-30}$$

Due to the presence of the terms $(\partial^2u^k/\partial u'^j\,\partial u'^i)\,(\partial\varphi/\partial u^k)$ in Eq. (5-30), the set of second partial derivatives does not transform like a tensor covariant of order two. It can be shown, however, that in an affine space A_n, the set of second partial derivatives forms an affine tensor, covariant of order two.

Equation (5-30) would seem to indicate that the derivative of a tensor is not, in general, a tensor. In order to verify this notion, let $f_i(u)$ be the covariant components of an arbitrary vector relative to the admissible coordinate system U. Under the admissible group of X_n, the partial derivatives $\partial f_i/\partial u^j$ transform according to

$$\frac{\partial f_i'}{\partial u'^j}=\frac{\partial}{\partial u'^j}\left(\frac{\partial u^k}{\partial u'^i}f_k\right)$$

$$=\frac{\partial u^m}{\partial u'^j}\frac{\partial u^k}{\partial u'^i}\frac{\partial f_k}{\partial u^m}+\frac{\partial^2u^k}{\partial u'^j\,\partial u'^i}f_k \tag{5-31}$$

Again, Eq. (5-31) is not the transformation law for a tensor unless the admissible group is restricted to the affine group, in which case the terms $(\partial^2 u^k/\partial u'^j \, \partial u'^i) \, f_k$ each vanish identically.

In order to construct a differential calculus for tensors, it is necessary to define an operation of differentiation in such a way that the derivative of a tensor is a tensor. This task is most easily accomplished in terms of certain combinations of the partial derivatives of the metric tensor, which are known as the Christoffel three index symbols. Let $g_{ij}(u)$ and $g^{ij}(u)$ be the covariant and contravariant representations of the metric tensor for the manifold X_n, relative to the coordinate system U. The Christoffel three index symbols of the first kind, relative to U, are defined to be the n^3 quantities

$$[ij, k] = \frac{1}{2}\left(\frac{\partial g_{ik}}{\partial u^j} + \frac{\partial g_{jk}}{\partial u^i} - \frac{\partial g_{ij}}{\partial u^k}\right), \quad i, j, k = 1, \ldots, n \tag{5-32}$$

The Christoffel three index symbols of the second kind are the inner products of the Christoffel symbols of the first kind with the contravariant components of the metric tensor

$$\begin{Bmatrix} k \\ ij \end{Bmatrix}_u = g^{km}(u) \, [i\,j, m]_u \tag{5-33}$$

Since the metric tensor is symmetric, it follows the Christoffel symbols of both kinds are symmetric with respect to the indices i and j.

$$[ij, k] = [ji, k]$$

$$\begin{Bmatrix} k \\ i\,j \end{Bmatrix} = \begin{Bmatrix} k \\ ji \end{Bmatrix}$$

One consequence of the symmetry is to reduce the number of distinct Christoffel symbols of each kind. There are only $\frac{1}{2}n^2(n+1)$ distinct symbols of each kind associated with each coordinate system of X_n.

Since the derivatives of a tensor are not, in general, tensors, we should not expect the Christoffel symbols to necessarily be tensors. In order to determine the tensorial character of these symbols, it is necessary to examine their transformation properties under the admissible group of X_n. Let $g_{ij}(u)$ be the covariant components of the metric tensor relative to the admissible coordinate system U. If U' is another admissible coordinate system of X_n, the covariant components of the metric tensor relative to U' are

$$g'_{ij}(u') = \frac{\partial u^m}{\partial u'^i} \frac{\partial u^n}{\partial u'^j} g_{mn}(u'; T_o; u) \tag{5-34}$$

Relative to the coordinate system U', the Christoffel symbols of the first kind are given by

$$[ij, k]_{u'} = \frac{1}{2}\left(\frac{\partial g'_{ik}}{\partial u'^j} + \frac{\partial g'_{jk}}{\partial u'^i} - \frac{\partial g'_{ij}}{\partial u'^k}\right) \tag{5-35}$$

Consider the term $\partial g'_{ij}/\partial u'^k$. Differentiating (5-34) with respect to u'^k we obtain

$$\frac{\partial g'_{ij}}{\partial u'^k} = g_{mn}\left(\frac{\partial^2 u^m}{\partial u'^k\, \partial u'^i}\frac{\partial u^n}{\partial u'^j} + \frac{\partial u^m}{\partial u'^i}\frac{\partial^2 u^n}{\partial u'^k\, \partial u'^j}\right) + \frac{\partial u^m}{\partial u'^i}\frac{\partial u^n}{\partial u'^j}\frac{\partial u^r}{\partial u'^k}\frac{\partial g_{mn}}{\partial u^r} \tag{5-36}$$

Similarly

$$\frac{\partial g'_{ij}}{\partial u'^k} = g_{mn}\left(\frac{\partial^2 u^m}{\partial u'^k\, \partial u'^i}\frac{\partial u^n}{\partial u'^j} + \frac{\partial^2 u^m}{\partial u'^k\, \partial u'^j}\frac{\partial u^n}{\partial u'^i}\right) + \frac{\partial u^m}{\partial u'^j}\frac{\partial u^r}{\partial u'^k}\frac{\partial u^n}{\partial u'^i}\frac{\partial g_{mn}}{\partial u^r} \tag{5-37}$$

and

$$\frac{\partial g'_{jk}}{\partial u'^i} = g_{nr}\left(\frac{\partial^2 u^n}{\partial u'^i\, \partial u'^j}\frac{\partial u^r}{\partial u'^j} + \frac{\partial^2 u^n}{\partial u'^i\, \partial u'^k}\frac{u^r}{\partial u'^j}\right) + \frac{\partial u^n}{\partial u'^j}\frac{\partial u^r}{\partial u'^k}\frac{\partial u^m}{\partial u'^i}\frac{\partial g_{nr}}{\partial u^m} \tag{5-38}$$

Combining Eqs. (5-36), (5-37), and (5-38), we obtain the transformation law

$$[ij, k]_{u'} = \frac{\partial u^m}{\partial u'^i}\frac{\partial u^n}{\partial u'^j}\frac{\partial u^r}{\partial u'^k}[mn, r]_u + \frac{\partial^2 u^m}{\partial u'^i\, \partial u'^j}\frac{\partial u^n}{\partial u'^k}g_{mn} \tag{5-39}$$

Thus, the Christoffel three index symbols of the first kind transform as tensors only in an affine space. In a similar fashion, it can be shown that the three index symbols of the second kind transform according to

$$\left\{\begin{matrix} i \\ jk \end{matrix}\right\}_{u'} = \frac{\partial u'^i}{\partial u^m}\frac{\partial u^n}{\partial u'^j}\frac{\partial u^r}{\partial u'^k}\left\{\begin{matrix} m \\ nr \end{matrix}\right\}_u + \frac{\partial^2 u^m}{\partial u'^j\, \partial u'^i}\frac{\partial u'^i}{\partial u^m} \tag{5-40}$$

It follows from (5-40) that the three index symbols of the second kind transform as tensors only in an affine space.

In any coordinate system U of X_n for which the $g_{ij}(u)$ are all constants, the Christoffel three index symbols of the first and second kind are all zero. However, if the g_{ij} are all constants for some admissible coordinate system of X_n, there exists some coordinate system U' of X_n such that the fundamental quadratic form is

$$ds = \sum_{i=1}^{n} du'^i\, du'^i$$

It can also be shown that the three index symbols of the second kind with respect to a given coordinate system all vanish identically only if the components of the metric tensor with respect to that system are all constants. Hence, a necessary and sufficient condition for the manifold X_n to be Euclidean is that the Christoffel three index symbols of the second kind vanish identically in some admissible coordinate system of X_n.

If a_{ij} is any symmetric tensor of order two which has the contravariant representation a^{ij}, it is possible to define sets of quantities with respect to the a_{ij} which are analogous to the Christoffel symbols. For example, if we define

$$[ij, k; a_{ij}]=\tfrac{1}{2}\left(\frac{\partial a_{ik}}{\partial u^j}+\frac{\partial a_{jk}}{\partial u^i}-\frac{\partial a_{ij}}{\partial u^k}\right)$$

we find that the symbols $[ij, k; a_{ij}]$ transform according to

$$[ij, k; a_{ij}]_{u'}=\frac{\partial u^m}{\partial u'^i}\frac{\partial u^n}{\partial u'^j}\frac{\partial u^r}{\partial u'^k}[mn, r; a_{mn}]_u+\frac{\partial^2 u^m}{\partial u'^i\,\partial u'^j}\frac{\partial u^n}{\partial u'^k}\,a_{mn}$$

under the admissible group in X_n. Similarly, we find that if we define symbols of the second kind as

$$\left\{\begin{matrix} k \\ ij;\, a_{ij}\end{matrix}\right\}=a^{km}[ij, m; a_{ij}]$$

these symbols transform according to

$$\left\{\begin{matrix} k \\ ij;\, a_{ij}\end{matrix}\right\}_{u'}=\frac{\partial u^m}{\partial u'^i}\frac{\partial u^n}{\partial u'^j}\frac{\partial u'^k}{\partial u^r}\left\{\begin{matrix} r \\ mn;\, a_{mn}\end{matrix}\right\}_u+\frac{\partial^2 u^m}{\partial u'^i\,\partial u'^j}\frac{\partial u'^k}{\partial u^r}\,a^{rn}a_{mn}$$

The two set of symbols defined with respect to the arbitrary symmetric tensor a_{ij} do not prove to be particularly useful in subsequent developments, so we shall not consider them further.

EXAMPLE 11. As we have previously seen, one of the admissible coordinate systems in the Euclidean space E_3 is the system of spherical coordinates (r, θ, φ), which are defined by

$$x=r\sin\theta\cos\varphi$$
$$y=r\sin\theta\sin\varphi$$
$$z=r\cos\theta$$

With respect to this coordinate system, the covariant and contravariant components of the metric tensor are

$$g_{11}=1,\ g_{12}=g_{21}=g_{13}=g_{31}=0,\ g_{22}=r^2,\ g_{23}=g_{32}=0,$$
$$g_{33}=r^2\sin^2\theta;$$
$$g^{11}=1,\ g^{12}=g^{21}=g^{13}=g^{31}=0,\ g^{22}=\frac{1}{r^2},$$
$$g^{23}=g^{32}=0,\ g^{33}=\frac{1}{r^2\sin^2\theta}$$

respectively. The Christoffel three index symbols relative to spherical coordinates can be calculated either from the definitions, Eqs. (5-32) and (5-33), or by the transformation laws, Eqs. (5-39) and (5-40). The transformation laws can be conveniently used since the three index symbols relative to the cartesian coordinate system are all zero.

We shall first carry out the direct calculation. Since the off diagonal elements of the metric tensor are zero, the Christoffel symbols with all indices distinct are zero. Further, g_{11} is constant, which implies that those symbols with a repeated index of 1 are zero. Finally, the components of the metric tensor are independent of φ, so that if the index 3 appears an odd number of times, the symbol is zero. This reduces the number of possible independent non-zero symbols of each kind to seven. We have

$$[12, 2]=[21, 2]=\tfrac{1}{2}\frac{\partial g_{22}}{\partial r}=r,\ [13, 3]=[31, 3]=\tfrac{1}{2}\frac{\partial g_{33}}{\partial r}=r\sin^2\theta$$

$$[22, 1]=-\tfrac{1}{2}\frac{\partial g_{22}}{\partial r}=-r,\ [22, 2]=\tfrac{1}{2}\frac{\partial g_{22}}{\partial\theta}=0$$

$$[23, 3]=[32, 3]=\tfrac{1}{2}\frac{\partial g_{33}}{\partial\theta}=r^2\sin\theta\cos\theta$$

$$[33, 1]=-\tfrac{1}{2}\frac{\partial g_{33}}{\partial r}=-r\sin^2\theta,\ [33, 3]=-\tfrac{1}{2}\frac{\partial g_{33}}{\partial\theta}=-r^2\sin\theta\cos\theta$$

and

$$\begin{Bmatrix}1\\22\end{Bmatrix}=g^{11}[22, 1]=-r,\ \begin{Bmatrix}1\\33\end{Bmatrix}=g^{11}[33, 1]=-r\sin^2\theta$$

$$\begin{Bmatrix}2\\12\end{Bmatrix}=\begin{Bmatrix}2\\21\end{Bmatrix}=g^{22}[12, 2]=1/r,\ \begin{Bmatrix}2\\22\end{Bmatrix}=g^{22}[22, 2]=0$$

$$\begin{Bmatrix}2\\33\end{Bmatrix}=g^{22}[33, 3]=-\sin\theta\cos\theta,\ \begin{Bmatrix}3\\13\end{Bmatrix}=\begin{Bmatrix}3\\31\end{Bmatrix}=g^{33}[13, 3]=1/r$$

$$\begin{Bmatrix}3\\23\end{Bmatrix}=\begin{Bmatrix}3\\32\end{Bmatrix}=g^{33}[23, 3]=\cot\theta$$

Since the Christoffel symbols are zero in cartesian coordinates, we have from Eq. (5-39) that

$$[ij, k]=\frac{\partial^2 x^m}{\partial u^i\,\partial u^j}\frac{\partial x^n}{\partial u^k}g_{mn}=\sum_{m=1}^{3}\frac{\partial^2 x^m}{\partial u^i\,\partial u^j}\frac{\partial x^m}{\partial u^k}$$

Then,

$$\begin{aligned}[12, 2]&=\frac{\partial^2 x}{\partial r\,\partial\theta}\frac{\partial x}{\partial\theta}+\frac{\partial^2 y}{\partial r\,\partial\theta}\frac{\partial y}{\partial\theta}+\frac{\partial^2 z}{\partial r\,\partial\theta}\frac{\partial z}{\partial\theta}\\&=(\cos\theta\cos\varphi)(r\cos\theta\cos\varphi)+\\&\quad(\cos\theta\sin\varphi)(r\cos\theta\sin\varphi)+(-\sin\theta)(-r\sin\theta)\end{aligned}$$

$$=r(\cos^2\theta\cos^2\varphi+\cos^2\theta\sin^2\varphi+\sin^2\theta)$$
$$=r$$

$$[13, 3]=\frac{\partial^2 x}{\partial r\,\partial\varphi}\frac{\partial x}{\partial\varphi}+\frac{\partial^2 y}{\partial r\,\partial\theta}\frac{\partial y}{\partial\varphi}+\frac{\partial^2 z}{\partial r\,\partial\varphi}\frac{\partial z}{\partial\varphi}$$
$$=(-\sin\theta\sin\varphi)(-r\sin\theta\sin\varphi)+(\sin\theta\cos\varphi)(r\sin\theta\cos\varphi)$$
$$=r\sin^2\theta$$

The calculation of the remaining non-zero three index symbols of the first kind from the transformation law is left as an exercise. Similarly, it follows from Eq. (5-40) that

$$\begin{Bmatrix}2\\12\end{Bmatrix}=\frac{\partial^2 x}{\partial r\,\partial\theta}\frac{\partial\theta}{\partial x}+\frac{\partial^2 y}{\partial r\,\partial\theta}\frac{\partial\theta}{\partial y}+\frac{\partial^2 z}{\partial r\,\partial\theta}\frac{\partial\theta}{\partial z}$$
$$=(\cos\theta\cos\varphi)(\cos\theta\cos\varphi/r)+$$
$$(\cos\theta\sin\varphi)(\cos\theta\sin\varphi/r)+(-\sin\theta)(-\sin\theta/r)$$
$$=1/r$$

The remaining three index symbols of the second kind follow in the same way.

5-6. Differentiation of Tensors. We now return to the problem of differentiating a tensor in such a way that the result is a tensor. We have seen that the transformation law for the ordinary derivative of the components of a tensor of order one differs from the transformation law for the components of a tensor of order two by a term which involves the second derivative. A term of this same form appears in the transformation law for the Christoffel three index symbols of the second kind.

We can solve Eq. (5-40) for the second derivative to obtain

$$\frac{\partial^2 u^n}{\partial u'^j\,\partial u'^k}=\frac{\partial u^n}{\partial u'^i}\begin{Bmatrix}i\\jk\end{Bmatrix}_{u'}-\frac{\partial u^m}{\partial u'^j}\frac{\partial u^r}{\partial u'^k}\begin{Bmatrix}n\\mr\end{Bmatrix}_u \tag{5-41}$$

Substituting this result in the expression for the j-th partial derivative of f_i,

$$\frac{\partial f'_i}{\partial u'^j}=\frac{\partial u^m}{\partial u^j}\frac{\partial u^n}{\partial u'^i}\frac{\partial f_n}{\partial u^m}+\frac{\partial^2 u^n}{\partial u'^j\,\partial u'^i}f_n$$
$$=\frac{\partial u^m}{\partial u'^j}\frac{\partial u^n}{\partial u'^i}\frac{\partial f_n}{\partial u^m}+\frac{\partial u^m}{\partial u'^k}\begin{Bmatrix}k\\ij\end{Bmatrix}_{u'}f_m-$$
$$\frac{\partial u^m}{\partial u'^j}\frac{\partial u^r}{\partial u'^i}\begin{Bmatrix}n\\mr\end{Bmatrix}_u f_n \tag{5-42}$$

Now, $(\partial u^n/\partial u'^k)\,f_n=f'_k$, so that after some rearrangement of terms, we may write (5-42) in the form

$$\left(\frac{\partial f'_i}{\partial u'^j}-\left\{\begin{matrix} k \\ ij \end{matrix}\right\}_{u'} f'_k\right)=\left(\frac{\partial f_n}{\partial u^m}-\left\{\begin{matrix} r \\ mn \end{matrix}\right\}_u f_r\right)\frac{\partial u^m}{\partial u'^j}\frac{\partial u^n}{\partial u'^i} \tag{5-43}$$

Each quantity in parentheses may be regarded as being one of a set of n^2 quantities defined in the U' and U coordinate systems respectively. Then, Eq. (5-43) is the transformation law for a tensor covariant of order two. We define

$$f_{i,j}=\frac{\partial f_i}{\partial u^j}-\left\{\begin{matrix} k \\ ij \end{matrix}\right\} f_k \tag{5-44}$$

to be the covariant derivative of the covariant vector f_i with respect to u^j. It is clear that this covariant derivative is a tensor covariant of order two. In an affine space, the Christoffel symbols vanish identically, and the covariant derivative of f_i reduces to the ordinary derivative.

Now, let f^i be a contravariant vector and consider the derivative of f^i with respect to u^j. The transformation law for this derivative under the admissible group is

$$\frac{\partial f'^i}{\partial u'^j}=\frac{\partial u'^i}{\partial u^m}\frac{\partial u^n}{\partial u'^j}\frac{\partial f^m}{\partial u^n}+\frac{\partial^2 u'^i}{\partial u^m\,\partial u^n}\frac{\partial u^n}{\partial u'^j}f^m \tag{5-45}$$

If we interchange the role of the primed and unprimed coordinates in (5-41), we have

$$\frac{\partial^2 u'^i}{\partial u^m\,\partial u^n}=\frac{\partial u'^i}{\partial u^r}\left\{\begin{matrix} r \\ mn \end{matrix}\right\}_u-\frac{\partial u'^j}{\partial u^m}\frac{\partial u'^k}{\partial u^n}\left\{\begin{matrix} i \\ jk \end{matrix}\right\}_{u'} \tag{5-46}$$

We substitute (5-46) for the second derivative term in (5-45) and rearrange terms to obtain

$$\frac{\partial f'^i}{\partial u'^j}+\left\{\begin{matrix} i \\ jk \end{matrix}\right\}_{u'} f'^k=\left(\frac{\partial f^m}{\partial u^n}+\left\{\begin{matrix} m \\ nr \end{matrix}\right\}_u f^r\right)\frac{\partial u^n}{\partial u'^j}\frac{\partial u'^i}{\partial u^m} \tag{5-47}$$

Thus, the set of n^2 quantities

$$\frac{\partial f^m}{\partial u^n}+\left\{\begin{matrix} i \\ jk \end{matrix}\right\}_u f^r$$

are the components of a mixed tensor covariant of order one and contravariant of order one. We define

$$f^i_{,j}=\frac{\partial f^i}{\partial u^j}+\left\{\begin{matrix} i \\ jk \end{matrix}\right\}_u f^k \tag{5-48}$$

to be the covariant derivative with respect to u^j of the contravariant vector f^i.

The extension of the definition of covariant differentiation to higher order tensors is not difficult. For example, the covariant derivative with respect to u^m of the mixed tensor $A^{i_1 \ldots i_r}_{j_1 \ldots j_s}$ is defined to be

$$A^{i_1\ldots i_r}_{j_1\ldots j_s,m}=\frac{\partial A^{i_1\ldots i_r}_{j_1\ldots j_s}}{\partial u^m}-\begin{Bmatrix} n \\ j_1 m \end{Bmatrix} A^{i_1\ldots i_r}_{n j_1\ldots j_s}-\ldots-$$

$$\begin{Bmatrix} n \\ j_1 m \end{Bmatrix} A^{i_1\ldots i_r}_{j_1\ldots j_{s-1} n}+\begin{Bmatrix} i_1 \\ nm \end{Bmatrix} A^{n i_2\ldots i_r}_{j_1\ldots j_s}+\ldots+$$

$$\begin{Bmatrix} i_r \\ nm \end{Bmatrix} A^{i_1\ldots i_{r-1} n}_{j_1\ldots j_s} \tag{5-49}$$

In order to complete the definition of covariant differentiation, we define the covariant derivative with respect to u^i of the scalar function ϕ to be the ordinary derivative

$$\phi,_{\,i}=\frac{\partial \phi}{\partial u^i} \tag{5-50}$$

The covariant derivatives of the sum and direct product of two tensors have the same form as the ordinary derivative of the sum and product of two scalar functions. If

$$S^{i_1\ldots i_r}_{j_1\ldots j_s}=A^{i_1\ldots i_r}_{j_1\ldots j_s}+B^{i_1\ldots i_r}_{j_1\ldots j_s}$$

then,

$$S^{i_1\ldots i_r}_{j_1\ldots j_s,\,m}=A^{i_1\ldots i_r}_{j_1\ldots j_s,\,m}+B^{i_1\ldots i_r}_{j_1\ldots j_s,\,m} \tag{5-51}$$

Similarly, if

$$P^{i_1\ldots i_r p_1\ldots p_t}_{j_1\ldots j_s q_1\ldots q_v}=A^{i_1\ldots i_r}_{j_1\ldots j_s}\times B^{p_1\ldots p_t}_{q_1\ldots q_v}$$

then the covariant derivative of the product is

$$P^{i_1\ldots i_r p_1\ldots p_t}_{j_1\ldots j_s q_1\ldots q_v,\,m}=A^{i_1\ldots i_r}_{j_1\ldots j_s,\,m}\times B^{p_1\ldots p_t}_{q_1\ldots q_v}+A^{i_1\ldots i_r}_{j_1\ldots j_s}\times B^{p_1\ldots p_t}_{q_1\ldots q_v,\,m} \tag{5-52}$$

We shall next show that the operations of covariant differentiation and contraction commute. In order to do this, it is first convenient to consider the covariant derivatives of the Kronecker delta. By definition

$$\delta^i_{j,\,k}=\frac{\partial \delta^i_j}{\partial u^k}-\begin{Bmatrix} m \\ jk \end{Bmatrix}\delta^i_m+\begin{Bmatrix} i \\ mk \end{Bmatrix}\delta^m_j=0 \tag{5-53}$$

Hence, the Kronecker delta behaves as a constant under covariant differentiation. The contracted tensor $A^{i_1\ldots i_{m-1} i_{m+1}\ldots i_r}_{j_1\ldots j_{n-1} j_{n+1}\ldots j_s}$ can be obtained as the direct product

$$A^{i_1\ldots i_{m-1} i_{m+1}\ldots i_r}_{j_1\ldots j_{n-1} j_{n+1}\ldots j_s}=A^{i_1\ldots i_r}_{j_1\ldots j_s}\,\delta^{j_n}_{i_m}$$

It then follows from (5-52) and (5-53) that the operation of contraction commutes with the operation of covariant differentiation.

It is not difficult to obtain expressions for the partial derivatives of the metric tensor g_{ij} in terms of the Christoffel three index symbols. It follows from (5-32) that

$$\frac{\partial g_{ij}}{\partial u^k}=[ik,\,j]+[jk,\,i] \tag{5-54}$$

In terms of the three index symbols of the second kind, the derivative is expressed as

$$\frac{\partial g_{ij}}{\partial u^k}=g_{mj}\begin{Bmatrix} m \\ ik \end{Bmatrix}+g_{mi}\begin{Bmatrix} m \\ jk \end{Bmatrix} \tag{5-55}$$

In order to obtain the analogous formulas for the derivatives of the contravariant components of the metric tensor, we differentiate the identity

$$g^{ij}\,g_{jm}=\delta^i_m$$

with respect to u^k to obtain

$$g_{jm}\,\frac{\partial g^{ij}}{\partial u^k}=-g^{ij}\,\frac{\partial g_{jm}}{\partial u^k}$$

If we multiply by g^{jn} and express the derivative terms on the right by either (5-54) or (5-55), we have

$$\frac{\partial g^{ij}}{\partial u^k}=-g^{ni}\,g^{jm}\,([nk, m]+[mk, n])$$

$$=-g^{im}\begin{Bmatrix} j \\ mk \end{Bmatrix}-g^{jm}\begin{Bmatrix} i \\ mk \end{Bmatrix} \tag{5-56}$$

Now, consider the covariant derivative of g_{ij} with respect to u^k

$$g_{ij,k}=\frac{\partial g_{ij}}{\partial u^k}-g_{mj}\begin{Bmatrix} m \\ ik \end{Bmatrix}-g_{mi}\begin{Bmatrix} m \\ jk \end{Bmatrix}=0 \tag{5-57}$$

It is left as an exercise to show that

$$g^{ij}_{,k}=0 \tag{5-58}$$

The results (5-57) and (5-58) are known as Ricci's Theorem. There are two immediate corollaries of Ricci's theorem which are of interest. First, since the elements of the metric tensor behave as constants under covariant differentiation, it follows that the operation of raising or lowering indices commutes with the operation of covariant differentiation,

$$(g_{im}\,A^m_{jk})_{,n}=g_{im}\,A^m_{jk,n}=A_{ijk,n}$$

$$(g^{ij}\,A_{jk})_{,n}=g^{ij}\,A_{jk,n}=A^i_{k,n}$$

The second corollary is that, if in a given coordinate system U of X_n, the Christoffel three index symbols of the second kind vanish identically, then the $g_{ij}(u)$ are all constants.

Since the covariant derivative of a tensor is a tensor, it can be covariantly differentiated a second time to form yet another tensor, which is called the second covariant derivative of the original tensor. Let A_{ij} be a tensor, covariant of order two, whose components are continuously twice differentiable functions of the coordinates (u^i), and consider the two second covariant derivatives $A_{ij,km}$ and $A_{ij,mk}$. Since

$$A_{ij,k}=\frac{\partial A_{ij}}{\partial u^k}-\begin{Bmatrix} m \\ jk \end{Bmatrix} A_{im}-\begin{Bmatrix} m \\ ik \end{Bmatrix} A_{mj}$$

we have

$$A_{ij,km}=\frac{\partial A_{ij,k}}{\partial u^m}-\begin{Bmatrix} r \\ im \end{Bmatrix} A_{rj,k}-\begin{Bmatrix} r \\ jm \end{Bmatrix} A_{ir,k}-\begin{Bmatrix} r \\ km \end{Bmatrix} A_{ij,r}$$

$$\begin{aligned} A_{ij,km}=\frac{\partial}{\partial u^m}\left(\frac{\partial A_{ij}}{\partial u^k}-\begin{Bmatrix} m \\ jk \end{Bmatrix} A_{im}-\begin{Bmatrix} m \\ ik \end{Bmatrix} A_{mj}\right)- \\ \begin{Bmatrix} r \\ im \end{Bmatrix}\left(\frac{\partial A_{rj}}{\partial u^k}-\begin{Bmatrix} n \\ jk \end{Bmatrix} A_{rn}-\begin{Bmatrix} n \\ rk \end{Bmatrix} A_{nj}\right)- \\ \begin{Bmatrix} r \\ jm \end{Bmatrix}\left(\frac{\partial A_{ir}}{\partial u^k}-\begin{Bmatrix} n \\ rk \end{Bmatrix} A_{in}-\begin{Bmatrix} n \\ ik \end{Bmatrix} A_{nr}\right)- \\ \begin{Bmatrix} r \\ km \end{Bmatrix}\left(\frac{\partial A_{ij}}{\partial u^r}-\begin{Bmatrix} n \\ jr \end{Bmatrix} A_{in}-\begin{Bmatrix} n \\ ir \end{Bmatrix} A_{nj}\right) \end{aligned} \tag{5-59}$$

Similarly,

$$\begin{aligned} A_{ij,mk}=\frac{\partial}{\partial u^k}\left(\frac{\partial A_{ij}}{\partial u^m}-\begin{Bmatrix} n \\ jm \end{Bmatrix} A_{in}-\begin{Bmatrix} n \\ im \end{Bmatrix} A_{nj}\right)- \\ \begin{Bmatrix} r \\ ik \end{Bmatrix}\left(\frac{\partial A_{rj}}{\partial u^m}-\begin{Bmatrix} n \\ jm \end{Bmatrix} A_{rn}-\begin{Bmatrix} n \\ rm \end{Bmatrix} A_{nj}\right)- \\ \begin{Bmatrix} r \\ jk \end{Bmatrix}\left(\frac{\partial A_{ri}}{\partial u^m}-\begin{Bmatrix} n \\ rm \end{Bmatrix} A_{in}-\begin{Bmatrix} n \\ im \end{Bmatrix} A_{nr}\right)- \\ \begin{Bmatrix} r \\ km \end{Bmatrix}\left(\frac{\partial A_{ij}}{\partial u^r}-\begin{Bmatrix} n \\ jr \end{Bmatrix} A_{in}-\begin{Bmatrix} n \\ ir \end{Bmatrix} A_{nj}\right) \end{aligned} \tag{5-60}$$

Forming the difference $A_{ij,km}$ and $A_{ij,mk}$ we obtain

$$\begin{aligned} A_{ij,km}-A_{ij,mk}=\frac{\partial}{\partial u^k}\left(\begin{Bmatrix} n \\ im \end{Bmatrix} A_{jn}+\begin{Bmatrix} n \\ im \end{Bmatrix} A_{nj}\right)- \\ \frac{\partial}{\partial u^m}\left(\begin{Bmatrix} n \\ jk \end{Bmatrix} A_{in}+\begin{Bmatrix} n \\ ik \end{Bmatrix} A_{nj}\right)+ \\ \begin{Bmatrix} r \\ ik \end{Bmatrix}\left(\frac{\partial A_{rj}}{\partial u^m}-\begin{Bmatrix} n \\ jm \end{Bmatrix} A_{rn}-\begin{Bmatrix} n \\ rm \end{Bmatrix} A_{nj}\right)- \\ \begin{Bmatrix} r \\ im \end{Bmatrix}\left(\frac{\partial A_{rj}}{\partial u^k}-\begin{Bmatrix} n \\ jk \end{Bmatrix} A_{rn}-\begin{Bmatrix} n \\ rk \end{Bmatrix} A_{nj}\right)+ \\ \begin{Bmatrix} r \\ jk \end{Bmatrix}\left(\frac{\partial A_{ir}}{\partial u^m}-\begin{Bmatrix} n \\ rm \end{Bmatrix} A_{in}-\begin{Bmatrix} n \\ im \end{Bmatrix} A_{nr}\right)- \\ \begin{Bmatrix} r \\ jm \end{Bmatrix}\left(\frac{\partial A_{ir}}{\partial u^k}-\begin{Bmatrix} n \\ rk \end{Bmatrix} A_{in}-\begin{Bmatrix} n \\ ik \end{Bmatrix} A_{nr}\right) \end{aligned}$$

If we carry out the indicated differentiations, we have, after some manipulation of indices

$$A_{ij,km}-A_{ij,mk}=A_{in}\left(\frac{\partial}{\partial u^k}\begin{Bmatrix} n \\ jm \end{Bmatrix}-\frac{\partial}{\partial u^m}\begin{Bmatrix} n \\ jk \end{Bmatrix}+\begin{Bmatrix} r \\ jm \end{Bmatrix}\begin{Bmatrix} n \\ rk \end{Bmatrix}-\begin{Bmatrix} r \\ jk \end{Bmatrix}\begin{Bmatrix} n \\ rm \end{Bmatrix}\right)+A_{nj}\left(\frac{\partial}{\partial u^k}\begin{Bmatrix} n \\ im \end{Bmatrix}-\frac{\partial}{\partial u^m}\begin{Bmatrix} n \\ ik \end{Bmatrix}+\begin{Bmatrix} r \\ im \end{Bmatrix}\begin{Bmatrix} n \\ rk \end{Bmatrix}-\begin{Bmatrix} r \\ ik \end{Bmatrix}\begin{Bmatrix} n \\ rm \end{Bmatrix}\right) \quad (5\text{-}61)$$

Since both $A_{ij,km}$ and $A_{ij,mk}$ are tensors covariant of order four, their difference is a tensor covariant of order four, and each term on the right of (5-61) is a tensor covariant of order four. Each term on the right of (5-61) is an inner product of the parenthesis with a tensor covariant of order two, and hence, by the quotient law, each parenthesis is a tensor covariant of order three and contravariant of order one. We define

$$R^n_{jkm}=\frac{\partial}{\partial u^k}\begin{Bmatrix} n \\ jm \end{Bmatrix}-\frac{\partial}{\partial u^m}\begin{Bmatrix} n \\ jk \end{Bmatrix}+\begin{Bmatrix} r \\ jm \end{Bmatrix}\begin{Bmatrix} n \\ rk \end{Bmatrix}-\begin{Bmatrix} r \\ jk \end{Bmatrix}\begin{Bmatrix} n \\ rm \end{Bmatrix} \quad (5\text{-}62)$$

to be the mixed representation of the Riemann-Christoffel tensor (also known as the second Riemann-Christoffel tensor). In terms of this tensor, the difference between the second covariant derivatives is

$$A_{ij,km}-A_{ij,mk}=A_{in}\,R^n_{jkm}+A_{nj}\,R^n_{ikm} \quad (5\text{-}63)$$

Equation (5-63) can be generalized to higher order tensors. In the general case, the difference between the second covariant derivatives is given by

$$A^{i_1\ldots i_r}_{j_1\ldots j_s,km}-A^{i_1\ldots i_r}_{j_1\ldots j_s,mk}=\sum_{n=1}^{s} A^{i_1\ldots i_r}_{j_1\ldots j_{n-1}\,n j_{n+1}\ldots j_s}\,R^n_{j_nkm}-\sum_{n=1}^{r} A^{i_1\ldots i_{n-1}\,n i_{n+1}\ldots i_r}_{j_1\ldots j_s}\,R^{i_n}_{nkm} \quad (5\text{-}64)$$

It is clear from Eq. (5-64) that in order to interchange the order of covariant differentiation, it is both necessary and sufficient that the Riemann-Christoffel tensor vanish identically. The Riemann-Christoffel tensor is identically zero if and only if the Christoffel three index symbols of the second kind all vanish in some coordinate system U of X_n. This latter condition is both necessary and sufficient for the manifold X_n to be a Euclidean manifold. Thus, a given manifold is Euclidean if and only if the Riemann-Christoffel tensor for X_n is a null tensor. Further, the second covariant derivative of a tensor is independent of the order of differentiation in a Euclidean manifold, and only in such a manifold.

It follows from the definition (5-62) that the mixed representation of the Riemann-Christoffel tensor can be formally written as

$$R^i_{jkm} = \begin{vmatrix} \frac{\partial}{\partial u^k} & \frac{\partial}{\partial n^m} \\ \begin{Bmatrix} i \\ jk \end{Bmatrix} & \begin{Bmatrix} i \\ jm \end{Bmatrix} \end{vmatrix} + \begin{vmatrix} \begin{Bmatrix} i \\ nk \end{Bmatrix} & \begin{Bmatrix} i \\ nm \end{Bmatrix} \\ \begin{Bmatrix} n \\ jk \end{Bmatrix} & \begin{Bmatrix} n \\ jm \end{Bmatrix} \end{vmatrix} \tag{5-65}$$

It is also convenient to consider the covariant representation of the Riemann-Christoffel tensor (which is also known as the Riemann-Christoffel tensor of the first kind),

$$R_{ijkm} = g_{in} R^n_{jkm} \tag{5-66}$$

If we substitute the explicit form of the mixed representation, Eq. (5-62), in (5-66), we obtain

$$R_{ijkm} = g_{in} \frac{\partial}{\partial u^k} \begin{Bmatrix} n \\ jm \end{Bmatrix} - g_{in} \frac{\partial}{\partial u^m} \begin{Bmatrix} n \\ jk \end{Bmatrix} + \begin{Bmatrix} r \\ jm \end{Bmatrix} [rk, i] - \begin{Bmatrix} r \\ jk \end{Bmatrix} [rm, i]$$

But,

$$g_{in} \frac{\partial}{\partial u^k} \begin{Bmatrix} n \\ jm \end{Bmatrix} = \frac{\partial}{\partial u^k} [jm, i] - \begin{Bmatrix} n \\ jm \end{Bmatrix} \frac{\partial g_{in}}{\partial u^k}$$

$$= \frac{\partial}{\partial u^k} [jm, i] - \begin{Bmatrix} n \\ jm \end{Bmatrix} [ik, n] - \begin{Bmatrix} n \\ jm \end{Bmatrix} [nk, i]$$

and hence,

$$R_{ijkm} = \frac{\partial}{\partial u^k} [jm, i] - \frac{\partial}{\partial u^m} [jk, i] + \begin{Bmatrix} n \\ jk \end{Bmatrix} [im, n] - \begin{Bmatrix} n \\ jm \end{Bmatrix} [ik, n] \tag{5-67}$$

Equation (5-67) is formally equivalent to the expansion of

$$R_{ijkm} = \begin{vmatrix} \frac{\partial}{\partial u^k} & \frac{\partial}{\partial u^m} \\ [jk, i] & [jm, i] \end{vmatrix} + \begin{vmatrix} \begin{Bmatrix} n \\ jk \end{Bmatrix} & \begin{Bmatrix} n \\ jm \end{Bmatrix} \\ [ik, n] & [im, n] \end{vmatrix}$$

So far in our discussion of covariant differentiation, we have considered only absolute tensors. The concept of covariant differentiation can be extended to include relative tensors of weight W. We shall extend the idea only to tensors covariant of order one, and indicate the more general result. Let f_i be a covariant vector of weight W. Then

$$\frac{\partial f'_i}{\partial u'^j} = \frac{\partial}{\partial u'^j} \left|\frac{\partial u'^m}{\partial u^n}\right|^W \frac{\partial u^k}{\partial u'^i} f_k$$

$$= \left|\frac{\partial u'^m}{\partial u^n}\right|^W \frac{\partial u^k}{\partial u'^i} \frac{\partial u^r}{\partial u'^j} \frac{\partial f_k}{\partial u^r} + \left|\frac{\partial u'^m}{\partial u^n}\right|^W \frac{\partial^2 u^k}{\partial u'^j \partial u'^i} f_k +$$

$$\frac{\partial}{\partial u'^j} \left|\frac{\partial u'^m}{\partial u^n}\right|^W \frac{\partial u^k}{\partial u'^i} f_k$$

Now,

$$\frac{\partial}{\partial u'^j}\left|\frac{\partial u'^m}{\partial u^n}\right|^W = W\left|\frac{\partial u'^m}{\partial u^n}\right|^{W-1}\frac{\partial u^r}{\partial u'^j}\frac{\partial}{\partial u^r}\left(\frac{\partial u'^m}{\partial u^n}\right)$$

and

$$\frac{\partial}{\partial u^r}\left(\frac{\partial u'^m}{\partial u^n}\right) = \frac{\partial^2 u'^p}{\partial u^r\,\partial u^q}\frac{\partial u^q}{\partial u'^p}\frac{\partial u'^m}{\partial u^n}$$

Hence,

$$\frac{\partial f'_i}{\partial u'^j} = \left|\frac{\partial u'^m}{\partial u^n}\right|^W \frac{\partial u^k}{\partial u'^i}\frac{\partial u^r}{\partial u'^j}\frac{\partial f_k}{\partial u^r} + \frac{\partial^2 u^k}{\partial u'^j\,\partial u'^i} f_k +$$

$$W\frac{\partial^2 u'^p}{\partial u^r\,\partial u^q}\frac{\partial u^q}{\partial u'^p}\frac{\partial u^r}{\partial u'^j}\frac{\partial u^k}{\partial u'^i} f_k$$

Substituting for the second derivatives from Eqs. (5-41) and (5-46), we obtain after a certain amount of manipulation of indices

$$\frac{\partial f'_i}{\partial u'^j} = \left|\frac{\partial u'^m}{\partial u^n}\right|^W \frac{\partial u^n}{\partial u'^i}\frac{\partial u^t}{\partial u'^j}\left(\frac{\partial f_n}{\partial u^t} - W\begin{Bmatrix} q \\ tq \end{Bmatrix}_u f_n - \begin{Bmatrix} k \\ tn \end{Bmatrix}_u f_k\right) +$$

$$\begin{Bmatrix} m \\ ij \end{Bmatrix}_{u'} f'_m + W\begin{Bmatrix} s \\ js \end{Bmatrix}_u f'_i$$

Hence, the quantity

$$\frac{\partial f_n}{\partial u^t} - W\begin{Bmatrix} q \\ tq \end{Bmatrix} f_n - \begin{Bmatrix} k \\ tn \end{Bmatrix} f_k$$

transforms according to the transformation law for a tensor of order two and weight W. We therefore define the covariant derivative with respect to u^j of the contravariant vector of weight W to be

$$f_{i,j} = \frac{\partial f_i}{\partial u^j} - W\begin{Bmatrix} k \\ jk \end{Bmatrix} f_i - \begin{Bmatrix} k \\ ij \end{Bmatrix} f_k \tag{5-68}$$

In general, if the tensor $A^{i_1\ldots i_r}_{j_1\ldots j_s}$ is of weight W, its covariant derivative with respect to u^k is

$$A^{i_1\ldots i_r}_{j_1\ldots j_s,k} = \frac{\partial}{\partial u^k} A^{i_1\ldots i_r}_{j_1\ldots j_s} - W\begin{Bmatrix} m \\ km \end{Bmatrix} A^{i_1\ldots i_r}_{j_1\ldots j_s} -$$

$$\begin{Bmatrix} m \\ j_1 k \end{Bmatrix} A^{i_1\ldots i_r}_{m j_2\ldots j_s} - \ldots - \begin{Bmatrix} m \\ j_s k \end{Bmatrix} A^{i_1\ldots i_r}_{j_1\ldots j_{s-1} m} +$$

$$\begin{Bmatrix} i_1 \\ mk \end{Bmatrix} A^{m i_2\ldots i_r}_{j_1\ldots j_s} + \ldots + \begin{Bmatrix} i_r \\ mk \end{Bmatrix} A^{i_1\ldots i_{r-1} m}_{j_1\ldots j_s} \tag{5-69}$$

If the tensor is absolute, $W=0$, and Eq. (5-69) reduces to (5-49).

It is possible to obtain a rather interesting geometrical interpretation of the covariant derivative. We shall restrict this interpretation to the covariant derivative of an absolute or polar vector in a three-dimensional Euclidean manifold. Let (u^1, u^2, u^3) be a generalized coordinate system in a three-dimensional Euclidean manifold which is covered by the orthogonal cartesian system (x^1, x^2, x^3). Let $P(u^1, u^2, u^3)$ and $Q(u^1+du^1, u^2+du^2, u^3+du^3)$ be two points which are infinitesimally separated. If $\mathbf{f}(u^1, u^2, u^3)$ is a vector field defined at the points P and Q, we shall calculate the change in $\mathbf{f}$ between the two points. In order to obtain a covariant description of $\mathbf{f}$, we introduce the unitary vectors $\mathbf{e}_i(P)$ and $\mathbf{e}_i(Q)$ as bases. At the point P, the vector $\mathbf{f}$ has the representation

$$\mathbf{f}(P)=f^i(P)\,\mathbf{e}_i(P)$$

and at the point Q

$$\mathbf{f}(Q)=f^i(Q)\,\mathbf{e}_i(Q)$$

The change in the vector $\mathbf{f}$ as we go from P to Q is induced in part by a change in the components f^i and in part by a change in the basis vectors $\mathbf{e}_i$. Neglecting the higher order differentials, the total change in $\mathbf{f}$ is given by

$$d\mathbf{f}=df^i\,\mathbf{e}_i+f^i\,d\mathbf{e}_i$$

Similarly, the rate of change of $\mathbf{f}$ with respect to the coordinate u^j is given by

$$\frac{\partial \mathbf{f}}{\partial u^j}=\frac{\partial f^i}{\partial u^j}\mathbf{e}_i+f^i\frac{\partial \mathbf{e}_i}{\partial u^j} \tag{5-70}$$

It is now necessary to obtain a representation of $\partial\mathbf{e}_i/\partial u^j$ in terms of the basis set $\mathbf{e}_k$. If $\mathbf{r}$ is the radius vector from an arbitrary origin to the point P, the unitary basis vectors are defined by

$$\mathbf{e}_i=\frac{\partial \mathbf{r}}{\partial u^i} \tag{5-71}$$

The covariant components of the metric tensor are defined to be the inner products of the set of unitary vectors

$$g_{ij}=\mathbf{e}_i\cdot\mathbf{e}_j$$

and hence,

$$\frac{\partial g_{ij}}{\partial u^k}=\mathbf{e}_i\cdot\frac{\partial \mathbf{e}_j}{\partial u^k}+\frac{\partial \mathbf{e}_i}{\partial u^k}\cdot\mathbf{e}_j$$

Similarly,

$$\frac{\partial g_{ik}}{\partial u^j}=\mathbf{e}_i\cdot\frac{\partial \mathbf{e}_k}{\partial u^j}+\frac{\partial \mathbf{e}_i}{\partial u^j}\cdot\mathbf{e}_k$$

$$\frac{\partial g_{jk}}{\partial u^i}=\mathbf{e}_j\cdot\frac{\partial \mathbf{e}_k}{\partial u^i}+\frac{\partial \mathbf{e}_j}{\partial u^i}\cdot\mathbf{e}_k$$

Then,

$$[ij, k]=\tfrac{1}{2}\left(\frac{\partial g_{ik}}{\partial u^j}+\frac{\partial g_{jk}}{\partial u^i}-\frac{\partial g_{ij}}{\partial u^k}\right)$$

$$=\tfrac{1}{2}\left[\mathbf{e}_i\cdot\left(\frac{\partial \mathbf{e}_k}{\partial u^j}-\frac{\partial \mathbf{e}_j}{\partial u^k}\right)+\mathbf{e}_j\cdot\left(\frac{\partial \mathbf{e}_k}{\partial u^i}-\frac{\partial \mathbf{e}_i}{\partial u^k}\right)+\mathbf{e}_k\cdot\left(\frac{\partial \mathbf{e}_i}{\partial u^j}+\frac{\partial \mathbf{e}_j}{\partial u^i}\right)\right]$$

However, it follows form (5-71) that $\partial\mathbf{e}_i/\partial u^j=\partial\mathbf{e}_j/\partial u^i$, and hence

$$\mathbf{e}_k\cdot\frac{\partial \mathbf{e}_i}{\partial u^j}=[ij, k]$$

For any vector **a**, $\mathbf{a}\cdot\mathbf{e}_k$ is the k-th covariant component of **a** in the representation $\mathbf{a}=\mathbf{a}_k\mathbf{e}^k$, where the $\mathbf{e}^k$ are the reciprocal unitary basis. Therefore,

$$\frac{\partial \mathbf{e}_i}{\partial u^j}=[ij, k]\ \mathbf{e}^k$$

and

$$\mathbf{e}^m\cdot\frac{\partial \mathbf{e}_i}{\partial u^j}=[ij, k]\ \mathbf{e}^k\cdot\mathbf{e}^m$$
$$=g^{mk}\ [ij, k]$$
$$=\begin{Bmatrix} m \\ ij \end{Bmatrix}$$

The quantity $(\partial\mathbf{e}_i/\partial u^j)\cdot\mathbf{e}^m$ is the m-th contravariant component of the vector $\partial\mathbf{e}_i/\partial u^j$, so that we may write

$$\frac{\partial \mathbf{e}_i}{\partial u^j}=\begin{Bmatrix} k \\ ij \end{Bmatrix}\mathbf{e}_k$$

If we substitute this result in Eq. (5-70), we find that

$$\frac{\partial \mathbf{f}}{\partial u^j}=\frac{\partial f^i}{\partial u^j}\mathbf{e}_i+f^i\begin{Bmatrix} k \\ ij \end{Bmatrix}\mathbf{e}_k$$
$$=\left(\frac{\partial f^i}{\partial u^j}+f^i\begin{Bmatrix} i \\ kj \end{Bmatrix}\right)\mathbf{e}_i \tag{5-72}$$

The bracketed quantity in Eq. (5-72) is the covariant derivative of f^i with respect to u^j, and we see that this is the i-th contravariant component of the vector $\partial\mathbf{f}/\partial u^j$. If we represent **f** in terms of its covariant components and the reciprocal unitary vectors $\mathbf{e}^i$, we obtain the similar result

$$\frac{\partial \mathbf{f}}{\partial u^j}=\left(\frac{\partial f^i}{\partial u^j}-f_k\begin{Bmatrix} k \\ ij \end{Bmatrix}\right)\mathbf{e}^i=f_{i,j}\mathbf{e}^i \tag{5-73}$$

EXAMPLE 12. In a three-dimensional Euclidean space, with the cartesian coordinates (x^1, x^2, x^3), the quantities dx^i/dt, where the parameter t represents the time, are the contravariant components of a vector, known as the velocity vector. On the other hand, the second time derivatives d^2x^i/dt^2 do not transform as the components of a vector. Under the admissible group in E_3, the second time derivatives of the coordinates transform according to

$$\frac{d^2u'^i}{dt^2}=\frac{\partial u'^i}{\partial u^m}\frac{d^2u^m}{dt^2}+\frac{\partial^2u'^i}{\partial u^m\partial u^n}\frac{du^m}{dt}\frac{du^n}{dt}$$

It is rather important, in many applications that we identify the second derivatives of the cartesian coordinates as the contravariant cartesian components of a vector, known as the acceleration. We, therefore, need to define a set of contravariant components a^i such that in cartesian coordinates

$$a^i=\frac{d^2x^i}{dt^2}$$

The term $\partial^2u'^i/\partial u^m\partial u^n$ can be expressed in terms of the three index symbols of the second kind as

$$\frac{\partial^2u'^i}{\partial u^m\partial u^n}=\left\{\begin{matrix} r \\ mn \end{matrix}\right\}_u\frac{\partial u'^i}{\partial u^r}-\left\{\begin{matrix} i \\ jk \end{matrix}\right\}_{u'}\frac{\partial u'^j}{\partial u^m}\frac{\partial u'^k}{\partial u^n}$$

Then,

$$\frac{d^2u'^i}{dt^2}=\frac{\partial u'^i}{\partial u^m}\frac{d^2u^m}{dt^2}+$$

$$\left(\left\{\begin{matrix} r \\ mn \end{matrix}\right\}_u\frac{\partial u'^i}{\partial u^r}-\left\{\begin{matrix} i \\ jk \end{matrix}\right\}_{u'}\frac{\partial u'^j}{\partial u^m}\frac{\partial u'^k}{\partial u^n}\right)\frac{du^m}{dt}\frac{du^n}{dt}$$

However, the first time derivatives transform as contravariant vectors, and hence,

$$\left\{\begin{matrix} i \\ jk \end{matrix}\right\}_{u'}\frac{\partial u'^j}{\partial u^m}\frac{\partial u'^k}{\partial u^n}\frac{du^m}{dt}\frac{du^n}{dt}=\left\{\begin{matrix} i \\ jk \end{matrix}\right\}_{u'}\frac{du'^j}{dt}\frac{du'^k}{dt}$$

Rearranging terms,

$$\frac{d^2u'^i}{dt^2}+\left\{\begin{matrix} i \\ jk \end{matrix}\right\}_{u'}\frac{du'^j}{dt}\frac{du'^k}{dt}=\left(\frac{d^2u^m}{dt^2}+\left\{\begin{matrix} m \\ nr \end{matrix}\right\}_u\frac{du^n}{dt}\frac{du^r}{dt}\right)\frac{\partial u'^i}{\partial u^m}$$

which is the transformation law of a contravariant vector. Further, since the three index symbols all vanish in a cartesian coordinate system,

$$\frac{d^2x^m}{dt^2}=\frac{d^2x^m}{dt^2}+\left\{\begin{matrix} m \\ nr \end{matrix}\right\}_x\frac{dx^n}{dt}\frac{dx^r}{dt}$$

Thus, the contravariant components of the acceleration vector can be defined to be

$$a^m=\frac{d^2u^m}{dt^2}+\left\{\begin{matrix} m \\ nr \end{matrix}\right\}_u\frac{du^n}{dt}\frac{du^r}{dt}=(du^m/dt),_t$$

The generalization of this result to the differentiation of an arbitrary vector with respect to a parameter is left as an exercise.

EXAMPLE 13. We have previously used the manifold defined by the co-latitude, θ and the longitude, φ, on the surface of a sphere of radius r_0 as an example of a non-Euclidean manifold. We can now show, by calculating the Riemann-Christoffel tensor that such a manifold is indeed non-Euclidean. On the surface of a sphere of radius r_0, we have the fundamental quadratic form

$$ds^2=r_0^2\,(d\theta)^2+r_0^2\sin^2\theta\,(d\varphi)^2$$

The calculation of the Riemann-Christoffel tensor is somewhat simplified by making a change in scale so that $r_0=1$. After this scale change, the components of the metric tensor are immediately identified as

$$g_{11}=1;\quad g_{12}=g_{21}=0;\quad g_{22}=\sin^2\theta$$

The only non-zero three index symbols are

$$[12, 2]=[21, 2]=\sin\theta\cos\theta;\quad [22, 1]=-\sin\theta\cos\theta$$

$$\begin{Bmatrix}2\\12\end{Bmatrix}=\begin{Bmatrix}2\\21\end{Bmatrix}=\cot\theta;\quad \begin{Bmatrix}1\\22\end{Bmatrix}=-\sin\theta\cos\theta$$

Then, the sixteen components of the first Riemann-Christoffel tensor are

$$\begin{array}{llll} R_{1111}=0, & R_{1112}=0, & R_{1121}=0, & R_{1122}=0\\ R_{1211}=0, & R_{1212}=\sin^2\theta & R_{1221}=-\sin^2\theta, & R_{1222}=0\\ R_{2111}=0, & R_{2112}=-\sin^2\theta, & R_{2121}=\sin^2\theta, & R_{2122}=0\\ R_{2211}=0, & R_{2212}=0, & R_{2221}=0, & R_{2222}=0 \end{array}$$

Then, there is no coordinate system in the manifold for which the three index symbols are all zero. It then follows that the metric tensor can never have all components equal to constants, and hence the manifold is non-Euclidean.

5-7. The Completely Anti-symmetric Unit Tensor. In this section, we shall introduce some special tensors which are extremely useful in the applications of the tensor formalism. Since the properties of these tensors depend on permutations of the indices, it is well to review some of the basic properties of permutations of the integers. If we have n integers arranged in a definite order, the arrangement is known as a permutation of the n integers. It can be shown that there are $n!$ different permutations of the set of n integers. We shall assume that the set of integers which we are considering are the consecutive integers starting with unity.

A transposition is defined as an operation in which two elements of a given permutation are interchanged. Any given permutation can be obtained from any other given permutation by carrying out one or more transpositions. For example, we can obtain the permutation 1423 from the permutation 4312 by the sequence of transpositions

$$4312 \quad 3412 \quad 1432 \quad 1423$$

The sequence of transpositions is not uniquely defined. In the example above, we could have equally well chosen the sequence

$$4312 \quad 4132 \quad 1432 \quad 1423$$

If the integers 1 to n appear in increasing order

$$1, 2, \ldots, n$$

we shall say that this is the normal permutation or natural ordering. An inversion in a given permutation occurs whenever two elements in the permutation appear in a different order than in the normal permutation. For example, the permutation 1243 has one inversion, whereas the permutation 1423 has two inversions. If a given permutation has an even number of inversions, it is called an even permutation, whereas a permutation having an odd number of inversions is called an odd permutation.

In any admissible coordinate system U of X_n, we define a set of n^n quantities

$$e_{i_1 \ldots i_n} = \begin{cases} +1; \; i_1 \ldots i_n \text{ an even permutation of } 1, \ldots, n \\ -1; \; i_1 \ldots i_n \text{ an odd permutation of } 1, \ldots, n \\ 0; \; \text{otherwise} \end{cases} \tag{5-74}$$

We similarly define

$$e^{i_1 \ldots i_n} = \begin{cases} +1; \; i_1 \ldots i_n \text{ an even permutation of } 1, \ldots, n \\ -1; \; i_1 \ldots i_n \text{ an odd permutation of } 1, \ldots, n \\ 0; \; \text{otherwise} \end{cases} \tag{5-75}$$

The set of quantities defined by either (5-74) or (5-75) is called an e-system or set of permutation symbols. It is clear from the defining equations that both $e_{i_1 \ldots i_n}$ and $e^{i_1 \ldots i_n}$ are anti-symmetric with respect to every pair of indices, i.e. are completely anti-symmetric. We shall see later that the $e_{i_1 \ldots i_n}$ are the components of a relative tensor of weight -1, and the set of quantities $e^{i_1 \ldots i_n}$ are the components of a relative tensor of weight $+1$. Since there are $n!$ permutations of the integers $1, \ldots, n$, there are $n!$ non-zero permutation symbols of each kind in an n-dimensional manifold.

In order to establish the tensorial character of the permutation symbols, we first need to review some of the properties of determinants of order n. Consider the third order determinant

$$\begin{vmatrix} a_1^1 & a_2^1 & a_3^1 \\ a_1^2 & a_2^2 & a_3^2 \\ a_1^3 & a_2^3 & a_3^3 \end{vmatrix} = a_1^1 a_2^2 a_3^3 + a_1^2 a_2^3 a_3^1 + a_1^3 a_2^1 a_3^2 - a_1^1 a_2^3 a_3^2 - a_1^3 a_2^2 a_3^1 - a_1^2 a_2^1 a_3^3$$

We have written the expansion of the determinant in such a way that the subscripts in each term are in natural order. We note in the expansion that a given term is positive if the subscripts are an even permutation of 1, 2, 3 and is negative if the subscripts are an odd permutation of 1, 2, 3. Thus, we may write

$$\det(a_j^i) = e_{ijk}\, a_1^i\, a_2^j a_3^k \tag{5-76}$$

On the other hand, if we expand det (a_j^i) so that the superscripts are the normal permutation,

$$\det(a^i) = a_1^1 a_2^2 a_3^3 + a_2^1 a_3^2 a_1^3 + a_3^1 a_1^2 a_2^3 - a_1^1 a_3^2 a_2^3 - a_3^1 a_2^2 a_1^3 - a_2^1 a_1^2 a_3^3$$

we see that a given term is positive if the subscripts are an even permutation of 1, 2, 3, and negative if the subscripts are an odd permutation of 1, 2, 3. Hence,

$$\det(a_j^i) = e^{ijk}\, a_i^1\, a_j^2\, a_k^3 \tag{5-77}$$

These results can be immediately generalized to the case of determinants of higher order

$$\begin{vmatrix} a_1^1 & \dots & a_n^1 \\ \cdot & \dots & \cdot \\ \cdot & \dots & \cdot \\ \cdot & \dots & \cdot \\ a_1^n & \dots & a_n^n \end{vmatrix} = e^{i_1\dots i_n} a_{i_1}^1 a_{i_2}^2 \dots a_{i_n}^n = e_{i_1\dots i_n}\, a_1^{i_1} \dots a_n^{i_n} \tag{5-78}$$

The essential result which we require from the theory of determinants is that an interchange of two rows (or two columns) in the determinant changes the sign of the determinant. The condition for the interchange of rows can be expressed as

$$e^{i_1\dots i_n} \det(a_c^r) = e^{j_1\dots j_n}\, a_{j_1}^{i_1} a_{j_2}^{i_2} \dots a_{j_n}^{i_n} \tag{5-79}$$

Similarly, the condition for the interchange of rows is expressed as

$$e_{j_1\dots j_n} \det(a_c^r) = e_{i_1\dots i_n}\, a_{j_1}^{i_1} a_{j_2}^{i_2} \dots a_{j_n}^{i_n} \tag{5-80}$$

Now, consider a transformation between two admissible coordinate systems of X_n,

$$u'^i = u'^i(u^1 \dots u^n) \tag{5-81}$$

The Jacobian of the transformation is

$$J(u', u) = \det(\partial u'^i / \partial u^j)$$

The Jacobian for the inverse transformation is

$$J'(u, u') = \det(\partial u^i / \partial u'^j)$$

The product of the two transformations is the identity transformation, and hence, $JJ'=1$, or

$$\frac{1}{J}=\left|\partial u^i/\partial u'^j\right|$$

In Eq. (5-79), let the determinant det (a^r_c) be the Jacobian of the transformation (5-81), so that

$$e^{i_1\ldots i_n}\,J=\frac{\partial u'^{i_1}}{\partial u^{j_1}}\cdots\frac{\partial u'^{i_n}}{\partial u^{j_n}}\,e^{j_1\ldots j_n}$$

or

$$e^{i_1\cdots i_n}=\left|\frac{\partial u^i}{\partial u'^j}\right|\frac{\partial u'^{i_1}}{\partial u^{j_1}}\cdots\frac{\partial u'^{i_n}}{\partial u^{j_n}}\,e^{j_1\cdots j_n}$$

which is the transformation law for a contravariant tensor of order n and weight $+1$. Similarly, it follows from (5-80) that

$$e_{i_1\ldots i_n}=\left|\frac{\partial u^i}{\partial n'^j}\right|^{-1}\frac{\partial u^{j_1}}{\partial u'^{i_1}}\cdots\frac{\partial u^{j_n}}{\partial u'^{i_n}}\,e_{j_1\cdots j_n}$$

which is the transformation law for a covariant tensor of order n and weight -1. Since the permutation symbols $e^{1\ldots n}$ are the components of a tensor, they are commonly known as the contravariant, completely anti-symmetric unit tensor in X_n. Similarly, the permutation symbols $e_{1\ldots n}$ define the covariant, completely anti-symmetric unit tensor of weight -1 in X_n.

We occasionally require a generalization of the Kronecker delta. This generalization is easily obtained in terms of the completely anti-symmetric unit tensors. In a three-dimensional manifold X_n, consider the direct product

$$e^{mnr}\times e_{ijk}=\begin{cases}+1;\ mnr\text{ an even permutation of }ijk\\-1;\ mnr\text{ an odd permutation of }ijk\\0;\ \text{otherwise}\end{cases}$$

We define the generalized Kronecker delta of order three in X_n to be

$$\delta^{mnr}_{ijk}=e^{mnr}\times e_{ijk} \tag{5-82}$$

Since e^{mnr} is a tensor of weight $+1$, and e_{ijk} is a tensor of weight -1, it is clear that δ^{mnr}_{ijk} is an absolute tensor covariant of order three and contravariant of order three. Now, contract δ^{mnr}_{ijk} to obtain

$$\delta^{mn}_{ij}=\delta^{mnk}_{ijk}=\delta^{mn1}_{ij1}+\delta^{mn2}_{ij2}+\delta^{mn3}_{ij3}$$

If $m=n$ or $i=j$, then each term of the sum vanishes. If we choose $i=1$ and $j=2$, the only non-zero term is δ^{mn3}_{123} which is $+1$ if mn is an even permutation of 12, and is -1 if mn is an odd permutation of 12. If we choose $i=1$, $j=3$, the only non-zero term is δ^{mn2}_{132}; this term has value $+1$ if $m=1$, $n=3$; and -1 if $m=3$, $n=1$. Similar results obtain if we choose $i=2, j=3$; $i=3, j=1$; $i=3, j=2$. In summary,

$$\delta_{ij}^{mn}=\begin{cases}+1; & m=i,\ n=j,\ i\neq j\\ -1; & m=j,\ n=i,\ i\neq j\\ \ \ 0; & \text{otherwise}\end{cases} \tag{5-83}$$

defines the second order generalized Kronecker delta in X_3. If we contract the second order Kronecker delta, we have

$$\delta_{ij}^{mj}=\delta_{i1}^{m1}+\delta_{i2}^{m2}+\delta_{i3}^{m3}$$

Setting $i=1$,

$$\delta_{1j}^{mj}=\delta_{12}^{m2}+\delta_{13}^{m3}=\begin{cases}2; & m=1\\ 0; & m\neq 1\end{cases}$$

Similar results are obtained by choosing $i=2$ and $i=3$. Hence, the usual Kronecker delta is obtained as one-half the singly contracted Kronecker delta of order two,

$$\delta_i^m=\tfrac{1}{2}\ \delta_{ij}^{mj} \tag{5-84}$$

In an n-dimensional manifold, we generalize the preceeding results and define the generalized Kronecker delta of order n to be

$$\delta_{j_1\ldots j_n}^{i_1\ldots i_n}=e^{i_1\ldots i_n}\times e_{j_1\ldots j_n} \tag{5-85}$$

The Kronecker deltas of order $m<n$ are obtained by repeated contraction of (5-85) and multiplication by a normalizing factor,

$$\delta_{j_1\ldots j_m}^{i_1\ldots i_m}=\frac{1}{(n-m)!}\ \delta_{j_1\ldots j_m j_{m+1}\ldots j_n}^{i_1\ldots i_m j_{m+1}\ldots j_n} \tag{5-86}$$

or more generally,

$$\delta_{j_1\ldots j_m}^{i_1\ldots i_m}=\frac{(n-k)!}{(n-m)!}\ \delta_{j_1\ldots j_m j_{m+1}\ldots j_k}^{i_1\ldots i_m j_{m+1}\ldots j_k} \tag{5-87}$$

5-8. Vector Analysis. One very useful application of the tensor formalism is in vector analysis. We have formulated the rules for the manipulation of vectors in terms of three-dimensional Euclidean spaces. Many of the concepts and operations of vector analysis can be extended by means of the tensor formalism, to non-Euclidean spaces of arbitrary dimensionality. Even in Euclidean three-dimensional space, many complicated analyses involving combinations of vectors and operations on vectors are simplified by the use of the tensor formalism.

Consider a three-dimensional Euclidean manifold which is covered by an orthogonal cartesian system (x^1, x^2, x^3). In any coordinate system (u^1, u^2, u^3) of the manifold, a given vector $\mathbf{a}$ is specified by either its covariant components a_i or its contravariant components a^i. In the coordinate system U, the two sets of components are related by

$$a_i = g_{ij}\, a^j;\; a^i = g^{ij}\, a_j$$

where g_{ij} and g^{ij} are the covariant and contravariant representations of the Euclidean metric tensor with respect to the coordinate system U. If (u'^1, u'^2, u'^3) is a set of coordinates obtained from (u^1, u^2, u^3) by an admissible transformation, the covariant and contravariant components of the vector **a**, relative to the system U' are obtained from the components of **a** relative to U through the transformations

$$a'_i(u') = \frac{\partial u^j}{\partial u'^i}\, a_j(u'; T_0; u)$$

and

$$a'^i(u') = \frac{\partial u'^i}{\partial u^j}\, a^j\,(u'; T_0; u)$$

respectively.

If **a** and **b** are two vectors whose components relative to the coordinate system U are known, then the components relative to U of their sum **c** are given by

$$c_i = a_i + b_i;\; c^i = a^i + b^i$$

It is clear that **c** is a vector of the same kind as **a** and **b**. We have previously defined the inner product of two vectors **a** and **b** to be the contracted tensor formed from the direct product of **a** and **b**.

$$\mathbf{a}\cdot\mathbf{b} = a^i b_i = a_i b^i = g_{ij}\, a^i b^j = g^{ij}\, a_i b_j$$

The cross product of the two vectors **a** and **b** is defined to be the relative tensor (pseudo-vector) whose components are given by

$$(\mathbf{a}\times\mathbf{b})^i = e^{ijk}\, a_j b_k \tag{5-88}$$

If we carry out the indicated sum on the right, we obtain the customary relation

$$(\mathbf{a}\times\mathbf{b})^i = a_j b_k - a_k b_j; \qquad ijk \text{ an even permutation of } 123$$

Similarly, the mixed triple product is defined as the relative scalar (pseudo-scalar)

$$(\mathbf{a}\cdot\mathbf{b}\times\mathbf{c}) = e^{ijk}\, a_i b_j c_k \tag{5-89}$$

The commutation and anti-commutation properties of the mixed triple product are immediately obvious from Eq. (5-89). The final algebraic operation we will define is the triple cross product. In terms of the tensor formalism,

$$[\mathbf{a}\times(\mathbf{b}\times\mathbf{c})]^i = \delta^{ij}_{km}\, a_j b^k c^m \tag{5-90}$$

This quantity is obviously an absolute or polar vector unless one or all of the factors are pseudo-vectors. Carrying out the indicated sums on the right, we have

$$[\mathbf{a}\times(\mathbf{b}\times\mathbf{c})]^i=(a_jc^j)\,b^i-(a_jb^j)\,c^i$$

or, in the usual three vector notation

$$\mathbf{a}\times(\mathbf{b}\times\mathbf{c})=(\mathbf{a}\cdot\mathbf{c})\,\mathbf{b}=(\mathbf{a}\cdot\mathbf{b})\,\mathbf{c}$$

To further illustrate the utility of the tensor formalism in vector algebra, consider the quantity

$$\begin{aligned}[(\mathbf{a}\times\mathbf{b})\times(\mathbf{c}\times\mathbf{d})]^i&=\delta^{ij}_{km}(\mathbf{a}\times\mathbf{b})_jc^kd^m\\&=(\mathbf{a}\times\mathbf{b})_jd^jc^i-(\mathbf{a}\times\mathbf{b})_jc^jd^i\\&=(e^{jkm}a_jb_kd_m)\,c^i-(e^{jkm}a_jb_kc_m)\,d^i\end{aligned}$$

or

$$(\mathbf{a}\times\mathbf{b})\times(\mathbf{c}\times\mathbf{d})=(\mathbf{a}\cdot\mathbf{b}\times\mathbf{d})\;\mathbf{c}-(\mathbf{a}\cdot\mathbf{b}\times\mathbf{c})\;\mathbf{d}$$

Similarly,

$$\begin{aligned}(\mathbf{a}\times\mathbf{b})\cdot(\mathbf{c}\times\mathbf{d})&=e^{ijk}(\mathbf{a}\times\mathbf{b})_ic_jd_k\\&=\delta^{jk}_{mn}a^mb^nc_jd_k\\&=a^mc_mb^nd_n-a^md_mb^nc_n\\&=(\mathbf{a}\cdot\mathbf{c})(\mathbf{b}\cdot\mathbf{d})-(\mathbf{a}\cdot\mathbf{d})(\mathbf{b}\cdot\mathbf{c})\end{aligned}$$

The reader may recall that the proof of these identities and the identity (5-90) by strictly vector methods is not so straightforward or simple.

We took as our model for the covariant transformation law, the transformation which takes the partial derivatives $\partial\phi/\partial u^i$ into the partial derivatives $\partial\phi'/\partial u'^j$,

$$\frac{\partial\phi'}{\partial u^j}=\frac{\partial u^i}{\partial u'^j}\frac{\partial\phi}{\partial u^i}$$

Since the partial derivatives define the components of the gradient, we have

$$(\text{grad }\phi)_i=\phi,_i \tag{5-91}$$

The contravariant components of the gradient of ϕ are given by

$$(\text{grad }\phi)=(g^{ij}\phi),_j$$

In the event that ϕ is a pseudo-scalar, the gradient of ϕ is the pseudo-vector whose components are given by (5-91).

If f^i is a contravariant vector defined in U, we define the divergence of f^i as the invariant contracted form

$$\text{div }\mathbf{f}=f^i,_i=(g^{ij}f_j),_i \tag{5-92}$$

We shall now show that Eq. (5-92) is the same as the more

customary representation of the divergence of a vector relative to a generalized coordinate system (u^1, u^2, u^3), in a three-dimensional Euclidean space, if we expand the covariant derivative,

$$\text{div } \mathbf{f} = \frac{\partial f^i}{\partial u^i} + \begin{Bmatrix} i \\ mi \end{Bmatrix} f^m$$

It can be shown that

$$\begin{Bmatrix} i \\ im \end{Bmatrix} = \frac{\partial \ln(\sqrt{g})}{\partial u^m} = \frac{1}{2g} \frac{\partial g}{\partial u^m}$$

where g is the determinant of the g_{ij}. Hence,

$$\text{div } \mathbf{f} = \frac{\partial f^i}{\partial u^i} + \frac{1}{2g} \frac{\partial g}{\partial u^m} f^m = \frac{1}{\sqrt{g}} \frac{\partial}{\partial u^i}(\sqrt{g}\, f^i)$$

which is customary expression for the divergence relative to the coordinate system U in a three-dimensional Euclidean space.

Let ψ be an absolute scalar. We may then combine Eqs. (5-91) and (5-92) in order to obtain an expression for the Laplacian of ψ

$$\nabla^2 \psi = \text{div grad } \psi = g^{ij} \psi_{,ij} \tag{5-93}$$

Except for relations and operations involving the cross product or the curl, there is no difficulty in generalizing vector analysis to n-dimensional non-Euclidean manifolds. In a fixed coordinate system U of X_n, we describe a vector $\mathbf{f}$ by either its covariant components f_i, or its contravariant components f^i, and the two representations are related through the metric tensor for the manifold. As before, the inner product of two vectors is defined as the invariant

$$\mathbf{a} \cdot \mathbf{b} = a^i b_i = a_i b^i = g_{ij} a^i b^j = g^{ij} a_i b_j$$

where the sums are from one to n.

In X_n, the differential operators "grad" and "div" have the form

$$(\text{grad } \phi)_i = \phi_{,i} \tag{5-94}$$

and

$$\text{div } \mathbf{f} = f^i_{,i} \tag{5-95}$$

As before, we can combine Eqs. (5-94) and (5-95) to obtain an expression for the Laplacian of the scalar ψ,

$$\nabla^2 \psi = g^{ij} \psi_{,ij} \tag{9-96}$$

Some difficulty is encountered when we attempt to generalize the concept of the cross product of two vectors. We may, however, define a set of quantities in X_n, which reduce to the components of the usual cross product in Euclidean three space. If $A_{i_1 \ldots i_m}$ is a completely anti-symmetric tensor of order m, then the relative tensor

$$A^{*i_{m+1} \ldots i_n} = \frac{1}{m!} e^{i_1 \ldots i_m i_{m+1} \ldots i_n} A_{i_1 \ldots i_m} \tag{5-97}$$

is defined to be the order $n-m$ dual of $A_{i_1\ldots i_m}$. If **a** and **b** are two vectors in X_n, we define their cross product to be the order $n-2$ dual

$$C^{*i_1\ldots i_{n-2}}=\tfrac{1}{2}\,e^{i_1\ldots i_{n-2}jk}\,(a_jb_k-a_kb_j) \tag{5-98}$$

Similarly, in X_n we define the curl of a vector **a** to be the $n-2$ order dual

$$C^{*i_1\ldots i_{n-2}}=\tfrac{1}{2}\,e^{i_1\ldots i_{n-2}jk}\,(a_{k,j}-a_{j,k}) \tag{5-99}$$

In Euclidean three space, Eqs. (5-98) and (5-99) reduce to

$$C^{*i}=\tfrac{1}{2}\,e^{ijk}(a_jb_k-a_kb_j)=e^{ijk}a_jb_k$$

and

$$C^{*i}=-e^{ijk}a_{,}b_k$$

respectively.

PROBLEMS

5-1. In E_3 show that the set of coordinates (u, v, w) where

$$x=\sqrt{uv}\cos w,\ y=\sqrt{uv}\sin w\ ,$$

$$z=\frac{u-v}{2}$$

$0\leq u<\infty$, $0\leq v<\infty$, $0\leq w<2\pi$, is an admissible coordinate system.

5-2. Many of the forces between two particles, which occur in nature, can be derived from a potential function of the form

$$U=a\,(x^2+y^2+z^2)^{-1/2}$$

where (x, y, z) are cartesian coordinates in E_3.

(i) Show that U is a scalar invariant in E_3.

(ii) Determine the form of U in terms of the spherical coordinates (r, θ, φ).

5-3. In terms of a cartesian coordinate system (x^1, x^2, x^3) in E_3, the velocity of a particle is given by the contravariant vector with components $(dx^1/dt, dx^2/dt, dx^3/dt)$. Use the transformation law T_2 to determine the components of the velocity vector in terms of the spherical coordinates (r, θ, φ).

5-4. In a cartesian coordinate system (x, y, z), the covariant components of the tensor A_{ij} are

$$A_{11}=x^2,\quad A_{12}=A_{21}=xy,\quad A_{13}=A_{31}=xz$$
$$A_{22}=y^2,\quad A_{23}=A_{32}=yz,\quad A_{33}=z^2$$

Find the components of A_{ij} relative to the parabolic coordinates (u, v, w) defined in Prob. 5-1.

5-5. Show that the sum and difference of two given tensors of the same kind as the given tensors

5-6. Let a_{ij} be the components of a covariant tensor of order two, and let A^{ij} be the cofactor of a_{ij} in $\det(a_{ij})$. Show that $A^{ij}/\det(a_{ij})$ are the components of a tensor contravariant of order two.

5-7. In the restricted theory of relativity, it is customary to introduce a pseudo-Euclidean manifold E_4' known as Minkowski four-space, and a pseudo-cartesian coordinate system (x^0, x^1, x^2, x^3) so that the fundamental quadratic form is

$$ds^2=(dx^0)^2-(dx^1)^2-(dx^2)^2-(dx^3)^2$$

(i) Show that the set of coordinates $(r, \theta, \varphi, \psi)$ defined by

$$x^0=r\cosh\psi$$
$$x^1=r\sin\theta\cos\varphi\sinh\psi$$
$$x^2=r\sin\theta\sin\varphi\sinh\psi$$
$$x^3=r\cos\theta\sinh\psi$$

is a set of admissible coordinates in E_4.

(ii) Determine the fundamental quadratic form relative to $(r, \theta, \varphi, \psi)$.

5-8. Calculate the metric coefficients for the manifold X_2 which describes the surface of the ellipsoid

$$\frac{x^2}{9}+\frac{y^2}{9}+\frac{z^2}{10}=1$$

Use the set of surface coordinates (u, v), where

$$x=3\sqrt{1-u^2}\cos v$$
$$y=3\sqrt{1-u^2}\sin v$$
$$z=\sqrt{10}\,u$$

5-9. Show that the Christoffel three index symbols of the second kind transform according to

$$\left\{ {k \atop ij} \right\}_{u'}=\frac{\partial u'^k}{\partial u^m}\frac{\partial u^n}{\partial u'^i}\frac{\partial u^r}{\partial u'^j}\left\{ {m \atop nr} \right\}_u+\frac{\partial^2 u^n}{\partial u'^i\,\partial u'^j}\frac{\partial u'^k}{\partial u^n}$$

5-10. If $g_{ij}=0$ for $i\neq j$, show that $\left\{ {k \atop ij} \right\}=0$ whenever i, j, and k are distinct.

5-11. Use the transformation laws to calculate the Christoffel three index symbols of both kinds in terms of the spherical coordinates (r, θ, φ).

5-12. Show that

$$\frac{\partial g_{ij}}{\partial u^k}-\frac{\partial g_{jk}}{\partial u^i}=[jk, i]-[ij, k]$$

5-13. In a coordinate system U of X_n, let g denote the determinant of the g_{ij}, and show that

$$\frac{\partial}{\partial u^i}(\ln\sqrt{g})=\left\{ {m \atop im} \right\}$$

5-14. Use the identity $g_{im}g^{mj}=\delta_i^j$ to show that $\delta^{ij}_{,k}=0$.

5-15. Show that

$$R_{ijkm}=\tfrac{1}{2}\left(\frac{\partial g_{im}}{\partial u^j\partial u^k}+\frac{\partial g_{jk}}{\partial u^i\partial u^m}-\frac{\partial g_{ik}}{\partial u^j\partial u^m}-\frac{\partial g_{jm}}{\partial u^i\partial u^k}\right)+$$
$$g^{nr}([jk, r][im, n]-[jm, r][ik, n])$$

5-16. Prove the Bianchi identity

$$R^i_{jkm,n}+R^i_{jmn,k}+R^i_{jnk,m}=0$$

5-17. Let $f^i(x^1, x^2, x^3, s)$ be a contravariant vector, defined in E_3, which depends on the parameter s. Determine a contravariant vector $f^i_{,s}$ whose cartesian representation is

$$(f^i_{,s})_x = \frac{df^i}{ds}$$

This vector can be interpreted as the covariant derivative of f^i with respect to the parameter s.

5-18. Calculate the components of the acceleration vector relative to the elliptic cylinder coordinates (μ, φ, z), where

$$x = a \cosh \mu \cos \varphi$$
$$y = a \sinh \mu \sin \varphi$$
$$z = z$$

5-19. Determine the second Riemann-Christoffel tensor for the X_2 manifold defined on the surface of a unit sphere.

5-20. Use the e-system $e_{i_1 \ldots i_n}$ to show that

(i) $\det(a^i_j) \det(b^i_j) = \det(c^i_j)$, $c^i_j = a^i_k b^k_j$

(ii) $\dfrac{\partial a}{\partial u^i} = \dfrac{\partial a^j_k}{\partial u^i} A^j_k$, $a = \det(a^i_j)$; A^k_j is the cofactor of a^k_j in a

5-21. Use the tensor formalism to find an explicit representation of the Laplacian of a vector in terms of the spheroidal coordinates (μ, θ, φ) in E_3, where

$$x = a \cosh \mu \cos \theta \cos \varphi$$
$$y = a \cosh \mu \cos \theta \sin \varphi$$
$$z = a \sinh \mu \sin \theta$$

CHAPTER 6

Vectorial Mechanics

In the previous chapters, we have obtained sufficient mathematical formalism that we are now in a position to apply vector and tensor methods to the analysis of many physical problems. Although such methods are applicable in almost all areas of physical and engineering analysis, we shall consider in detail only two of the important areas of application. In this chapter we consider the classical mechanics of a single particle, a system of particles, and rigid bodies. In the next chapter, we consider the properties of the classical electromagnetic field.

6-1. Mechanics of a Particle. In order to discuss the behaviour of a particle under the influence of one or more forces which may act on the particle, it is necessary to first introduce some frame of reference. By a frame of reference, we shall mean some method of describing the spatial location of the mass particle and the time at which the particle occupied that location. We shall consider the time only as a convenient parameter for the description of the spatial motion. One very useful specification of a reference frame is to introduce an arbitrary origin and a set of orthogonal cartesian coordinate (x^1, x^2, x^3) whose coordinate axes are constant in time. In terms of this frame of reference, the location of a mass particle at time t is specified by the instantaneous value of the radius vector $\mathbf{r}(t)$ from the arbitrary origin to the particle as shown in Fig. 6-1. A complete history of the behaviour of the particle is given by specifying the values of $\mathbf{r}(t)$ for all values of the time t. The principal problem of the mechanics of a mass particle can be stated in the following way: Given a system of forces acting on a particle of mass m, which occupies the location $\mathbf{r}(t_0)$ at some initial time t_0, determine the vector $\mathbf{r}(t)$ for all $t \geqslant t_0$. The curve traced by the tip of the vector $\mathbf{r}(t)$ for $t \geqslant t_0$, is known as the trajectory of the particle.

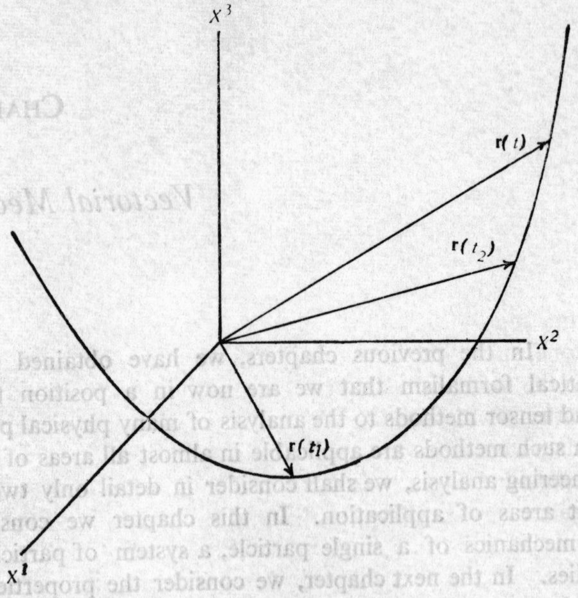

Figure 6-1.

We shall assume that the vector $\mathbf{r}(t)$ has a continuous second derivative with respect to the parameter t. Although there is no mathematical justification for this assumption, it is reasonable on physical grounds. Motions for which $\mathbf{r}(t)$ does not have a continuous second derivative are physically possible, and do occur in many important situation. However, the analysis of such motions is extremely complicated, and is beyond the scope of this work. We shall denote the time derivative of $\mathbf{r}(t)$ by $\dot{\mathbf{r}}(t)$ and define the velocity of the mass particle to be

$$\mathbf{u}(t)=\dot{\mathbf{r}}(t)=\frac{d\mathbf{r}}{dt} \tag{6-1}$$

If the motion of the particle is such that it has a constant velocity

$$\mathbf{u}(t)=\mathbf{u}_0$$

then the particle trajectory is the straight line

$$\mathbf{r}(t)=\mathbf{r}(t_0)+\mathbf{u}_0\, t$$

Such motion is called uniform rectilinear motion.

The second time derivative of $\mathbf{r}(t)$, denoted by $\ddot{\mathbf{r}}(t)$ is defined to be the acceleration of the particle

$$\mathbf{a}(t)=\frac{d\mathbf{u}}{dt}=\frac{d^2\mathbf{r}}{dt^2}=\ddot{\mathbf{r}}(t) \tag{6-2}$$

If the acceleration of the particle is constant in time, the trajectory is given by

$$\mathbf{r}(t)=\mathbf{r}(t_0)+\mathbf{u}(t_0)\,t+\tfrac{1}{2}\,\mathbf{a}t^2$$

where $\mathbf{r}(t_0)$ and $\mathbf{u}(t_0)$ represent the location and velocity of the particle at some initial time t_0, and a is the constant acceleration. If the acceleration is not constant in time, the trajectory is obtained by integrating (6-2) twice with respect to time and expressing the constants of integration in terms of an initial location and velocity at reference time t_0. Except for very simple cases, it is not generally possible to carry out this programme in closed form. Although it is usually possible to define higher derivatives of $\mathbf{r}(t)$ with respect to t, this is not usually done, since the second derivative or acceleration is directly related to the forces acting on the particle.

A quantity which is related to the velocity, and which completely characterizes the motion of a material particle is the linear momentum. For a particle of mass m, moving with velocity $\mathbf{u}$, the linear momentum is defined as the vector quantity

$$\mathbf{p}=m\mathbf{u} \tag{6-3}$$

The study of the motion of a particle or system of particles is greatly simplified if we can determine quantities which characterize the motion that are constant in time. Such quantities are known as constants of the motion, and depend, naturally, on the system under consideration and the forces acting on the system. As we shall see later, for any system there are a relatively large number of constants of the motion, but only a very limited number of these are of particular physical interests. The conditions under which quantities describing the motion of a particle or system of particles are constants of the motion are generally contained in conservation theorems or laws. The conservation laws for the physically significant constants of the motion are consequences of certain assumptions concerning the nature of space and time. The assumption that space is homogeneous leads to a conservation law which was first stated by Newton on experimental grounds. If $\mathbf{F}$ represents the vector sum of all forces acting on a particle of mass m, then

$$\mathbf{F}=\frac{d\mathbf{p}}{dt}=\dot{\mathbf{p}} \tag{6-4}$$

The conservation law, written in the form of Eq. (6-4) is known as Newton's second law of motion. It also expresses:

The Principle of Conservation of Linear Momentum: *The*

linear momentum of a mass particle is a constant of the motion, unless the particle is acted upon by an externally applied force.

Equation (6-4) enables us to determine the complete history of the mass particle if all of the external forces acting on the particle are known functions of the time. In this case, we can resolve Eq. (6-4) into cartesian components and obtain the system of ordinary second order linear differential equations

$$m\,\frac{d^2x^1}{dt^2}=F^1(t),\quad m\,\frac{d^2x^2}{dt^2}=F^2(t),\quad m\,\frac{d^2x^3}{dt^2}=F^3(t) \tag{6-5}$$

The solution of the system of differential equations (6-5) which satisfies prescribed conditions at some given time t_0 defines the complete history of the particle of mass m.

If we describe the motion of the particle in terms of a generalized system of coordinates (u^1, u^2, u^3), the expressions relating the force components and the acceleration components are more complicated than in the case of cartesian components due to the fact that the unitary or reciprocal unitary vectors are not constant.

EXAMPLE 1. The polar spherical coordinates (r, θ, φ) are related to the cartesian coordinates (x^1, x^2, x^3) by

$$x^1=r\sin\theta\cos\varphi,\quad x^2=r\sin\theta\sin\varphi,\quad x^3=r\cos\theta$$

Any force $\mathbf{F}$ acting on the particle has the representation

$$\mathbf{F}=F^r\,\mathbf{e}_r+F^\theta\,\mathbf{e}_\theta+F^\varphi\,\mathbf{e}_\varphi$$

where $(\mathbf{e}_r, \mathbf{e}_\theta, \mathbf{e}_\varphi)$ are the unitary vectors associated with the coordinate system. The instantaneous radius vector to the particle is

$$\mathbf{r}=r\mathbf{e}_r$$

and the velocity of the particle is

$$\mathbf{u}=\frac{dr}{dt}\,\mathbf{e}_r+\frac{d\theta}{dt}\,\mathbf{e}_\theta+\frac{d\varphi}{dt}\,\mathbf{e}_\varphi$$

The difficulty arises when we calculate the acceleration components relative to the spherical coordinates, since the quantities d^2x^i/dt^2 do not transform as the contravariant components of the acceleration vector. We must, therefore, calculate the new components from the general definition of the acceleration vector,

$$a^m=\frac{d^2u^m}{dt^2}+\begin{Bmatrix} m \\ nr \end{Bmatrix}\frac{du^n}{dt}\frac{du^r}{dt} \tag{6-6}$$

In spherical coordinates, the only non-zero three index symbols of the second kind are

$$\begin{Bmatrix} 1 \\ 22 \end{Bmatrix}=-r\,;\quad \begin{Bmatrix} 1 \\ 33 \end{Bmatrix}=-r\sin^2\theta$$

$$\left\{{2 \atop 12}\right\}=\left\{{2 \atop 21}\right\}=\frac{1}{r}\;;\;\left\{{2 \atop 33}\right\}=-\sin\theta\cos\theta$$

$$\left\{{3 \atop 13}\right\}=\left\{{3 \atop 31}\right\}=\frac{1}{r}\;;\;\left\{{3 \atop 23}\right\}=\left\{{3 \atop 32}\right\}=\cot\theta$$

It then follows that the acceleration components are

$$a^r=\frac{d^2r}{dt^2}-r\left(\frac{d\theta}{dt}\right)^2-r\sin^2\theta\left(\frac{d\varphi}{dt}\right)^2$$

$$a^\theta=\frac{d^2\theta}{dt^2}+\frac{2}{r}\frac{dr}{dt}\frac{d\theta}{dt}-\sin\theta\cos\theta\left(\frac{d\varphi}{dt}\right)^2$$

$$a^\varphi=\frac{d^2\varphi}{dt^2}+\frac{2}{r}\frac{dr}{dt}\frac{d\varphi}{dt}+2\cot\theta\frac{d\theta}{dt}\frac{d\varphi}{dt}$$

The integration of the resulting equations of motion is a formidable undertaking.

We now seek other quantities which may characterize the motion of a particle and conservation laws for these quantities. Consider a particle with linear momentum **p**, and let **r**(a) be the radius vector from an arbitrary point a to the particle. The angular momentum of the particle about the point a is the axial vector

$$\mathbf{l}(a)=\mathbf{r}(a)\times\mathbf{p} \tag{6-7}$$

It is generally more convenient to refer the angular momentum to the origin of coordinates. Let **a** be the radius vector from the origin to the point a, so that $\mathbf{r}(a)=\mathbf{r}-\mathbf{a}$, where **r** is the radius vector from the origin to the particle and **a** is a constant vector. Then,

$$\mathbf{l}(a)=(\mathbf{r}-\mathbf{a})\times\mathbf{p}=\mathbf{r}\times\mathbf{p}-\mathbf{a}\times\mathbf{p}$$

We define the first term to be the angular momentum of the particle about the origin, denoted by **l**. Let us now consider the time derivative of the angular momentum of a particle about the origin,

$$\begin{aligned}\frac{d\mathbf{l}}{dt}&=\frac{d}{dt}(\mathbf{r}\times\mathbf{p})=m\frac{d}{dt}(\mathbf{r}\times\dot{\mathbf{r}})\\&=m(\dot{\mathbf{r}}\times\dot{\mathbf{r}}+\mathbf{r}\times\ddot{\mathbf{r}})\\&=\mathbf{r}\times\mathbf{F}\end{aligned} \tag{6-8}$$

where the term $\dot{\mathbf{r}}\times\dot{\mathbf{r}}$ vanishes since it is the cross product of a vector with itself, and we have equated the mass times the acceleration of the particle to the total force acting on the particle. The quantity

$$\mathbf{N}=\mathbf{r}\times\mathbf{F}$$

is known as the moment of force or torque on the particle about the origin. If we define the angular momentum of a particle about an arbitrary point a, we may also define the torque about the point a

$$\mathbf{N}(a)=\mathbf{r}(a)\times\mathbf{F}$$

With respect to the point a, the analogue of Eq. (6-8) is

$$\dot{\mathbf{l}}(a)=\mathbf{N}(a) \tag{6-9}$$

Equation (6-9) expresses:

The Principle of Conservation of Angular Momentum: *If the total externally applied torque about any point, acting on a particle is zero, then the angular momentum of the particle about the same point is a constant of the motion.*

The principle of conservation of angular momentum is a consequence of the assumption of the isotropy of space.

If the linear momentum of particle is a constant, clearly the angular momentum is conserved. On the other hand, it is quite possible for the angular momentum to be constant, and the linear momentum not to be a conserved quantity for the given motion of the particle.

EXAMPLE 2. Consider a particle of mass m moving with constant speed u in a circle of radius r as shown in Fig. 6-2. If we

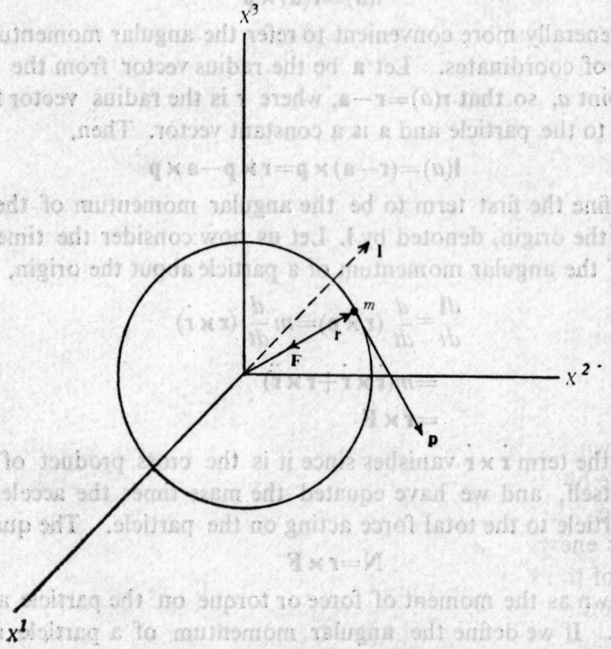

Figure 6-2.

choose the centre of the circle as the origin, the angular momentum about this point is

$$\mathbf{l}=\mathbf{r}\times\mathbf{p}$$

Now, $\mathbf{p}$ is a vector of magnitude mu and is always tangent to the circle at the location of the particle, whereas $\mathbf{r}$ is of magnitude r along a radial line from the origin to the particle. Hence the axial vector has the constant magnitude mur and a constant direction out of the plane of the paper. On the other hand, the linear momentum $\mathbf{p}$ is of constant magnitude, but is continually changing direction and hence is not conserved. We may also see the conservation of angular momentum by calculating the total torque about the origin which acts on the particle m. The only force acting on the particle is the centripetal force

$$\mathbf{F}=-\frac{mu^2\,\mathbf{r}}{r^2}$$

Hence, the net torque about the origin which acts on the particle is

$$\mathbf{N}=\mathbf{r}\times\mathbf{F}=0$$

and the angular momentum about the origin is a constant of the motion.

If the net force $\mathbf{F}$ on a particle acts through a differential distance $d\mathbf{r}$, the differential element of work which the force does on the particle is defined to be the inner product

$$dW=\mathbf{F}\cdot d\mathbf{r} \tag{6-10}$$

Assuming the mass of the particle to be constant, we may write

$$dW=m\ddot{\mathbf{r}}\cdot d\mathbf{r}$$

$$=m\dot{\mathbf{r}}\cdot d\dot{\mathbf{r}}=d(\tfrac{1}{2}m\,|\,\dot{\mathbf{r}}\,|^2)=dT$$

The quantity

$$T=\tfrac{1}{2}\,m\,|\,\dot{\mathbf{r}}\,|^2=\frac{|\,\mathbf{p}\,|^2}{2m} \tag{6-11}$$

is known as the kinetic energy of the particle. The work done on the particle by the force F produces a change in the kinetic energy of the particle. If there are no net forces acting on the particle, then the kinetic energy is a constant of the motion. Similarly, if the only effect of the forces acting on the particle is to produce a change in the direction of the motion, then the kinetic energy is again conserved.

In many physical situations, the force field $F(\mathbf{r}, t)$ which acts on a particle is a conservative field in the sense that the line integral

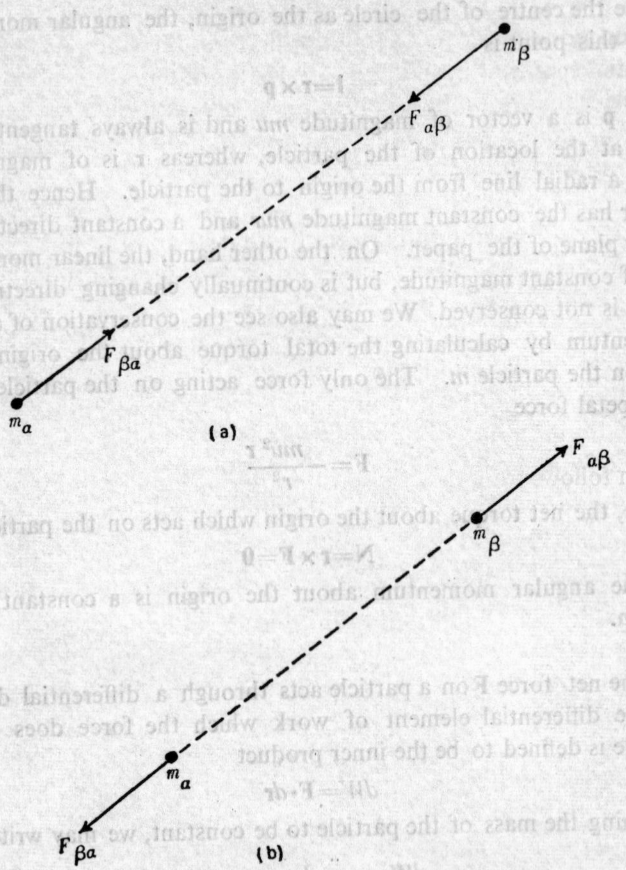

Figure 6-3. (a) Attractive force; (b) Repulsive force.

$$\int_{\mathbf{r}_1}^{\mathbf{r}_2} \mathbf{F}(\mathbf{r}, t) \cdot d\mathbf{r}$$

is independent of the path joining the end points $\mathbf{r}_1$ and $\mathbf{r}_2$. As we have seen in Sec. 2-7, a necessary and sufficient condition that $\mathbf{F}(\mathbf{r}, t)$ be a conservative field is that $\mathbf{F}$ be derived as the gradient of a scalar. We therefore, write

$$\mathbf{F}(\mathbf{r}, t) = -\text{grad } U(\mathbf{r}, t) \tag{6-12}$$

Then, the work done by the force field $F(r, t)$ in moving the particle from $\mathbf{r}_1$ to $\mathbf{r}_2$ for fixed time t is

$$\int_{\mathbf{r}_1}^{\mathbf{r}_2} \mathbf{F}(\mathbf{r}, t) \cdot d\mathbf{r} = U(\mathbf{r}_2, t) - U(\mathbf{r}_1, t)$$

The function $U(\mathbf{r}, t)$ defined by Eq. (6-12) is, therefore, known as the potential energy of the particle, and the work done by $\mathbf{F}$ in moving the particle from $\mathbf{r}_1$ to $\mathbf{r}_2$ is the difference in potential energy at $\mathbf{r}_2$ and $\mathbf{r}_1$.

Now, suppose that $\mathbf{F}$ and consequently U do not explicitly depend on the time t, and consider

$$\frac{d}{dt}(T+U)=\frac{dT}{dt}+\frac{dU}{dt}$$

Now,

$$\frac{dT}{dt}=\mathbf{F}\cdot\dot{d\mathbf{r}}$$

and

$$\frac{dU}{dt}=\dot{d\mathbf{r}}\cdot\text{grad } U+\frac{\partial U}{\partial t}$$

It then follows from (6-12) that

$$\frac{d}{dt}(T+U)=\frac{\partial U}{\partial t}=0$$

since U is not an explicit function of the time. We thus obtain:

The Principle of Conservation of Energy: *If the forces acting on a particle are derivable from a time independent potential function, then the sum of the kinetic and potential energies of the particle is conserved. This conservation law stems from the assumption of the homogeneity of time.*

6-2. Systems of Particles. We now turn our attention to the mechanics of systems of particles. Much, but not all, of the mechanics of a single particle can be applied to a system of particles. Suppose that we are considering a system of N particles, where the α-th particle has a mass m_α and is located in space by the radius vector $\mathbf{r}_\alpha$. In this section and the balance of this chapter, we shall "tag" the particles with Greek indices and continue to label the coordinates with Latin indices. The specification of radius vector $\mathbf{r}_\alpha$ for each particle of the system at a particular time defines the instantaneous configuration of the system. Then, the chief problem of the mechanics of a system of particles is the determination of the configuration of the system for all $t \geqslant t_0$, given the initial configuration at time t_0, and the set of forces acting on the system.

In specifying the forces acting on the particles of a system, it is necessary to consider any possible forces of interaction among the particles as well as any externally applied forces. If the particles do not exert forces on one another, i.e. are non-interacting of free particles, each particle moves independently of the others under the action of the externally applied forces. The instantaneous configuration is obtained by considering the instantaneous locations of the N independent particles. The mechanics of each particle is described in Sec. 6-1.

In the more usual case of a system of particles, there are forces of interaction between the particles of the system. The nature of these forces varies from the rather simple force which occur when two non-interacting particles collide to fairly complex interactions which take place between charged particles. A complete description of the behaviour of systems under the influence of the various interactions would consume much more space than is available here. However, there are many properties of systems of particles which do not depend on the exact nature of the interaction forces. It is to these properties that we shall give our attention.

Consider two interacting particles m_α and m_β. We shall denote the force exerted on m_α by m_β as $\mathbf{F}_{\beta\alpha}$ and the force exerted on m_β by m_α as $\mathbf{F}_{\alpha\beta}$. The direction of these forces will usually, but not always be either attractive or repulsive. In any case, there is abundant physical evidence to support the assertion that

$$\mathbf{F}_{\beta\alpha}=-\mathbf{F}_{\alpha\beta} \tag{6-13}$$

Equation (6-13) is generally known as Newton's third law of motion or the law of action and reaction. This is illustrated in Fig. 6-3 for both attractive and repulsive forces. If the total force $\mathbf{F}_\alpha$ on the particle m_α due to interaction with the other particles of the system is of the form

$$\mathbf{F}_\alpha=\sum_{\beta=1}^{N}{}' \mathbf{F}_{\beta\alpha} \tag{6-14}$$

where the prime on the summation indicates that the term $\beta=\alpha$ is omitted, the forces of interaction are called two body forces. These two body forces are independent in the sense that $\mathbf{F}_{\beta\alpha}$ depends only on the presence of the two particles m_β and m_α.

We may express the total force $\mathbf{F}_\alpha$ acting on the particle m as the vector sum of the total interaction force on m_α and any externally applied forces on m_α.

$$\mathbf{F}_\alpha=\mathbf{F}_\alpha^{\text{int}}+\mathbf{F}_\alpha^{\text{ext}}$$

If the interaction forces are two body forces, $\mathbf{F}_\alpha^{\text{int}}$ is given by (6-14). The total force acting on the system of particles is

$$\mathbf{F}=\Sigma\, \mathbf{F}_\alpha$$

where the sum extends over all particles in the system. If the total externally applied force vanishes for all configuration of the system, then the system is said to be isolated.

The total liner momentum of a system of N mass particles is defined to be

$$\mathbf{P}=\sum_{\alpha=1}^{N} \mathbf{p}_\alpha$$

where $\mathbf{p}_\alpha = m_\alpha \mathbf{u}_\alpha$ is the linear momentum of the α-th particle. From Newton's second law, the time rate of change of momentum of the α-th particle is equal to the net force $\mathbf{F}_\alpha$ which acts on the particle,

$$\mathbf{F}_\alpha=\frac{d}{dt}\mathbf{p}_\alpha$$

Summing over all of the particles in the system, we obtain

$$\mathbf{F}=\sum_{\alpha=1}^{N} \mathbf{F}_\alpha=\frac{d}{dt}\sum_{\alpha=1}^{N} \mathbf{p}_\alpha=\frac{d\mathbf{P}}{dt} \tag{6-15}$$

Hence, the total linear momentum of an isolated system is a constant of the motion. This does not, however, mean that the momenta of the individual particles are conserved. The latter fact is easily confirmed experimentally.

EXAMPLE 3. Consider an isolated system consisting of two mass particles m_1 and m_2 with $m_1=m_2$, which interact only by direct collision (Fig. 6-4). If m_2 is initially at rest and m_1 moves toward it with velocity $\mathbf{u}$, the initial momenta are

$$\mathbf{p}_1=m_1\,\mathbf{u},\ \mathbf{p}_2=0,\ \mathbf{P}=m_1\,\mathbf{u}$$

After the collision of the two particles, the particle m_2 is observed to move off with velocity $\mathbf{u}$, whereas the particle m_1 remains at rest. Thus, the momenta after the collision are

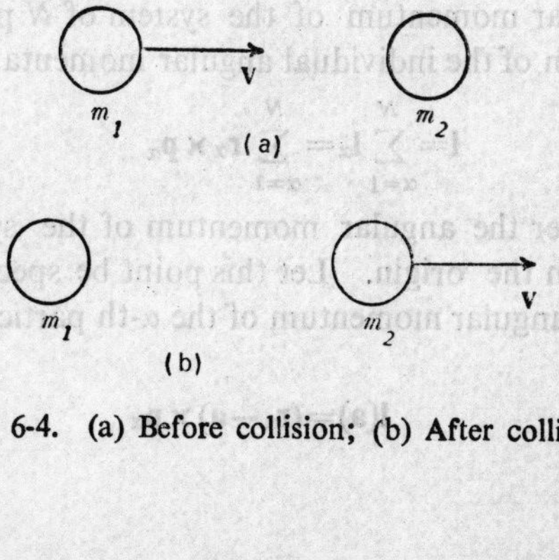

Figure 6-4. (a) Before collision; (b) After collision.

$$\mathbf{p}_1=0,\ \mathbf{p}_2=m_2\,\mathbf{u}=m_1\,\mathbf{u},\ \mathbf{P}=m_2\,\mathbf{u}=m_1\,\mathbf{u}$$

It is clear that the individual momenta of the two particles are not conserved during the collision, whereas the total momentum of the isolated system is constant.

If the particles of a system interact with two body forces, the total interaction force on the α-th particle can be written

$$\mathbf{F}_\alpha^{\text{int}}=\mathbf{F}_{\beta\alpha}+\sum_{\gamma=1}^{N}{}' \mathbf{F}_{\gamma\alpha}$$

where the prime on the sum indicates that the two terms $\gamma=\alpha$ and $\gamma=\beta$ are to be omitted. Similarly, the total interaction force exerted on the β-th particle is given by

$$\mathbf{F}_\beta^{\text{int}}=\mathbf{F}_{\alpha\beta}+\sum_{\gamma=1}^{N}{}' \mathbf{F}_{\gamma\beta}=-\mathbf{F}_{\beta\alpha}+\sum_{\gamma=1}^{N}{}' \mathbf{F}_{\gamma\beta}$$

Then,

$$\mathbf{F}_\alpha^{\text{int}}+\mathbf{F}_\beta^{\text{int}}=\sum_{\gamma=1}^{N}{}' (\mathbf{F}_{\gamma\alpha}+\mathbf{F}_{\gamma\beta})$$

When we consider the total interaction force, the terms cancel in pairs, and this establishes the result that for a system interacting with two body forces, only the externally applied forces contribute to the total force acting on the system.

As we have previously seen, the angular momentum of an isolated particle is defined to be

$$\mathbf{l}=\mathbf{r}\times\mathbf{p}$$

where $\mathbf{r}$ is the radius vector to the particle which has linear momentum $\mathbf{p}$. If $\mathbf{r}_\alpha$ and $\mathbf{p}_\alpha$ are the radius vector and linear momentum respectively of the α-th particle of a system of particles, we define the angular momentum of this particle about the origin to be

$$\mathbf{l}_\alpha=\mathbf{r}_\alpha\times\mathbf{p}_\alpha \tag{6-16}$$

The total angular momentum of the system of N particles, about the origin is the sum of the individual angular momenta,

$$\mathbf{l}=\sum_{\alpha=1}^{N}\mathbf{l}_\alpha=\sum_{\alpha=1}^{N}\mathbf{r}_\alpha\times\mathbf{p}_\alpha$$

Let us consider the angular momentum of the system about some point other than the origin. Let this point be specified by the radius vector $\mathbf{a}$. The angular momentum of the α-th particle about the point $\mathbf{a}$ is

$$\mathbf{l}(\mathbf{a})=(\mathbf{r}_\alpha-\mathbf{a})\times\mathbf{p}_\alpha$$

$$=\mathbf{r}_\alpha \times \mathbf{p}_\alpha - \mathbf{a} \times \mathbf{p}_\alpha$$

The total angular momentum about the point a is

$$\mathbf{l}(\mathbf{a}) = \sum_{\alpha=1}^{N} \mathbf{r}_\alpha \times \mathbf{p}_\alpha = \mathbf{a} \times \sum_{\alpha=1}^{N} \mathbf{p}_\alpha = \mathbf{l}(0) - \mathbf{a} \times \mathbf{P}$$

where $\mathbf{l}(0)$ is the total angular momentum of the system about the origin and $\mathbf{P}$ is the total linear momentum of the system. Thus, if the total linear momentum of a system of particles is zero, then the angular momentum of the system is the same about any point.

If the interaction forces for a system of particles are all two body forces which are directed along the lines joining pairs of particles, we may express the force acting on the α-th particle due to the presence of the β-th particle in the form

$$\mathbf{F}_{\beta\alpha} = |\mathbf{F}_{\beta\alpha}| \frac{\mathbf{r}_\alpha - \mathbf{r}_\beta}{|\mathbf{r}_\alpha - \mathbf{r}_\beta|} \tag{6-17}$$

By Newton's third law, the force on the β-th particle due to the presence of the α-th particle is

$$\mathbf{F}_{\alpha\beta} = -|\mathbf{F}_{\beta\alpha}| \frac{\mathbf{r}_\alpha - \mathbf{r}_\beta}{|\mathbf{r}_\alpha - \mathbf{r}_\beta|} \tag{6-18}$$

The time rate of change of the total angular momentum of a system about the origin is

$$\dot{\mathbf{l}} = \frac{d}{dt} \sum_{\alpha=1}^{N} \mathbf{r}_\alpha \times \mathbf{p}_\alpha = \sum_{\alpha=1}^{N} m_\alpha (\dot{\mathbf{r}}_\alpha \times \dot{\mathbf{r}}_\alpha + \mathbf{r}_\alpha \times \ddot{\mathbf{r}}_\alpha)$$

where we assume that the mass of each particle is constant. The product $m_\alpha \ddot{\mathbf{r}}_\alpha$ may be replaced by the total force acting on the α-th particle, and we may write

$$\dot{\mathbf{l}} = \sum_{\alpha=1}^{N} \mathbf{r}_\alpha \times \mathbf{F}_\alpha^{\text{ext}} + \sum_{\alpha=1}^{N} \sum_{\beta=1}^{N} \frac{|\mathbf{F}_{\beta\alpha}|\, \mathbf{r}_\alpha}{|\mathbf{r}_\alpha - \mathbf{r}_\beta|} \times (\mathbf{r}_\alpha = \mathbf{r}_\beta)$$

In the second sum, the terms of the form $\mathbf{r}_\alpha \times \mathbf{r}_\alpha$ are naturally zero, and in the double sum, the remaining terms cancel in pairs as a consequence of (6-17) and (6-18). Now, $\mathbf{r}_\alpha \times \mathbf{F}_\alpha^{\text{ext}}$ is the torque on the α-th particle due to external forces, and we may define the total externally applied torque on the system to be

$$\mathbf{N}^{\text{ext}} = \sum_{\alpha=1}^{N} \mathbf{N}^{\text{ext}} = \sum_{\alpha=1}^{N} \mathbf{r}_\alpha \times \mathbf{F}_\alpha$$

Then,

$$\dot{\mathbf{l}} = \mathbf{N}^{\text{ext}}$$

and we see that the total angular momentum of a system of particles is conserved if the total torque exerted on the system due to externally applied forces vanishes.

The kinetic energy of a system of particles is defined to be the sum of the kinetic energies of the individual particles,

$$T=\sum_{\alpha=1}^{N} T_\alpha=\tfrac{1}{2}\sum_{\alpha=1}^{N} m_\alpha \dot{\mathbf{r}}_\alpha^2$$

In the case of a single particle, we saw that the kinetic energy is a constant of the motion provided that the total force acting on the particle vanishes. For a system of more than one particle, this is no longer necessarily true. The work done by the force $\mathbf{F}_\alpha$ in moving through the distance $d\mathbf{r}_\alpha$ is

$$dW_\alpha=\mathbf{F}_\alpha\cdot d\mathbf{r}_\alpha$$

The total work done by the net force

$$\mathbf{F}=\sum_{\alpha=1}^{N}\mathbf{F}_\alpha$$

in moving each particle a differential distance is

$$dW=\sum_{\alpha=1}^{N} dW_\alpha=\sum_{\alpha=1}^{N}\mathbf{F}_\alpha\cdot d\mathbf{r}_\alpha$$

This latter sum is equal to $\mathbf{F}\cdot d\mathbf{r}$ only if each particle is moved through the same distance $d\mathbf{r}$ by the total force acting on it.

If the integral of the force $\mathbf{F}_\alpha$, acting on a mass particle m_α is such that the line integral

$$\int \mathbf{F}_\alpha(\mathbf{r}, t)\cdot d\mathbf{r}=0$$

for every closed path, the force is said to be conservative and

$$\mathbf{F}_\alpha=-\text{grad}_\alpha U(\mathbf{r}_\alpha, \mathbf{r}_\beta, \ldots) \qquad (6\text{-}19)$$

where "grad$_\alpha$" operates only on the coordinates of the α-th particle. If the total force exerted on each particle of the system can be expressed in the form (6-19), then U is the potential function for the system of particles. If U exists, and is not explicit function of the time, then the sum

$$\sum_{\alpha=1}^{N} T_\alpha+U$$

is a constant of the motion. If U is an explicit fuction of the time, then

$$\frac{d}{dt}\left(\sum_{\alpha=1}^{N} T_\alpha+U\right)=\frac{\partial U}{\partial t}$$

In the case where the interaction forces within the system are all two body forces, the internal forces are conservative and

$$\mathbf{F}_\alpha^{\text{int}} = -\text{grad}_\alpha U$$

where

$$U = \sum_{\alpha,\,\beta=1}^{N}{}' \, U_{\alpha\beta}(\mathbf{r}_\alpha, \mathbf{r}_\beta)$$

and the prime on the summation indicates the ommision of the term $\alpha=\beta$.

In order to determine in detail the behaviour of the N-particle system, we must consider the $3N$ second order differential equations of the form

$$m_\alpha \ddot{x}_\alpha^i = F_\alpha^i(r_\alpha, r_\beta, \ldots, t) \tag{6-20}$$

where the F_α^i are specified functions of the coordinates of all the particles and the time. The solutions of the set of Eq. (6-20) satisfy a set of $3N$ given initial conditions. A complete solution of the problem has never been achieved for $N>2$.

In many cases we are not interested in the detailed behaviour of the individual particles of the system. For these cases a great deal of information about the behaviour of the aggregate of particles is obtained by considering the motion of the centre of mass. For a system of N particles, the total mass and the radius vector $\mathbf{R}$ to the centre of mass of the system are defined by

$$M = \sum_{\alpha=1}^{N} m_\alpha; \quad \mathbf{R} = \frac{1}{M}\sum_{\alpha=1}^{N} m_\alpha \mathbf{r}_\alpha \tag{6-21}$$

We shall now obtain certain conservation laws which govern the motion of the centre of mass of a system of particles. We shall suppose for convenience that the interaction forces for the system are all two body forces. From (6-21), we define the linear momentum of the centre of mass by

$$\mathbf{P} = M\dot{\mathbf{R}} = \sum_{\alpha=1}^{N} m_\alpha \dot{\mathbf{r}}_\alpha$$

and the time rate of change of the linear momentum of the centre of mass is

$$\dot{P} = \sum_{\alpha=1}^{N} m_\alpha \ddot{\mathbf{r}}_\alpha = \sum_{\alpha=1}^{N} \dot{\mathbf{p}}_\alpha$$

$$= \sum_{\alpha=1}^{N} (\mathbf{F}_\alpha^{\text{ext}} + \mathbf{F}_\alpha^{\text{int}}) = \mathbf{F}^{\text{ext}} \tag{6-22}$$

The second term in the sum vanishes since the internal forces cancel in pairs, and the first term is the total external force acting on the system. Hence, for an isolated system, the linear momentum of the centre of mass of the system is constant of the motion. In the case of a system which is not isolated, (6-22) is the equation of motion for the centre of mass.

Denote the radius vector of the α-th particle relative to the centre of mass by $\mathbf{r}'_\alpha$, so that

$$\mathbf{r}_\alpha = \mathbf{R} + \mathbf{r}'_\alpha$$

The linear momentum of the α-th particle relative to the centre of mass is $\mathbf{p}'_\alpha = m_\alpha \dot{\mathbf{r}}'_\alpha$, and the angular momentum of the entire system about the centre of mass is

$$\mathbf{l}' = \sum_{\alpha=1}^{N} \mathbf{r}'_\alpha \times \mathbf{p}'_\alpha = \sum_{\alpha=1}^{N} m_\alpha \mathbf{r}'_\alpha \times \dot{\mathbf{r}}'_\alpha$$

We obtain for the time rate of change of angular momentum,

$$\dot{\mathbf{l}}' = \sum_{\alpha=1}^{N} m_\alpha \mathbf{r}'_\alpha \times \ddot{\mathbf{r}}'_\alpha$$

In terms of the cartesian components of $\mathbf{r}_\alpha$, $\mathbf{R}$, and $\mathbf{r}'_\alpha$ this can be written

$$(\dot{\mathbf{l}}')^3 = \sum_{\alpha=1}^{N} m_\alpha \{(x'_\alpha)^1(\ddot{x}^2_\alpha - \ddot{x}'^2_\alpha) - (x'_\alpha)^2(\ddot{x}^1_\alpha - \ddot{x}'^1_\alpha)\}$$

$$= -\sum_{\alpha=1}^{N} m_\alpha (x'_\alpha)^1 \ddot{x}^2 + \sum_{\alpha=1}^{N} m_\alpha (x'_\alpha)^2 \ddot{x}^1 +$$

$$\sum_{\alpha=1}^{N} [(x'_\alpha)^1 F^2_\alpha - (x'_\alpha)^2 F^1_\alpha] \qquad (6\text{-}23)$$

where $\mathbf{F}_\alpha$ is the total force acting on the α-th particle. The first two terms vanish by the definition of the centre of mass. In the last term, the internal forces cancel in pairs, and the resulting sum is just N^3, where $\mathbf{N}$ is the total torque exerted on the system by the external forces.

It is clear that the total momentum of a system of particles with respect to the centre of mass is zero, and hence in a set of coordinates with origin at the centre of mass, the angular momentum of the system is the same about any point.

6-3. Variational Principles and Lagrange's Equations. The general problem of mechanics, even for the single particle case, is complicated by the fact that for many physical systems, the motion is in some way constrained. The conditions of constraint will normally involve externally applied forces of constraint which are not a priori known. In order to solve the Newtonian equations of motion it is necessary to know all of the external forces, and the forces of constraint are to be determined as part of the solution. Hence, in order to consider constrained systems, it is desirable to reformulate mechanics in such a way that the forces of constraint do not appear in the equations of motion.

The constraints on the motion of a system of particles can be classified in several ways. One particularly useful classification is the following:

(i) If the constraints can be expressed as relations between the coordinates of the particles and possibly the time in the form

$$f(r_1, r_2, \ldots, r_N, t)=0$$

then the constraints are said to be holonomic. One very common type of holonomic constraint is the requirement that the motion of a given system takes place on a specified surface. Another, and even simpler, example of holonomic constraints occurs when the particles of a system are required to remain a fixed distance from one another. In this case, the equations of constraint have the simple form

$$|\mathbf{r}_\alpha-\mathbf{r}_\beta|^2-C^2_{\alpha\beta}=0, \quad \alpha\neq\beta$$

where the $C_{\alpha\beta}$ are constants.

(ii) Constraints which are not holonomic are called non-holonomic. The classic example of a non-holonomic constraint is the constraint on a sphere which is to roll on a rough surface without slipping. This constraint is a condition on the velocity of the point of contact, i.e. the linear velocity of this point is zero, rather than a condition on the coordinates of the particles of the sphere.

In addition to the difficulties introduced from the unknown forces of constraint, the conditions of constraint also introduce the difficulty that the coordinates of the particles of the system are no longer independent. This follows since the coordinates, or their derivatives are related by the conditions of constraint. It then follows that the equations of motion

$$\dot{\mathbf{p}}_\alpha=\mathbf{F}_\alpha, \quad \alpha=1, \ldots, N \tag{6-24}$$

are no longer independent. In the case of holonomic constraints, the lack of independence of the equations of motion is frequently a great advantage. For an unconstrained system consisting of N particles, there are $3N$ independent coordinates and consequently $3N$ equations of the form (6-24). If there are now imposed k constraints on the system, the $3N$ coordinates are related by k equations of constraint and hence the system is described by $3N-k$ independent coordinates. The number of independent coordinates required to describe a given mechanical system is called the number of degrees of freedom of the system.

EXAMPLE 4. Consider two particles which are constrained to have a unit distance between them. Without the constraint, it requires three independent coordinates to specify the location of each particle; a total of six independent variables. However, the condition of constraint

$$|\mathbf{r}_1-\mathbf{r}_2|^2=1$$

implies that if five of the coordinates are known, the sixth is determined. Thus, the system has five degrees of freedom. A convenient set of independent quantities to describe the motion of this system might be the three cartesian coordinates of the location of one of the particles, say (x_1^1, x_1^2, x_1^3) and the two angles (θ, φ) which define the orientation of the vector $\mathbf{r}_1-\mathbf{r}_2$ (see Fig. 6-5). The Newtonian equations of motion for the three cartesian coordinates will be relatively simple, but the equations of motion determining the angular coordinates are somewhat more complicated.

It is possible to describe the motion of a system of N particles subject to k holonomic constraints in terms of a set of $3N-k$ independent variables q^i. The quantities q^i are known as the generalized coordinates of the system, and are introduced by the transformation

$$\begin{aligned} x_1^1&=x_1^1\,(q^1,\ldots,q^{3N},t)\\ x_1^2&=x_1^2\,(q^1,\ldots,q^{3N},t)\\ x_1^3&=x_1^3\,(q^1,\ldots,q^{3N},t)\\ &\vdots\\ x_{3N}^3&=x_{3N}^3\,(q^1,\ldots,q^{3N},t)\end{aligned}$$

and the conditions of constraint, expressed as

$$\begin{aligned} q^1&=C^1\\ &\vdots\\ q^k&=C^k\end{aligned}$$

where the C^i are constants. Since the equations expressing non-holonomic constraints are either inequalities or non-integrable differential forms, they cannot be used to eliminate dependent coordinates. Thus, although problems involving holonomic constraints are,

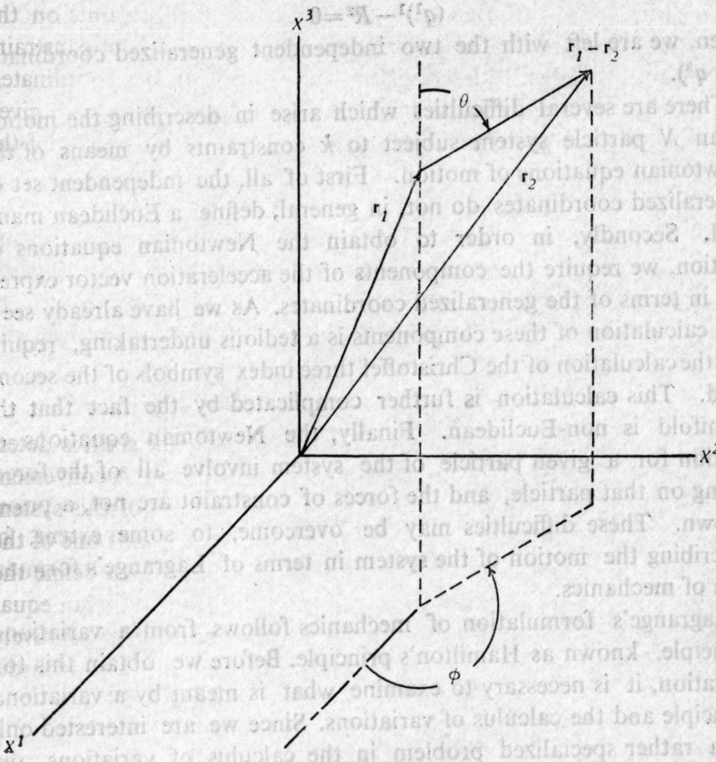

Figure 6-5.

in principle, always solvable, there is no general method of attack for problems involving non-holonomic constraints.

EXAMPLE 5. A particle is constrained to move on the surface of a sphere of radius R. In terms of the cartesian coordinates of the particle, the condition of constraint is expressed as

$$(x^1)^2+(x^2)^2+(x^3)^2=R^2$$

We introduce spherical polar coordinates by the usual transformation

$$x^1=q^1 \sin q^2 \cos q^3$$
$$x^2=q^1 \sin q^2 \sin q^3$$
$$x^3=q^1 \cos q^2$$

The generalized coordinates (q^1, q^2, q^3) as defined by the transformation equations arenot independent. However, q^1 can be eliminated by expressing the constraint condition as

$$(q^1)^1 - R^2 = 0$$

Then, we are left with the two independent generalized coordinates (q^2, q^3).

There are several difficulties which arise in describing the motion of an N particle system subject to k constraints by means of the Newtonian equations of motion. First of all, the independent set of generalized coordinates do not, in general, define a Euclidean manifold. Secondly, in order to obtain the Newtonian equations of motion, we require the components of the acceleration vector expressed in terms of the generalized coordinates. As we have already seen, the calculation of these components is a tedious undertaking, requiring the calculation of the Christoffel three index symbols of the second kind. This calculation is further complicated by the fact that the manifold is non-Euclidean. Finally, the Newtonian equations of motion for a given particle of the system involve all of the forces acting on that particle, and the forces of constraint are not, a priori, known. These difficulties may be overcome, to some extent, by describing the motion of the system in terms of Lagrange's formulation of mechanics.

Lagrange's formulation of mechanics follows from a variational principle, known as Hamilton's principle. Before we obtain this formulation, it is necessary to examine what is meant by a variational principle and the calculus of variations. Since we are interested only in a rather specialized problem in the calculus of variations, our discussion will not be in general terms. We first consider the one-dimensional variational problem, which can be stated as follows: Determine the curve $y(x)$ such that the integral

$$I = \int_{x_1}^{x_2} f(y, dy/dx, x)\, dx \tag{6-25}$$

has an extremal value for fixed end points (x_1, x_2) and a given function f. Since there is an infinite number of curves passing through the two fixed end points, it is necessary to distinguish the various curves by use of a continuous indexing parameter α. It is convenient to choose the value $\alpha=0$ to correspond to the curve which gives the integral (6-25) its extremal value. Then, the various curves, y which are to be considered can be represented by

$$y(x, \alpha) = y(x, 0) + \alpha\eta(x)$$

where $\eta(x)$ is an arbitrary function which vanishes at $x=x_1$, and $x=x_2$. Equation (6-25) can then be written as

$$I(\alpha) = \int_{x_1}^{x_2} f\left[y(x, \alpha), \frac{d}{dx} y(x, \alpha), x\right] dx$$

and the condition that the integral have an extremal value is

$$\left.\frac{\partial I}{\partial \alpha}\right|_{\alpha=0}=0$$

Since the limits of integration are fixed, we interchange differentiation and integration to obtain

$$\frac{\partial I}{\partial \alpha}=\int_{x_1}^{x_2}\left[\frac{\partial f}{\partial y}\frac{\partial y}{\partial \alpha}+\frac{\partial f}{\partial(dy/dx)}\frac{\partial}{\partial \alpha}(dy/dx)\right]dx$$

The second term is integrated by parts, and

$$\frac{\partial I}{\partial \alpha}=\frac{\partial f}{\partial(dy/dx)}\frac{\partial y}{\partial \alpha}\Bigg|_{x_1}^{x_2}+\int_{x_1}^{x_2}\left[\frac{\partial f}{\partial y}-\frac{d}{dx}\left(\frac{\partial f}{\partial(dy/dx)}\right)\right]\frac{\partial y}{\partial \alpha}\,dx$$

$$=\frac{\partial f}{\partial(dy/dx)}\eta(x)\Bigg|_{x_1}^{x_2}+\int_{x_1}^{x_2}\left[\frac{\partial f}{\partial y}-\frac{d}{dx}\left(\frac{\partial f}{\partial(dy/dx)}\right)\right]\eta(x)\,dx \qquad (6\text{-}26)$$

Since $\eta(x_1)=\eta(x_2)=0$, the integrated term vanishes, and the extremal condition is satisfied if

$$\left.\frac{\partial I}{\partial \alpha}\right|_{\alpha=0}=\int_{x_1}^{x_2}\left[\frac{\partial f}{\partial y}-\frac{d}{dx}\left(\frac{\partial f}{\partial(dy/dx)}\right)\right]\eta(x)\,dx=0$$

The right-hand side is independent of α, and $\eta(x)$ is an arbitrary function. Hence, a necessary and sufficient condition for the equality is

$$\frac{d}{dx}\left(\frac{\partial f}{\partial(dy/dx)}\right)-\frac{\partial f}{\partial y}=0 \qquad (6\text{-}27)$$

Equation (6-27) is the Euler-Lagrange equation for the variational problem specified by (6-25). Since f is a specified function of y, dy/dx, and x, (6-27) is a second order differential equation whose solution is the desired curve $y(x)$.

There is an alternative derivation of the Euler-Lagrange equation which involves the more customary notation of the calculus of variations. We multiply the first equality in (6-26) by $d\alpha$ and evaluate the derivatives with respect to α at $\alpha=0$. Define

$$\left(\left.\frac{\partial I}{\partial \alpha}\right|_{\alpha=0}\right)d\alpha=\delta I, \quad \left.\frac{\partial y}{\partial \alpha}\right|_{\alpha=0}d\alpha=\delta y$$

where δI and δy are known as the first variation of I and y respectively. Then,

$$\delta I=\frac{\partial f}{\partial(dy/dx)}\delta y\Bigg|_{x_1}^{x_2}+\int_{x_1}^{x_2}\left[\frac{\partial f}{\partial y}-\frac{d}{\partial x}\left(\frac{\partial f}{\partial(dy/dx)}\right)\right]\delta y\,dx$$

The quantity δy represents an arbitrary variation of the curve $y(x)$ from the curve which extremizes the integral, subject to the condition that $\delta y=0$ at $x=x_1$ and $x=x_2$. In terms of this notation, the condition for an extremum is $\delta I=0$, and we again obtain the Euler-Lagrange equation.

EXAMPLE 6. The classic example of the calculus of variations in one dimension is the brachistochrone problem. Given two points (x_1, y_1) and (x_2, y_2) in a vertical plane such that they do not lie on the same vertical line, find the curve joining them such that a particle starting from rest, without friction, will traverse this curve between the two points in minimum time.

It follows from the principle of conservation of energy that

$$\tfrac{1}{2}mv^2=mg\,(y-y_1)$$

Now,

$$v=\frac{ds}{dt}=\sqrt{1+(dy/dx)^2}\,\frac{dx}{dt}$$

and hence,

$$dt=\frac{\sqrt{1+(dy/dx)^2}}{\sqrt{2g\,(y-y_1)}}\,dx$$

The time of descent along the curve $y(x)$ is given by the integral

$$T=I=\int_{x_1}^{x_2}\sqrt{\frac{1+(dy/dx)^2}{2g\,(y-y_1)}}\,dx$$

The determination of the Euler-Lagrange equation and its solution is left as an exercise.

Let us now consider the N-dimensional variational problem, in which we wish to determine a set of independent functions $y_1(x), \ldots, y_N(x)$ such that the integral

$$I=\int_{x_1}^{x_2} f(y_1, \ldots, y_N, dy_1/dx, \ldots, dy_N/dx, x)\,dx \tag{6-28}$$

has an extremal value for fixed end points x_1 and x_2, and a specified function f. Taking the first variation of the integral, we obtain

$$\delta I=\int_{x_1}^{x_2}\sum_{i=1}^{N}\left[\frac{\partial f}{\partial y_i}\,\delta y_i+\frac{\partial f}{\partial(dy_i/dx)}\,\delta\left(\frac{dy_i}{dx}\right)\right]dx$$

In the second term, we shall assume that we can interchange the order of operations and write

$$\delta\left(\frac{dy_i}{dx}\right)=\frac{d(\delta y_i)}{dx}$$

We then integrate by parts to obtain

$$\delta I=\sum_{i=1}^{N}\frac{\partial f}{\partial(dy_i/dx)}\,\delta y_i\Big|_{x_1}^{x_2}+$$

$$\int_{x_1}^{x_2} \sum_{i=1}^{N} \left[\frac{\partial f}{\partial y_i} - \frac{d}{dx}\left(\frac{\partial f}{\partial (dy_i/dx)} \right) \right] \delta y_i \, dx$$

Since the arbitrary variations δy_i vanish at the end points x_1 and x_2, the extremal condition is

$$\delta I = \int_{x_1}^{x_2} \sum_{i=1}^{N} \left[\frac{\partial f}{\partial y_i} - \frac{d}{dx}\left(\frac{\partial f}{\partial (dy_i/dx)} \right) \right] \delta y_i \, dx = 0$$

The δy_i are independent, so that the extremal condition is satisfied if and only if each coefficient of δy_i is separately zero. We thus obtain the system of Euler-Lagrange equations

$$\frac{d}{dx}\left(\frac{\partial f}{\partial (dy_i/dx)} \right) - \frac{\partial f}{\partial y_i} = 0, \; i = 1, 2, \ldots, N \tag{6-29}$$

The system of Eqs. (6-29) is a set of N coupled second order differential equations whose solutions $y_i(x)$, $i=1, \ldots, N$, extremizes the integral (6-28).

Let us now consider a mechanical system consisting of N particles subject to k holonomic constraints. As we have already seen, the k conditions of constraint may be used, in connection with the $3N$ coordinates of the particles, to define a set of $3N-k$ generalized coordinates q^i, which are independent. The set of values of these $3N-k$ generalized coordinates defines the configuration of the system. Hence, we seek a set of differential equations which express the q^i's as functions of the time. These differential equations follow from a variational principle known as Hamilton's principle: *The change in the configuration of a mechanical system between times t_1 and t_2 is such that*

$$S = \int_{t_2}^{t_2} L(q^i, \dot{q}^i, t) \, dt \tag{6-30}$$

has an extremal value. The function $L(q^i, \dot{q}^i, t)$ is known as the Lagrangian of the system, and the integral is known as the action. In the case of a conservative system, the Lagrangian is given by

$$L(q^i, \dot{q}^i, t) = T - U \tag{6-31}$$

where T is the total kinetic energy and U is the potential function for the system.

Hamilton's principle is clearly a special case of the multidimensional variation problem, which is solved by the solutions of the $3N-k$ Euler-Lagrange equations

$$\frac{d}{dt}\left(\frac{\partial L}{\partial \dot{q}^i} \right) - \frac{\partial L}{\partial q^i} = 0, \; i = 1, \ldots, 3N-k \tag{6-32}$$

The Euler-Lagrange Eqs. (6-32) are known as the Lagrange equations of motion for the system. Their solutions, as functions of the time, specify the configuration of the system at all times. Note that the forces of constraint do not appear in the Lagrange equations and are restricted only by the fact that they must be conservative. We have thus obtained a set of equations of motion which do not explicitly involve the forces of constraint. On the other hand, if we are interested in the forces of constraint, the Lagrangian formulation, as given here, provides no mechanism for their determination.

EXAMPLE 7. The Atwood machine consists of two masses m_1 and m_2 connected by an inextensible string of negligible mass over a frictionless pulley as shown in Fig. 6-6. The string is of length l, and the two masses are free to move under the influence of gravity in the vertical plane containing the masses, the string, and the pulley. We wish to determine the location of the two masses

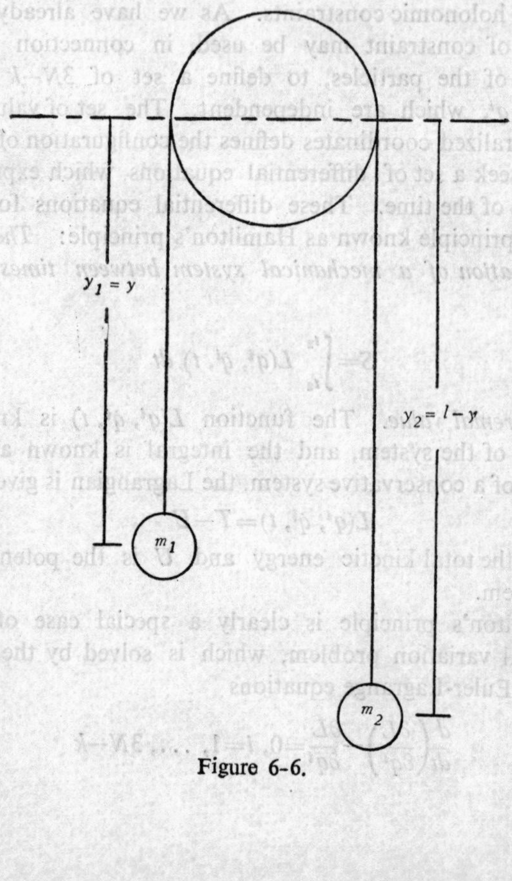

Figure 6-6.

relative to some fixed line as a function of time. If we choose the reference line as a horizontal line through the centre of the pulley and denote the distances of m_1 and m_2 from this line by y_1 and y_2 respectively, then the equation of constraint is

$$y_2 = l - y_1$$

The unknown force of constraint is the tension in the string. Since there is one equation of constraint, the system configuration is described by a single generalized coordinate y, which we take as the location of m_1, i.e. $y = y_1$, $y_2 = l - y$. If g denotes the acceleration due to gravity, the potential energy of the system relative to the chosen reference line is

$$\begin{aligned} U &= -m_1 g y - m_2 g(l - y) \\ &= -(m_1 - m_2)\, g y - m_2 g l \end{aligned}$$

The kinetic energy of the system is

$$\begin{aligned} T &= \tfrac{1}{2} m_1 \dot{y}_1^2 + \tfrac{1}{2} m_2 \dot{y}_2^2 \\ &= \tfrac{1}{2}(m_1 + m_2)\dot{y}^2 \end{aligned}$$

and the Lagrangian for the system is

$$\begin{aligned} L &= T - U \\ &= \tfrac{1}{2}(m_1 + m_2)\dot{y}^2 + (m_1 - m_2)\, g y + m_2 g l \end{aligned}$$

Since there is only one independent generalized coordinate, we obtain only one Lagrange equation of motion, namely

$$(m_1 + m_2)\ddot{y} - (m_1 - m_2)\, g = 0$$

The equation of motion has the solution

$$y(t) = \tfrac{1}{2}\, \frac{m_1 - m_2}{m_1 + m_2}\, g\, t^2 + \dot{y}_0 t + y_0$$

where $\dot{y}_0$ and y_0 are the velocity and position of m_1 at some initial time t_0. Although this problem, due to its simplicity, can also be solved in the Newtonian formulation, this is not the case for more complicated problems. However, in this simple case, the Newtonian equations of motion will also yield the tension in the string.

Let us consider an N particle system subject to k holonomic and m non-holonomic constraints. We shall suppose that we have used the k holonomic constraints to define a set of $3N - k$ generalized coordinates q^i. Since there are also m non-holonomic constraints, there are only $3N - k - m$ of the q^i which are independent. We shall consider only the case where the non-holonomic constraints are of the form

$$a_{ij} dq^j + a_{it} dt,\ i = 1, 2, \ldots, m,\ j = 1, 2, \ldots\ 3N - k$$

In the variational problem, the δq^i will not be independent because of the constraint condition among the differentials dq^i, but are related by m equations of the form

$$a_{ij}\delta q^j=0, i=1, 2, \ldots, m \tag{6-33}$$

We can use the non-holonomic constraints to eliminate the dependent coordinates by introducing the m Lagrange undetermined multipliers λ_i. If the system of Eqs. (6-33) holds, then it is equally true that

$$\lambda_i(a_{ij}\delta q^j)=0, i=1, \ldots, m \tag{6-34}$$

where the λ_i are as yet undetermined constants. From Hamilton's principle,

$$\int_{t_1}^{t_2} \sum_{j=1}^{3N-k} \left[\frac{\partial L}{\partial q^j}-\frac{d}{dt}\left(\frac{\partial L}{\partial \dot{q}^j}\right)\right] \delta q^j \, dt=0$$

where, as we have pointed out, the δq^j are not independent. However, if we choose the λ_i's so that

$$\frac{\partial L}{\partial \dot{q}^j}-\frac{d}{dt}\left(\frac{\partial L}{\partial \dot{q}^j}\right)+\sum_{i=1}^{m} \lambda_i a_{ij}=0, j=3N-k-m, \ldots, 3N-k \tag{6-35}$$

we can eliminate the dependent variations δq^j. Using (6-35), Hamilton's principle can be written in the form

$$\int_{t_1}^{t_2} \sum_{j=1}^{3N-k-m} \left\{\frac{\partial L}{\partial q^j}-\frac{d}{dt}\left(\frac{\partial L}{\partial \dot{q}^j}\right)+\sum_{i=1}^{m} \lambda_i a_{ij}\right\} \delta q^j \, dt=0$$

where the δq^j are now independent. Then,

$$\frac{\partial L}{\partial q^j}-\frac{d}{dt}\left(\frac{\partial L}{\partial \dot{q}^j}\right)+\sum_{i=1}^{m} \lambda_i a_{ij}=0, j=1, \ldots, 3N-k-m \tag{6-36}$$

We may combine Eqs. (6-35) and (6-36) to obtain the set of $3N-k$ equations of motion

$$\frac{\partial L}{\partial q^j}-\frac{d}{dt}\left(\frac{\partial L}{\partial \dot{q}^j}\right)+\sum_{i=1}^{m} \lambda_i a_{ij}=0, j=1, \ldots, 3N-k$$

However, we must now determine the $3N-k$ coordinates q^j, and the m constants λ_i, so that we need m additional equations. These are obtained by differentiating the equations of constraint with respect to t,

$$a_{ij}\, \dot{q}^j+a_{it}=0, i=1, \ldots, m$$

The method of undetermined multipliers can also be used in the case of holonomic constraints whenever it is inconvenient to define a set of independent generalized coordinates.

EXAMPLE 8. Consider the somewhat trivial example of a

hoop rolling, without slipping, down an inclined plane. Although the constraint is, in this case, holonomic, it is still convenient to use the method of undetermined multipliers. A convenient choice of generalized coordinates are the distance the hoop has moved from its starting point to the point of contact, and the total angle through which the hoop has turned. If the hoop has a radius r, and s denotes the distance from the start to the point of contact, the condition of constraint can be expressed as

$$r\,d\theta = ds$$

where θ is the total angle through which the hoop has turned. This condition of constraint is of the form

$$a_\theta d\theta + a_s ds = 0$$

with $a_\theta = r$, and $a_s = -1$.

The kinetic energy can be expressed as the sum of the kinetic energy of the centre of mass and the kinetic energy of the motion about the centre of mass. The first of these terms is

$$T_1 = \tfrac{1}{2} M \dot{s}^2$$

and the second is

$$T_2 = \tfrac{1}{2} I \dot{\theta}^2 = \tfrac{1}{2} M r^2 \dot{\theta}^2$$

where $I = Mr^2$ is the moment of inertia of the hoop about its centre. The potential energy of the system relative to the bottom of the inclined plane is

$$U = Mg\,(l - s) \sin \varphi$$

where l is the length of the incline, and φ is the angle of inclination. Combining these results, we find that the Lagrangian is

$$L = \tfrac{1}{2} M \dot{s}^2 + \tfrac{1}{2} M r^2 \dot{\theta}^2 - Mg\,(l - s) \sin \varphi$$

Since there is only one condition of constraint, there will be only one undetermined multiplier λ. It then follows that the equations of motion are

$$M\ddot{s} - Mg \sin \varphi + \lambda = 0$$

$$Mr^2 \ddot{\theta} - r\lambda = 0$$

These two equations along with the equivalent equation of constraint

$$r\dot{\theta} = \dot{s}$$

can be solved for s, θ, and λ.

Differentiating the equation of constraint with respect to the time,

$$r\ddot{\theta} = \ddot{s}$$

Substituting this result in the equation of motion for θ,

$$M\ddot{s}=\lambda$$

and this result substituted in the first equation of motion yields

$$M\ddot{s}-Mg\sin\varphi+M\ddot{s}=0$$

or

$$\ddot{s}=\frac{g\sin\varphi}{2}$$

Back-substitution of this result yields

$$\lambda=M\ddot{s}=\frac{Mg\sin\varphi}{2}$$

$$\ddot{\theta}=\frac{\ddot{s}}{r}=\frac{g\sin\varphi}{2r}$$

Thus, the hoop rolls down the plane with half the acceleration it would have in sliding down a frictionless plane.

To complete the Lagrangian formulation of mechanics, we note that if part of the forces acting on a given system are non-conservative, then Hamilton's principle is still valid if the Lagrangian for the system is of the form

$$L=T-U+W \tag{6-37}$$

In Eq. (6-37), U is the potential function for the conservative forces on the system, and

$$W=\sum_{\alpha=1}^{N}\mathbf{F}_{\alpha}\cdot d\mathbf{r}_{\alpha}$$

where F_{α} is the total non-conservative force acting on the α-th particle. If we substitute the Lagrangian as given by (6-37) in Hamilton's principle, and carry out the indicated variation, we obtain the system of Lagrange equations

$$\frac{d}{dt}\left(\frac{\partial L}{\partial \dot{q}^{j}}\right)-\frac{\partial L}{\partial q^{j}}=Q_{j} \tag{6-38}$$

where the Q_j are known as the generalized forces. These generalized forces are related to the actual forces by

$$Q_{j}=\sum_{\alpha=1}^{N}\mathbf{F}_{\alpha}\cdot\frac{\partial r_{\alpha}}{\partial q^{j}}$$

Although it is possible to obtain the Lagrange equations of motion directly from the Newtonian equations via D'Alembert's principle of virtual work, there are several advantages to the variational formulation. The principal advantage is that the variational technique may be extended to treat systems which are not mechanical in the Newtonian sense, such as the electromagnetic field.

6-4. Symmetries and Conservation Laws. In the motion of a mechanical system with f degrees of freedom, the $2f$ quantities q^i and $\dot{q}^i$, $i=1,\ldots,f$ which specify the configuration of the system, generally vary with the time. There are, however, certain functions of these $2f$ quantities which are constant in time, and whose values depend only on the initial configuration of the system. If the system is isolated, there are $2f-1$ of these constants or integrals of the motion. It is easy to prove this last assertion. Since the equations of motion are second order differential equations, their general solution contains $2f$ arbitrary constants. For an isolated system, the the equations of motion do not explicitly contain the time, and hence the choice of the time origin is completely arbitrary. Hence, one of the constants in the general solution of the equations of motion can be taken as an arbitrary additive time constant t_0. If we then eliminate $t+t_0$ from the $2f$ equations

$$q^i=q^i\,(t+t_0,\, C_1,\, C_2,\ldots C_{2f-1})$$

$$\dot{q}^i=\dot{q}^i\,(t+t_0,\, C_1,\, C_2,\ldots,C_{2f-1})$$

we can express the $2f-1$ arbitrary constants $(C_1, C_2,\ldots,C_{2f-1})$ as functions of the generalized coordinates and their time derivatives. These functions will be constants of the motion.

Only those constants of the motion which have the property of being additive, i.e. their values for a system composed of several non-interacting parts are equal to the sums of their values for the individual parts, are of any physical interest. We have already discussed these constants of the motion from an elementary point of view and have indicated that their conservation theorems derive from fundamental assumptions concerning the symmetry of space-time. We can now derive these conservation theorems from the relevant symmetries.

The assumption of homogeneity of time implies that the Lagrangian of an isolated system does not depend explicitly on the time. Then, the total time derivative of the Lagrangian is

$$\frac{dL}{dt}=\frac{\partial L}{\partial q^i}\,\dot{q}^i+\frac{\partial L}{\partial \dot{q}^i}\,\ddot{q}^i$$

However, from the equations of motion

$$\frac{\partial L}{\partial q^i}=\frac{d}{dt}\left(\frac{\partial L}{\partial \dot{q}^i}\right)$$

and

$$\frac{dL}{dt}=\dot{q}^i\;\frac{d}{dt}\left(\frac{\partial L}{\partial \dot{q}^i}\right)+\ddot{q}^i\;\frac{\partial L}{\partial \dot{q}^i}=\frac{d}{dt}\left(\dot{q}^i\;\frac{\partial L}{\partial \dot{q}^i}\right)$$

Rearranging terms,

$$\frac{d}{dt}\left(\dot{q}^i\,\frac{\partial L}{\partial \dot{q}^i}-L\right)=0$$

and we see that the quantity

$$H=\dot{q}^i\,\frac{\partial L}{\partial \dot{q}^i}-L \tag{6-39}$$

is a constant of the motion for an isolated system. We identify this constant of the motion as the total energy of the system, and its additivity follows immediately from the additivity of the Lagrangian, since it is a linear function of the Lagrangian. The law of conservation of H is valid not only for isolated systems, but also for systems in a time-independent external field. This follows since the only property of the Lagrangian which was used in obtaining the conservation theorem is its time independence.

For an isolated system, or a system in a constant external field, the Lagrangian is of the form

$$L=T\,(q^i,\,\dot{q}^i)-U\,(q^i)$$

where T is a homogeneous quadratic function of the velocities. It then follows from Euler's theorem on homogeneous functions that

$$\dot{q}^i\,\frac{\partial L}{\partial \dot{q}^i}=\dot{q}^i\,\frac{\partial T}{\partial \dot{q}^i}=2T \tag{6-40}$$

Substituting (6-40) in (6-39),

$$H=2T-T+U=T+U$$

and we see that our identification of H as the total energy of the system is correct.

The homogeneity of space implies that the mechanical properties, and hence the Lagrangian, of an isolated system are translationally invariant. Consider an isolated system, whose configuration is described by the orthogonal cartesian system (x^1, x^2, x^3), under a translation through the infinitesimal translation $\boldsymbol{\epsilon}$ with cartesian components $(\varepsilon^1, \varepsilon^2, \varepsilon^3)$. Such a transformation leaves the velocities of the particles of the system unchanged. The change in the Lagrangian induced by the infinitesimal translation is

$$\delta L=\sum_{\alpha=1}^{N}\varepsilon^i\,\frac{\partial L}{\partial x_\alpha^i}$$

where the sum extends over all of the particles in the system. Since the infinitesimal translation is arbitrary, the translational invariance of the Lagrangian is equivalent to

$$\sum_{\alpha=1}^{N}\frac{\partial L}{\partial x_{\alpha}^{i}}=0$$

From the equations of motion for an isolated system

$$\sum_{\alpha=1}^{N}\frac{\partial L}{\partial x_{\alpha}^{i}}=\sum_{\alpha=1}^{N}\frac{d}{dt}\left(\frac{\partial L}{\partial \dot{x}_{\alpha}^{i}}\right)=\frac{d}{dt}\sum_{\alpha=1}^{N}\frac{\partial L}{\partial \dot{x}_{\alpha}^{i}}=0$$

and we conclude that the vector with cartesian components

$$P_i=\sum_{\alpha=1}^{N}\frac{\partial L}{\partial \dot{x}_{\alpha}^{i}} \tag{6-41}$$

is a constant of the motion. We shall now show that vector P_i is the total linear momentum of the system. For an isolated system, the Lagrangian has the explicit form

$$L=\tfrac{1}{2}\sum_{\alpha=1}^{N} m_\alpha (\dot{x}_\alpha^i)^2-U(x_\alpha^i)$$

If we carry out the differentiation indicated in (6-41), we see that P_i is the total linear momentum of the system

$$P_i=\sum_{\alpha=1}^{N} m_\alpha\,\dot{x}_{i\alpha}$$

where we have again made use of the fact that there is no distinction between covariant and contravariant components in an orthogonal cartesian system.

The additivity of the linear momentum is obvious. However, unlike the energy, the linear momentum is additive whether or not the particles interact.

Although the three components of linear momentum are conserved only in the absence of an external field, if the potential energy of the field does not depend on one of the cartesian coordinates, say x^k, then the component of linear momentum p_k is conserved.

If the motion of the mechanical system is described by generalized coordinates q^i, the derivatives of the Lagrangian $L(q^i, \dot{q}^i)$ with respect to the generalized velocities

$$p_i=\frac{\partial L}{\partial \dot{q}^i} \tag{6-42}$$

are called generalized momenta. The derivatives of the Lagrangian with respect to the generalized coordinates

$$Q_i=\frac{\partial L}{\partial q^i} \tag{6-43}$$

are called generalized forces. In terms of the generalized forces and momenta, the equations of motion are

$$\frac{dp_i}{dt} = \dot{p}_i = Q_i \tag{6-44}$$

We have already seen that, in cartesian coordinates, the components of the linear momentum are the cartesian components of the vector $\mathbf{P}_\alpha$. The generalized momenta $p_{i\alpha}$ are the components of the vector $\mathbf{P}_\alpha$ relative to the generalized coordinate system. However, the generalized momenta are linear homogeneous functions of the generalized velocities $\dot{q}^i$, and do not, in general, reduce to the product of mass and the generalized velocities.

The linear momentum of an isolated mechanical system is different in different inertial frames of reference. If a reference frame K' moves relative to a reference frame K with velocity components v^i, then the cartesian components of the velocities $\dot{x}_{i\alpha}$ and $\dot{x}'_{i\alpha}$ of the α-th particle relative to the two frames are related by

$$\dot{x}_{i\alpha} = \dot{x}'_{i\alpha} + v_i$$

The total momentum of the system relative to the two different reference frames is

$$\begin{aligned} P_i &= \sum_{\alpha=1}^{N} m_\alpha \dot{x}_{i\alpha} = \sum_{\alpha=1}^{N} m_\alpha \dot{x}'_{i\alpha} + v_i \sum_{\alpha=1}^{N} m_\alpha \\ &= P'_i + M v_i \end{aligned} \tag{6-45}$$

where M is the total mass of the system.

In particular, there is a reference frame K' in which the total momentum is zero. The relative velocity of this frame can be immediately calculated from Eq. (6-45)

$$v_i = \frac{P_i}{M} = \frac{1}{M} \sum_{\alpha=1}^{N} m_\alpha \dot{x}_{i\alpha} \tag{6-46}$$

If the total momentum of a system is zero in a given reference frame K', we say that the system is at rest in the frame K'. Similarly, the velocity v_i given by (6-46) is the "velocity" of the "motion as a whole" of a system whose linear momentum is not zero. In this way, conservation of linear momentum leads to a natural definition of rest and velocity for a mechanical system treated as a distinct entity.

In terms of the coordinates of the centre of mass of the system, we can write (6-46) as

$$v_i = \dot{\bar{x}}_i$$

and interpret the law of conservation of linear momentum as stating that for an isolated system, the centre of mass is in uniform rectilinear motion. Thus, in considering the mechanical properties of an isolated system, it is natural to use a coordinate system (centre of mass system) in which the centre of mass is at rest. Although this ignores any uniform rectilinear motion of the entire system, this is not a serious matter, since such a motion is of little or no physical interest.

The energy of a mechanical system, at rest as a whole, is called the internal energy of the system, E_i. This internal energy includes the kinetic energy of the relative motion of the particles and their potential energy of interaction. If the system is in uniform rectilinear motion, as a whole, with relative velocity V, the total energy is

$$E=\tfrac{1}{2}MV^2+E_i$$

The conservation of angular momentum is a direct consequence of assuming that space is isotropic. Spatial isotropy implies that the mechanical properties of an isolated system are unchanged when the system, as a whole, is rotated in space. Hence, we consider an infinitesimal rotation of the system, and derive necessary and sufficient conditions for the system Lagrangian to be invariant.

Once again, we shall assume that the system configuration is described in terms of orthogonal cartesian coordinates, and let x^i_α be the displacement vector of the α-th particle. The infinitesimal rotation is described by an infinitesimal vector $\delta\phi^i$, whose magnitude is equal to the infinitesimal angle of rotation, and whose direction is that of the axis of rotation, with the positive sense defined by a right-hand rule. Then, the infinitesimal change in the displacement vector of each particle is

$$\delta x^i_\alpha=e^i_{jk}\,\delta\phi^j\,x^k_\alpha \tag{6-47}$$

where e^i_{jk} is the mixed representation of the completely anti-symmetric unit tensor of order three. Similarly, relative to a fixed system of coordinates, the components of the particle velocities are changed by an amount

$$\delta\dot{x}^i_\alpha=e^i_{jk}\,\delta\phi^j\,\dot{x}^k_\alpha \tag{6-48}$$

The requirement that the Lagrangian be invariant under the infinitesimal rotation is expressed as

$$\delta L=\sum_{\alpha=1}^{N}\left(\frac{\partial L}{\partial x^i_\alpha}\,\delta x^i_\alpha+\frac{\partial L}{\partial \dot{x}^i_\alpha}\,\delta\dot{x}^i\right)$$

$$=\sum_{\alpha=1}^{N}\left(\frac{\partial L}{\partial x^i_\alpha}\,e^i_{jk}\,\delta\phi^j x^k+\frac{\partial L}{\partial \dot{x}^i_\alpha}\,e^i_{jk}\,\delta\phi^j\dot{x}^k\right)=0$$

However, $\partial L/\partial \dot{x}^i_\alpha = p_{i\alpha}$ and $\partial L/\partial x^i_\alpha = \dot{p}_{i\alpha}$ so that

$$\sum_{\alpha=1}^{N} (\dot{p}_{i\alpha}\, e^i_{jk}\, \delta\phi^j\, x^k_\alpha + p_{i\alpha}\, e^i_{jk}\, \delta\phi^j\, \dot{x}^k_\alpha) = 0$$

Permuting factors and taking the common term outside the sum,

$$\delta\phi^i \sum_{\alpha=1}^{N} (e^j_{jk} x^j_\alpha \dot{p}^k_\alpha + e^i_{jk} \dot{x}^j_\alpha p^k_\alpha) = \delta\phi^i \frac{d}{dt} \sum_{\alpha=1}^{N} (e^j_{jk} x^j_\alpha p^k_\alpha)$$
$$= 0$$

Then, since the $\delta\phi^i$ are arbitrary, it follows that

$$\frac{d}{dt} \sum_{\alpha=1}^{N} (e^i_{jk} x^j_\alpha p^k_\alpha) = 0$$

and the regular momentum

$$l^i = \sum_{\alpha=1}^{N} e^i_{jk} x^j_\alpha p^k_\alpha$$

is a constant of the motion for an isolated system. Like the linear momentum, the angular momentum is additive whether or not the particles in the system interact. There are no other additive constants of the motion. Hence, every isolated system has seven additive constants of the motion: total energy, three components of linear momentum, and three components of angular momentum.

The definition of the angular momentum involves the coordinates of the particles and consequently depends on the choice of origin in general. The coordinates x^i and x'^i of a given point relative to two different origins which are separated by the vector a^i are related by

$$x^i = x'^i + a^i.$$

Hence,

$$l^i = \sum_{\alpha=1}^{N} e^i_{jk} x^j_\alpha p^k_\alpha$$
$$= \sum_{\alpha=1}^{N} e^i_{jk} x'^j_\alpha p^k_\alpha + e^i_{jk} a^j \sum_{\alpha=1}^{N} p^k_\alpha$$
$$= l'^i + e^i_{jk} a^j p^k \qquad (6\text{-}49)$$

We see from (6-49) that the angular momentum is independent of the choice of origin only when the system as a whole is at rest. This indeterminacy does not effect the law of conservation of angular momentum, since the linear momentum is also conserved for an isolated system.

Now consider two inertial frames K and K' such that the latter moves with velocity v^i relative to the former, and suppose that the origins of the two frames coincide at a given instant. At that instant, the coordinates of the particles in the two frames are the same, and their velocities are related by

$$\dot{x}^i = \dot{x}'^i + v^i$$

Hence,

$$l^i = \sum_{\alpha=1}^{N} m_\alpha e^i_{jk} x^j_\alpha \dot{x}^k_\alpha$$

$$= \sum_{\alpha=1}^{N} m_\alpha e^i_{jk} x^j_\alpha \dot{x}'^k_\alpha + \sum_{\alpha=1}^{N} m_\alpha e^i_{jk} x^j_\alpha v^k$$

The first sum is the angular momentum l'^i in the frame K'. In the second sum, we introduce the centre of mass coordinates X^i to obtain

$$l^i = l'^i + M e^i_{jk} X^j v^k$$

If the frame K' is that in which the system is at rest, as a whole, then v^i is the velocity of the centre of mass, Mv^i the total momentum of the system relative to the frame K, and

$$l^i = l'^i + e^i_{jk} X^j p^k$$

Thus, the angular momentum of a mechanical system consists of two parts: the intrinsic angular momentum of the system in its rest frame, and the angular momentum due to the motion of the system, as a whole.

Although conservation of all three components of angular momentum holds only for isolated systems, one or more components may be conserved for a system in an external field. Suppose that the external field has some axis of symmetry. Then, relative to an origin on this axis, the component of angular momentum along the symmetry axis is conserved since the mechanical properties of the system are invariant under rotations about such an axis.

The most important such case is that of a centrally symmetric field, i.e. a field in which the potential energy depends only on the distance from some particular point, known as the centre of the field. It is clear that the component of angular momentum along any axis through the centre is conserved during motion in such a field. This means that the angular momentum vector **l** defined with respect to the centre of the field is conserved.

Another important case is the homogeneous field along the x^3-axis. In such a field, the component l^3 of angular momentum is conserved independently of which point on the axis is taken as the origin. In circular cylinder coordinates (r, ϕ, z), the Lagrangian is given by

$$L=\tfrac{1}{2}\sum_{\alpha=1}^{N} m_\alpha(\dot{r}_\alpha^2+r_\alpha^2\dot{\phi}_\alpha^2+\dot{z}_\alpha^2)-U(r,\phi,z)$$

It follows from the proof of the conservation of angular momentum that

$$l^3=\sum_{\alpha=1}^{N}(\partial L/\partial\dot{\phi}_\alpha)=\sum_{\alpha=1}^{N} m_\alpha r_\alpha^2\dot{\phi}$$

6-5. Rigid Body Motion. One of the simplest examples of an N particle system with holonomic constraints is the rigid body. We shall define a rigid body to be a set of three or more non-colinear mass particles whose locations are specified by the vectors $\mathbf{r}_\alpha$, subject to the set of holonomic constraints

$$|\mathbf{r}_\alpha-\mathbf{r}_\beta|=C_{\alpha\beta},\ \alpha\neq\beta \tag{6-50}$$

The set of constraints (6-50) cannot all be independent, since it turns out that the number of constraints is greater than the number of cartesian coordinates required to specify the unconstrained configuration of N particles. In order to see this, we shall count the number of constraints between N particles. For each fixed value of α there are $N-1$ possible values of β; and there are N possible values of α. However, $N(N-1)$ overcounts the number of constraints by a factor of two, since the constraint equations are symmetric with respect to α and β. Thus, there are $N(N-1)/2$ constraint equations of the form (6-50). However, $N(N-1)/2$ is greater than $3N$ for $N\geqslant 7$. Hence, the $N(N-1)/2$ equations (6-50) are not independent. This being the case, we must now determine the number of independent constraint conditions.

From a given point in space, we can construct three and only three linearly independent, non-coplanar vectors. Select any three non-colinear points in the system and denote the position of these points by $\mathbf{r}_1$, $\mathbf{r}_2$, and $\mathbf{r}_3$. Then, for each other particle in the system, the vectors $\mathbf{r}_\alpha-\mathbf{r}_1$, $\mathbf{r}_\alpha-\mathbf{r}_2$, and $\mathbf{r}_\alpha-\mathbf{r}_3$ are the only linearly independent vectors which can be drawn from the particle at $\mathbf{r}_\alpha$. Then, the only independent rigid body constraints on the particle at $\mathbf{r}$ are

$$[(x_\alpha^1-x_i^1)^2+(x_\alpha^2-x_i^2)^2+(x_\alpha^3-x_i^3)^2]^{1/2}=C'_{\alpha i},\ i=1,2,3$$

Now, there are $N-3$ particles in the system, each of which is subject to three constraints of this form. This implies $3(N-3)=3N=9$ independent constraints on the system. However, the three non-colinear particles at $\mathbf{r}_1$, $\mathbf{r}_2$, and $\mathbf{r}_3$ are subject to the three constraints

$$[(x_i^1-x_j^1)^2+(x_i^2-x_j^2)^2+(x_i^3-x_j^3)^2]^{1/2}=C'_{ij},\ i,j=1,2,3\ i\neq j$$

This leaves $3N-6$ independent constraints which can be imposed on the system of particles under rigid body constraints. Thus, the number of degrees of freedom of a rigid body is six independently of the number of particles in the rigid body.

Since a rigid body has exactly six degrees of freedom, we need six independent coordinates to specify the configuration of the rigid body. One method of choosing such an independent set of variables is the following: Choose the reference point $\mathbf{r}_1$ as an arbitrary point in the rigid body. Then, with respect to a cartesian set of coordinates which are fixed in space, the point $\mathbf{r}_1$ has the instantaneous cartesian components x^i. Now choose the reference points $\mathbf{r}_2$ and $\mathbf{r}_3$ in such a way that

$$(\mathbf{r}_2-\mathbf{r}_1)\cdot(\mathbf{r}_3-\mathbf{r}_1)=0$$

It is clear that the points $\mathbf{r}_1$, $\mathbf{r}_2$, and $\mathbf{r}_3$ are non-colinear. We can now define a set of cartesian axes, fixed in the rigid body, with origin at $\mathbf{r}_1$ such that the x'^1-axis is in the direction $(\mathbf{r}_2-\mathbf{r}_1)$, the x'^2-axis is in the direction $(\mathbf{r}_3-\mathbf{r}_1)$, and the x'^3-axis is along $(\mathbf{r}_2-\mathbf{r}_1)\times(\mathbf{r}_3-\mathbf{r}_1)$. This is an orthogonal cartesian system, which is known as the set of body coordinates. The construction of this set of coordinate axes is shown in Fig. 6-7. The location and orientation of the rigid body is determined by the cartesian coordinates (x_0^1, x_0^2, x_0^3) of the origin of the body axes, and the direction angles of the body coordinate axes relative to the fixed set of space axes.

Since both the space axes and the body axes are orthogonal cartesian systems, for any particular point the body coordinates (x'^1, x'^2, x'^3) are related to the space coordinates (x^1, x^2, x^3) at any instant of time by the orthogonal transformation

$$x'^i = a^i_j x^j + x^i_0\,;\ \det(a^i_j) = +1$$

where the x^i_0 are the components, relative to the space coordinates of the origin of the body system. The set of space coordinates are fixed, whereas the body coordinates are changing in time. Hence, the matrix of the transformation $A=(a^i_j)$ and the components x^i_0 are functions of the time. If at some initial instant, $t=0$, the body axes and the space axes are coincident, the initial condition on the system is expressed as

$$A(0)=I;\ x^i_0(0)=0$$

Although the choice of origin for the system of body coordinates is arbitrary, there are certain choices which prove to be particularly useful in considering the dynamics of rigid body motion. If the rigid body is constrained to move with one point fixed, as in the case of the

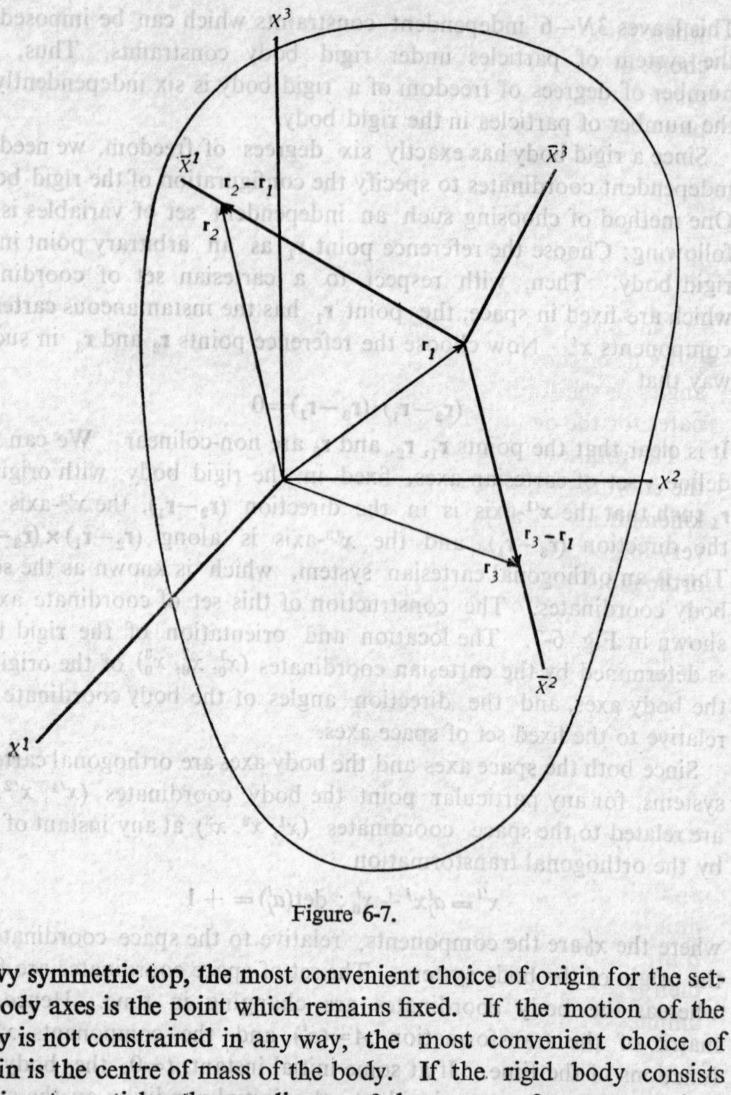

Figure 6-7.

heavy symmetric top, the most convenient choice of origin for the set of body axes is the point which remains fixed. If the motion of the body is not constrained in any way, the most convenient choice of origin is the centre of mass of the body. If the rigid body consists of discrete particles, the coordinates of the centre of mass are given by Eq. (6-21). For the case of a continuous distribution of mass with volume density $\rho(\mathbf{r}')$, the coordinates of the centre of mass are

$$X^i = \frac{\int \rho(\mathbf{r}')\, x'^i\, dv}{\int \rho(\mathbf{r}')\, dv}$$

where both integrals extend over the entire rigid body. For either choice of origin, the equations of motion of the origin of the body system are independent of the equations of motion for the rotational part of the transformation.

The coefficients a^i_j of the time dependent linear transformation which describes the motion of a rigid body are not a suitable choice of generalized coordinates, since they are connected by orthogonality relations, and hence are not independent. However, the orthogonal transformation may be described in terms of either the Euler angles or the Cayley-Klein parameters. Since either of these descriptions is in terms of three independent quantities, we may use either the Euler angles or the Cayley-Klein parameters as a set of generalized coordinates for the description of the rotational motion of a rigid body.

The rotations which we considered in Chapter 4 were rotations of the coordinate system through some finite angle of rotation. In the kinematical theory of rigid body motion, infinitesimal rotations play a very important role. By an infinitesimal rotation, we mean a proper orthogonal transformation of the form

$$x'^i=x^i+\epsilon^i_j\, x^j=(\delta^i_j+\epsilon^i_j)\, x^j \tag{6-51}$$

where the quantities ϵ^i_j are infinitesimals. The set of nine ϵ^i_j are the elements of what we shall call the infinitesimal matrix ϵ. In the usual matrix notation, the infinitesimal transformation (6-51) is expressed as

$$x'=(I+\epsilon)\, x \tag{6-52}$$

where I is the identity matrix.

One property which the infinitesimal orthogonal transformations do not have in common with their finite counterparts is that their multiplication is commutative. We have previously seen that if A and B are matrices representing two finite orthogonal transformations, then $AB\neq BA$ in general. Now, let $I+\epsilon_1$ and $I+\epsilon_2$ represent two infinitesimal rotations. Then

$$\begin{aligned}(I+\epsilon_1)\,(I+\epsilon_2)&=I^2+\epsilon_1 I+I\epsilon_2+\epsilon_1\epsilon_2\\ &=I+(\epsilon_1+\epsilon_2)\end{aligned}$$

neglecting higher order infinitesimals. On the other hand,

$$\begin{aligned}(I+\epsilon_1)\,(I+\epsilon_2)&=I+(\epsilon_1+\epsilon_2)\\ &=I+(\epsilon_2+\epsilon_1)\\ &=(I+\epsilon_2)\,(I+\epsilon_1)\end{aligned}$$

since matrix addition is commutative.

Now, consider the product of the two infinitesimal transformations $I+\epsilon$ and $I-\epsilon$,

$$(I+\epsilon)(I-\epsilon)=I+(\epsilon-\epsilon)=I$$

Hence, if the matrix A represents the infinitesimal orthogonal transformation $I+\epsilon$, then

$$A^{-1}=I-\epsilon$$

However, the inverse of an orthogonal matrix is equal to the transpose, so that

$$A^{-1}=I-\epsilon=A^T=(I+\epsilon)^T=I+\epsilon^T$$

from which

$$\epsilon=-\epsilon^T \tag{6-53}$$

Equation (6-53) defines an anti-symmetric matrix. In terms of the matrix elements of ϵ, Eq. (6-53) can be written in the form

$$\epsilon^i_j=-\epsilon^j_i$$

and we may write, without loss of generality

$$\epsilon=\begin{pmatrix} 0 & d\theta_3 & -d\theta_2 \\ -d\theta_3 & 0 & d\theta_1 \\ d\theta_2 & d\theta_1 & 0 \end{pmatrix} \tag{6-54}$$

where the differentials $d\theta_1$, $d\theta_2$, and $d\theta_3$ are related to the three independent parameters which specify the infinitesimal rotation.

Although we may associate with each finite rotation a direction, (the direction of the axis of rotation), and a magnitude, (the angle of rotation), it is not possible to associate a vector with such a rotation. This results from the fact that the matrices representing these finite rotations are not commutative under multiplication, which would imply that the corresponding vectors would not be commutative under addition. However, the infinitesimal matrices corresponding to infinitesimal rotations are commutative under multiplication, and this offers the possibility of representing infinitesimal transformations of the form $I+\epsilon$ by vectors. The effect of the infinitesimal transformation $I+\epsilon$ on the vector $\mathbf{r}$ with components (x^1, x^2, x^3) is to produce a vector $\mathbf{r}'$ which differs from $\mathbf{r}$ by only an infinitesimal amount,

$$\mathbf{r}'=(I+\epsilon)\,\mathbf{r}=\mathbf{r}+d\mathbf{r}$$

from which

$$d\mathbf{r}=\epsilon\mathbf{r}$$

In component form, this latter relation is equivalent to

$$\begin{aligned} dx^1&=x^2\,d\theta_3-x^3\,d\theta_2 \\ dx^2&=x^3\,d\theta_1-x^1\,d\theta_3 \\ dx^3&=x^1\,d\theta_2-x^2\,d\theta_1 \end{aligned} \tag{6-55}$$

The set of Eqs. (6-55) can be summarized in the single vector equation

$$d\mathbf{r} = \mathbf{r} \times d\boldsymbol{\theta}$$

where $d\boldsymbol{\theta}$ is a vector with cartesian components $(d\theta_1, d\theta_2, d\theta_3)$. In fact, $d\boldsymbol{\theta}$ must be an axial vector, since both $\mathbf{r}$ and $d\mathbf{r}$ are polar vectors and the cross product of any two polar vectors must be an axial vector.

One of the principal uses of the concept of infinitesimal rotations in rigid body mechanics is the description of the time rate of change of vectors as seen in the body set of axes and the space set of axes. Let $\mathbf{F}$ be an arbitrary vector. In the time interval dt, the change in $\mathbf{F}$ as seen from the body coordinates differs from the change in $\mathbf{F}$ as seen in the space coordinates. This effect results from the rotation of the body coordinates relative to the space coordinates during the time interval dt. The change in $\mathbf{F}$ as viewed in terms of the two sets of axes is given by

$$(d\mathbf{F})_B - (d\mathbf{F})_S = (d\mathbf{F})_{\text{Rot}}$$

The rotation of the body set of axes relative to the space set during the time interval dt is an infinitesimal rotation, and hence,

$$(d\mathbf{F})_{\text{Rot}} = -d\boldsymbol{\theta} \times \mathbf{F}$$

It then follows that

$$(d\mathbf{F})_S = (d\mathbf{F})_B + d\boldsymbol{\theta} \times \mathbf{F}$$

and the time rate of change of $\mathbf{F}$ is given by

$$\left(\frac{d\mathbf{F}}{dt}\right)_S = \left(\frac{d\mathbf{F}}{dt}\right)_B + \boldsymbol{\omega} \times \mathbf{F} \tag{6-56}$$

where

$$\boldsymbol{\omega} = \frac{d\boldsymbol{\theta}}{dt}$$

is defined to be the angular velocity of the rigid body.

If we choose the origin of the system of body axes to be the centre of mass of the rigid body, the motion of the body is described by ix independent coordinates which separate naturally into two groups. The first group consists of the three cartesian coordinates of the centre of mass relative to the fixed set of space axes. The second group of coordinates may be taken as the set of Euler angles which describe the orientation of the body axes relative to the fixed space axes. A similar separation of the kinetic energy can be made, and we may write

$$T = \tfrac{1}{2} Mv^2 + T'(\varphi, \theta, \psi)$$

where the first term is the kinetic energy due to the motion of the centre of mass, and is the same as the kinetic energy of a particle of mass equal to the total mass of the rigid body, and moving with the centre of mass. The second term is the kinetic energy of the rotational motion of the rigid body about the centre of mass.

If a similar separation of the potential function U can be made, then the Lagrangian can be divided into two parts, one involving only the translational coordinates of the centre of mass, and the other involving only the rotational coordinates (φ, θ, ψ). In this case, which is generally the only solvable one, the mechanical problem is separated into two independent problems. The mechanical problem of the motion of the centre of mass is the problem of a single mass particle which we have already considered. In the problem of the rotational motion, we consider the centre of mass to be fixed, and look at the rotational motion about this point.

When the centre of mass of the rigid body is regarded as stationary, the angular momentum of the rigid body about the centre of mass is

$$\mathbf{l} = \int \rho(\mathbf{r})\, \mathbf{r} \times \mathbf{u}(\mathbf{r})\, dv$$

where $\rho(\mathbf{r})$ is the mass density at the point $\mathbf{r}$ relative to the centre of mass; $\mathbf{u}(\mathbf{r})$ is the velocity of this mass element relative to the centre of mass; and the integral extends over the entire volume of the rigid body. The velocity $\mathbf{u}$ relative to the fixed set of space axes is due entirely to the rotational motion, and it follows from (6-56) that

$$\mathbf{u}(\mathbf{r}) = \boldsymbol{\omega} \times \mathbf{r}$$

Hence,

$$\begin{aligned} \mathbf{l} &= \int \rho(\mathbf{r})\, \mathbf{r} \times (\boldsymbol{\omega} \times \mathbf{r})\, dv \\ &= \int \rho(\mathbf{r})\, \boldsymbol{\omega} r^2 - (\boldsymbol{\omega} \cdot \mathbf{r})\, \mathbf{r}\, dv \end{aligned} \tag{6-57}$$

where $r = |\mathbf{r}|$. If we resolve (6-57) into cartesian components relative to the set of body axes, we obtain the system of linear equations

$$\begin{aligned} l^1 - \omega^1 \int \rho\, (r^2 - x^1 x_1)\, dv - \omega^2 \int \rho\, x^1 x_2\, dv - \omega^3 \int \rho\, x^1 x_3\, dv \\ l^2 = -\omega^1 \int \rho\, x^2 x_1\, dv + \omega^2 \int \rho (r^2 - x^2 x_2)\, dv - \omega^3 \int \rho\, x^2 x_3\, dv \\ l^3 = -\omega^1 \int \rho\, x^3 x_1\, dv - \omega^2 \int \rho\, x^3 x_2\, dv + \omega^3 \int \rho\, (r^2 - x^3 x_3)\, dv \end{aligned} \tag{6-58}$$

We define

$$I^i_i = \int \rho (r^2 - x^i x_i)\, dv \text{ (no sum on } i\text{)}$$

$$I^i_j = -\int \rho\, x^i\, x_j\, dv \tag{6-59}$$

and rewrite the set of Eqs. (6-58) in the form

$$l^i = I^i_j\, \omega^j$$

The diagonal coefficients I^i_i are known as the moments of inertia, and the off-diagonal coefficients I^i_j are called the products of inertia of the rigid body relative to the centre of mass.

The set of quantities I^i_j are the mixed components of an affine tensor of order two. In order to prove this, consider the affine transformation

$$x'^i = a^i_j\, x^j, \ \det(a^i_j) \neq 0$$

It then follows that the covariant components of **r** relative to the primed coordinates are

$$x'_i = a'^j_i\, x_j, \ A' = A^{-1}$$

Then, from (6-59),

$$I'^i_j = -\int \rho\, x'^i\, x'_j\, dv = -\int \rho\, a^i_k\, x^k\, a'^m_j\, x_m\, dv$$

$$= a^i_k\, I^k_m\, a'^m_j$$

which is the transformation law for the mixed components of an affine tensor of order two. Since the diagonal components are defined in a slightly different way, we must also investigate their behaviour under the affine group. Now,

$$I'^i_i = \int \rho\, (x'^j\, x'_j - x'^i\, x'_i)\, dv \qquad \text{(no sum on } i)$$

$$= \int \rho\, (a^j_k\, x^k\, a'^m_j\, x_m - a^i_k\, x^k\, a'^m_i\, x_m)\, dv \qquad \text{(no sum on } i)$$

$$= a^i_m\, I^m_m\, a'^m_i \qquad \text{(no sum on } i)$$

Since $a^j_k\, a'^m_j = \delta^m_k$. The tensor II with components I^i_j is known as the inertia tensor, and is characteristic of the rigid body.

In terms of the inertia tensor, II, the angular momentum of the system about the centre of mass is expressed by

$$\mathbf{l} = II \cdot \boldsymbol{\omega}$$

where $II \cdot \boldsymbol{\omega}$ is the affine vector with contravariant components $I^i_j\, \omega^j$.

The kinetic energy of the rigid body about the centre of mass is given by

$$T = \tfrac{1}{2} \int \rho(\mathbf{r})\, u^2(\mathbf{r})\, dv$$

However, since $\mathbf{u(r)}=\boldsymbol{\omega}\times\mathbf{r}$, it follows that

$$T=\tfrac{1}{2}\int \rho(\mathbf{r})\,[\mathbf{u}\cdot(\boldsymbol{\omega}\times\mathbf{r})]\,dv$$

$$=\frac{\boldsymbol{\omega}}{2}\cdot\int \rho(\mathbf{r})\,(\mathbf{r}\times\mathbf{u})\,dv$$

$$T=\frac{\boldsymbol{\omega}\cdot\mathbf{l}}{2}=\frac{\boldsymbol{\omega}\cdot\mathbf{II}\cdot\boldsymbol{\omega}}{2}$$

We may express the vector $\boldsymbol{\omega}$ as $\omega\mathbf{n}$, where $\omega=|\,\boldsymbol{\omega}\,|$ and $\mathbf{n}$ is a unit vector in the direction of $\boldsymbol{\omega}$. We may also define a scalar quantity I as

$$I=\mathbf{n}\cdot\mathbf{II}\cdot\mathbf{n}=\int \rho(\mathbf{r})\,[r^2-(\mathbf{r}\cdot\mathbf{n})^2]\,dv$$

The scalar I is known as the moment of inertia of the rigid body about the axis of rotation which is specified by the unit vector $\mathbf{n}$. In terms of the angular velocity $\boldsymbol{\omega}$ and the moment of inertia about the axis of rotation, the kinetic energy due to rotation is

$$T=\frac{I\omega^2}{2}$$

The preceeding analysis shows that the rotational dynamics of rigid body motion is very much dependent on the properties of the inertial tensor. It is, therefore, worthwhile to examine the properties of this tensor and its associated matrix. It is clear from the defining relation that the matrix representing II is a real symmetric matrix

$$I^i_j=I^j_i$$

The components of the inertial tensor depend on both the origin and the set of body axes chosen. Our expressions for the angular momentum and the rotational kinetic energy may be greatly simplified if, for a given origin, we can find an orthogonal set of body axes, (x'^1, x'^2, x'^3) such that the inertial tensor with respect to this set of body axes has the simple form

$$I'^i_j=I'^i_i\,\delta^i_j, \quad \text{(no sum on } i)$$

If this can be done, we may write the matrix representation of the inertial tensor as the diagonal matrix

$$II'=\begin{bmatrix} I_1 & 0 & 0 \\ 0 & I_2 & 0 \\ 0 & 0 & I_3 \end{bmatrix}$$

Now, the set of body axes (x'^1, x'^2, x'^3) are derived from a given set of body axes (x^1, x^2, x^3) by an orthogonal transformation A. Hence,

$$II' = A\ II\ A^{-1} = A^{-1}\ II\ A \tag{6-60}$$

since II' is a diagonal matrix, and II is a real symmetric matrix. Thus, the diagonal form of the inertial tensor is obtained from the original form through a similarity transformation by the orthogonal matrix A. The set of body axes

$$x'^i = a^i_j\ x^j$$

are known as the principal axes of the rigid body.

The determination of the principal axes for a given rigid body, and the consequent determination of the diagonal form of the inertial tensor is another example of the eigenvalue problem, we previously considered in connection with Euler's theorem on proper orthogonal transformations. We shall state here, some general results, without proof, concerning the eigenvalues and eigenvectors for a real symmetric matrix. The results we require are:

(i) The three eigenvalues of a real, symmetrix 3×3 matrix are all real numbers.

(ii) The eigenvectors corresponding to different eigenvalues are mutually orthogonal. If two of the eigenvalues have the same value (a doubly degenerate eigenvalue), the corresponding eigenvectors are linearly independent, and define a plane orthogonal to the eigenvector corresponding to the distinct eigenvalue. Further, any vector which lies in this plane is an eigenvector corresponding to the degenerate eigenvalue. Hence, there always exists a pair of orthogonal vectors in the plane which are eigenvectors of the given matrix. If all three eigenvalues have the same value (triply degenerate), the corresponding set of eigenvectors are linearly independent, but not necessarily orthogonal. However, in this case, any three-dimensional vector is an eigenvector, and it is thus possible to always find a mutually orthogonal set of eigenvectors for the tensor corresponding to the given 3×3 matrix.

(iii) The columns of the matrix A which diagonalizes the given 3×3 real symmetric matrix by a similarity transformation, are given by the elements of the three orthogonal eigenvectors corresponding to the given matrix.

It follows from the above properties, that there always exists a set of mutually orthogonal principal axes for a given rigid body. We now turn our attention to the actual determination of this set of axes. The elements of the diagonal form of the inertial tensor, known as the principal moments of inertia, are the eigenvalues of the matrix II, and are the roots of the determinantal equation

$$\begin{vmatrix} I_1^1-I & I_2^1 & I_3^1 \\ I_1^2 & I_2^2-I & I_3^2 \\ I_1^3 & I_2^3 & I_3^3-I \end{vmatrix}=0$$

The set of eigenvectors, or the principal axis of the rigid body, are obtained by solving the set of simultaneous equations

$$\begin{aligned} (I_1^1-I)x^1+I_2^1x^2+I_3^1x^3&=0 \\ I_1^2x^1+(I_2^2-I)x^2+I_3^2x^3&=0 \\ I_1^3x^1+I_2^3x^2+(I_3^3-I)x^3&=0 \end{aligned} \qquad (6\text{-}61)$$

for the appropriate eigenvalue I. Since the solution of (6-61) is unique only to within multiplicative constants, it is usually convenient to require the set of eigenvectors of II be unit vectors. This condition is sufficient to fix the multiplicative constants. If two of the principal moments of inertia, say I_1 and I_2, are the same, then the corresponding eigenvectors, $\mathbf{E}_1$ and $\mathbf{E}_2$, are not necessarily orthogonal. In this case, we choose any two orthogonal vectors in the plane determined by $\mathbf{E}_1$ and $\mathbf{E}_2$ as the principal axes. The remaining eigenvector $\mathbf{E}_3$ is orthogonal to both $\mathbf{E}_1$ and $\mathbf{E}_2$, and consequently is orthogonal to the pair of principal axes. If all three of the principal moments of inertia are the same, any three mutually orthogonal axes will serve as the principal axes of the rigid body.

A great simplification in the dynamical description of the rigid body results from choosing the principal axes of the rigid body as the set of body coordinates. In terms of this set of body axes, the components of the angular momentum relative to the centre of mass are given by the relatively simple expressions

$$l_1=I_1\omega_1,\ l_2=I_2\omega_2,\ l_3=I_3\omega_3$$

where ω_1, ω_2, and ω_3 are the components of the angular velocity vector relative to the principal axes.

Similarly, in terms of the principal axes, the kinetic energy has the simple form

$$T=\tfrac{1}{2}(I_1\omega_1^2+I_2\omega_2^2+I_3\omega_3^2)$$

It follows that the Lagrangian for the rotational part of the motion can be written in the form

$$L=T-U=\tfrac{1}{2}(I_1\omega_1^2+I_2\omega_2^2+I_3\omega_3^2)-U(\varphi, \theta, \psi)$$

where $U(\varphi, \theta, \psi)$ is the potential function of the rotational motion expressed as a function of the Euler angles. The components of the angular velocity with respect to the principal axes can be expressed in terms of the Euler angles and their time derivative. In this way, the Lagrangian can be expressed completely in terms of Euler angles and their time derivatives. The set of Lagrange equations then determine the Euler angles as functions of time, and in turn, these angles specify the orientation of the principal axes relative to a fixed set of space axes as functions of the time. Finally, these angles combined with the equations of motion for the centre of mass give a complete dynamical description of the motion of given rigid body.

PROBLEMS

6-1. Consider a particle of mass m acted on by a force $\mathbf{F}$ whose description in the circular cylinder coordinate system (ρ, φ, z) is

$$\mathbf{F}=F^{\rho}\,\mathbf{e}_{\rho}+F^{\varphi}\,\mathbf{e}_{\varphi}+F^{z}\,\mathbf{e}_{z}$$

Obtain the component form of the Newtonian equations of motion for the particle.

6-2. Show that if the force field acting on a particle of mass m depends on the velocity of the particle, then the force cannot be derived from a scalar potential function.

6-3. The escape velocity of a particle from the earth's surface is defined as the minimum velocity which a particle must have at the surface of the earth in order to escape from the earth's gravitational field. If the atmospheric resistance is ignored, the forces acting on such a particle are conservative. Ignoring the presence of any other material bodies, show that the escape velocity is 11.19 km/sec.

6-4. Consider a system of N mass particles with position vectors $\mathbf{r}_{\alpha}$, each acted on by a total force $\mathbf{F}_{\alpha}$.

(i) Show that

$$\frac{d}{dt}\sum_{\alpha=1}^{N}\mathbf{p}_{\alpha}\cdot\mathbf{r}_{\alpha}=2T+\sum_{\alpha=1}^{N}\mathbf{F}_{\alpha}\cdot\mathbf{r}_{\alpha}$$

(ii) Take the time average of the result of (i) to show that

$$\overline{T}=-\tfrac{1}{2}\,\overline{\sum_{\alpha=1}^{N}\mathbf{F}_{\alpha}\cdot\mathbf{r}_{\alpha}}$$

where the bar indicates the time average. This result is known as the virial theorem, and plays an important role in the kinetic theory of gases.

(iii) If the forces $\mathbf{F}_{\alpha}$ are derivable from a potential function U which has the form

$$U=a\,|\,r_{\alpha\beta}\,|^{-1}$$

show that

$$\overline{T}=-\tfrac{1}{2}\overline{U}$$

6-5. A neutron of mass 1.67×10^{-23} gm and velocity of 10^{9} cm/sec undergoes a collision with a stationary proton of mass 1.67×10^{-23} gm.

(i) If the particles stick together, what is the velocity of the composite dueteron?

(ii) If the two particles do not stick and there is no net energy loss in the collision, what is the velocity of each particle after the collision?

6-6. Show that the path between two points in a Euclidean plane which has an extremal arc length is the straight line between the two points.

6-7. Complete the brachistochrone problem by finding and solving the Euler-Lagrange equations.

6-8. A double pendulum consists of two mass points m_1 and m_2 connected to a fixed point 0 and to one another by weightless rigid rods of lengths l_1 and l_2 as shown in Fig. 6-8.

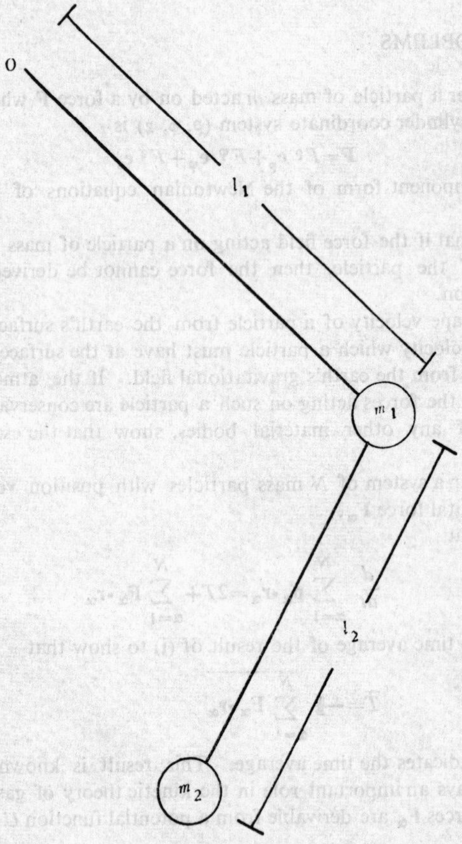

Figure 6-8.

(i) Obtain a set of generalized coordinates for the description of the system. Assume that the system is constrained to move in a vertical plane.

(ii) Find a set of Lagrange equations of motion for the system with respect to the generalized coordinates of part (i).

6-9. Find the Lagrangian equations of motion for a particle moving under the influence of the force

$$\mathbf{F}=F^r\,\mathbf{e}_r+F^\theta\,\mathbf{e}_\theta+F^\varphi\,\mathbf{e}_\varphi$$

in terms of the spherical coordinates (r, θ, φ).

6-10. Two particles of mass m are connected by a rigid, weightless rod o length l, whose centre is held fixed. Set up the Lagrange equations of motion for each particle.

6-11. A simple pendulum of mass m has its point of support in motion on a vertical circle with constant angular frequency ω. Determine the equations of motion in Lagrangian form.

6-12. A particle of mass m moves with velocity $\mathbf{v}_1$ in a half-space where its potential energy is U_1. It leaves this half-space, and enters a half-space where its potential energy has the different value U_2. Calculate the change in the direction of motion of the particle.

6-13. Calculate the cartesian components and the magnitude of the angular momentum of a particle in terms of the cylindrical coordinates (ρ, φ, z).

6-14. Calculate the cartesian components and the magnitude of the angular momentum of a particle in terms of the spherical coordinates (r, θ, φ).

6-15. Consider the cube bounded by the planes $x=0$, $x=1$, $y=0$, $y=1$, $z=0$, $z=1$, which has the uniform mass density ρ. Find the set of principal axes through the centre of mass, and compute the principal moments of inertia.

6-16. Calculate the principal moments of inertia of a tetratomic molecule such that the three atoms of mass m_1 lie at the vertices of an equilateral triangle of side a, and the single atom of mass m_2 lies a distance h from the plane of this triangle.

6-17. Determine the kinetic energy of a homogeneous cone rolling on a plane with its vertex fixed at the origin.

6-18. A uniform lamina in the shape of an equilateral triangle 10 cm on the side is spinning on a vertical axis through the perpendicular bisector of one of the sides, with vertex fixed. Set up the Lagrange equations of motion for the lamina.

CHAPTER 7

The Electromagnetic Field

By the term electromagnetic field, we shall mean the set of values of the four vector functions **E**, **D**, **B**, **H**, which are known as the electric intensity, electric displacement, magnetic induction, and magnetic intensity respectively. In general, we shall assume that these vectors are bounded functions of the points of space and the time, except possibly at isolated points, or in isolated regions. Further, with the exception of regions where the physical properties of space change in a discontinuous manner, the field vectors and their derivatives of all orders are continuous throughout space.

The first problem of electromagnetic theory is to relate the four field vectors, their derivatives, the physical sources of the electromagnetic field, and any parameters which are required to describe the properties of space. The most general set of such relations is the set of equations derived by Maxwell from a number of experimentally determined laws. The second and more difficult problem is to determine the electromagnetic field produced in a bounded or unbounded region of space by a given distribution of sources, where the field vectors or their derivatives are specified on the boundaries of the region. We shall consider the first problem in Sec. 7–1, and then devote the remainder of this chapter to the second problem with particular emphasis on the case of time independent fields.

7-1. The Field Equations. The fundamental equations of electromagnetic theory, i.e. Maxwell's equations, are based on four experimental laws which were discovered in the eighteenth and nineteenth centuries. Historically, the first of these laws is the Coulomb force law, which states that the force exerted on an electrically charged body q_1 due to the presence of an electrically charged body q_2 is proportional to the product of the charges, and inversely to the square of their separation, the constant of proportionality being different for different media. This electrostatic or Coulomb force is a two body force as long as the charges are stationary. Symbolically, the Coulomb law may be written

$$\mathbf{F}_1 = k \frac{q_1 q_2}{|\mathbf{r}_1 - \mathbf{r}_2|^3} (\mathbf{r}_1 - \mathbf{r}_2) \tag{7-1}$$

where $\mathbf{r}_1$ and $\mathbf{r}_2$ are the positions of the charges of q_1 and q_2 and the constant k depends on both the properties of the medium and the units used to measure the charge. If the charges are measured in Coulombs, their separation in metres, and the force in newtons, we define the so-called rationalized MKS units by taking the constant k to be

$$k = \frac{1}{4\pi\epsilon}$$

where ϵ is known as the permittivity of the medium, and has the physical dimensions of coul^2/newton-metre^2. In vaccum on free space, the constant $1/(4\pi\epsilon_0)$ has the numerical value 9.987×10^9 newton-metre^2/coulomb^2. The Coulomb force may be either attractive, if the charges are of different kinds, or repulsive, if the charges are of the same kind.

The electric field intensity E at the location of the charge q_1 is defined to be

$$\mathbf{E}(\mathbf{r}_1) = \lim_{q_1 \to 0} \mathbf{F}_1$$

if this limit exists. If the field at $\mathbf{r}_1$ is produced by a charge q_2 located at $\mathbf{r}_2$, it follows from (7-1) that

$$\mathbf{E}(\mathbf{r}_1) = \frac{1}{4\pi\epsilon} \frac{q_2}{|\mathbf{r}_1 - \mathbf{r}_2|^3} (\mathbf{r}_1 - \mathbf{r}_2) \tag{7-2}$$

Since the Coulomb force is a two body interaction for stationary charges, the electric intensity field due to a collection of N charges q_i at the locations $\mathbf{r}_i$ is

$$\mathbf{E}(\mathbf{r}) = \frac{1}{4\pi\epsilon} \sum_{i=1}^{N} \frac{q_i}{|\mathbf{r} - \mathbf{r}_i|^3} (\mathbf{r} - \mathbf{r}_i) \tag{7-3}$$

Although it is well known that the elements of electric charge occur in quantized amounts, it is usually possible in macroscopic electrodynamics to consider continuous charge distributions. If $\rho(\mathbf{r}')$ represents the charge per unit volume at the point $\mathbf{r}'$ within a bounded region V, then the electric intensity field at any point $\mathbf{r}$ outside V is given by

$$\mathbf{E}(\mathbf{r}) = \frac{1}{4\pi\epsilon} \int_V \frac{\rho(\mathbf{r}') (\mathbf{r} - \mathbf{r}')}{|\mathbf{r} - \mathbf{r}'|^3} \, dv \tag{7-4}$$

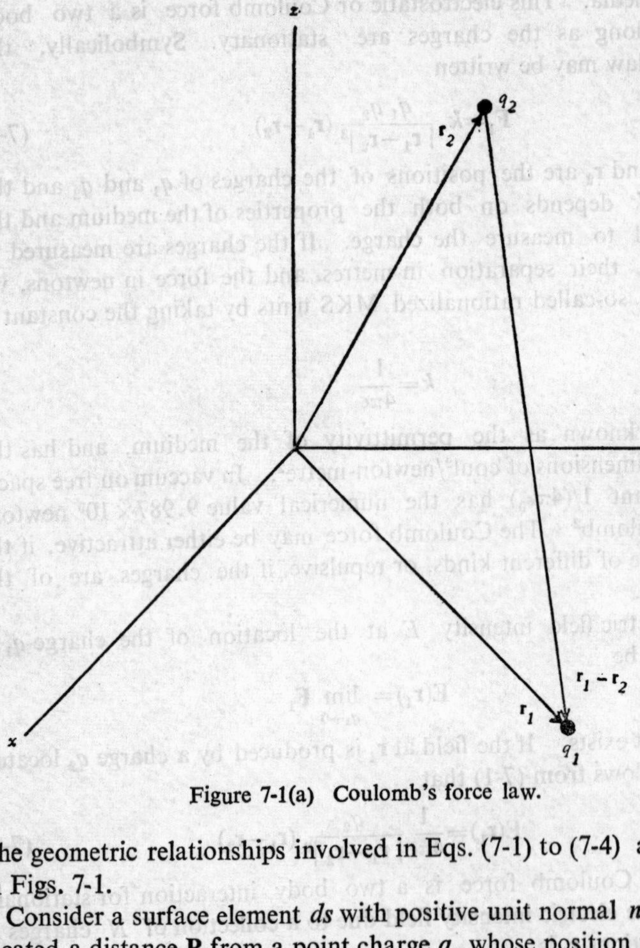

Figure 7-1(a) Coulomb's force law.

The geometric relationships involved in Eqs. (7-1) to (7-4) are shown in Figs. 7-1.

Consider a surface element *ds* with positive unit normal *n* which is located a distance **R** from a point charge *q*, whose position is specified by **r′** as shown in Fig. 7-2. It follows from (7-4) that

$$\mathbf{E}\cdot\mathbf{n}\, ds = \frac{q}{4\pi\epsilon}\frac{\mathbf{R}\cdot\mathbf{n}\, ds}{|\mathbf{R}|^3}$$

where **E** is the electric intensity measured on the surface element *ds*. The quantity

$$\frac{\mathbf{R}\cdot\mathbf{n}\, ds}{|\mathbf{R}|^3}$$

is the element of solid angle $d\Omega$ subtended about the point **r′** by the surface element *ds*. Hence,

$$\int_S \mathbf{E}\cdot\mathbf{n}\, ds = \frac{q}{4\pi\epsilon}\int_S d\Omega$$

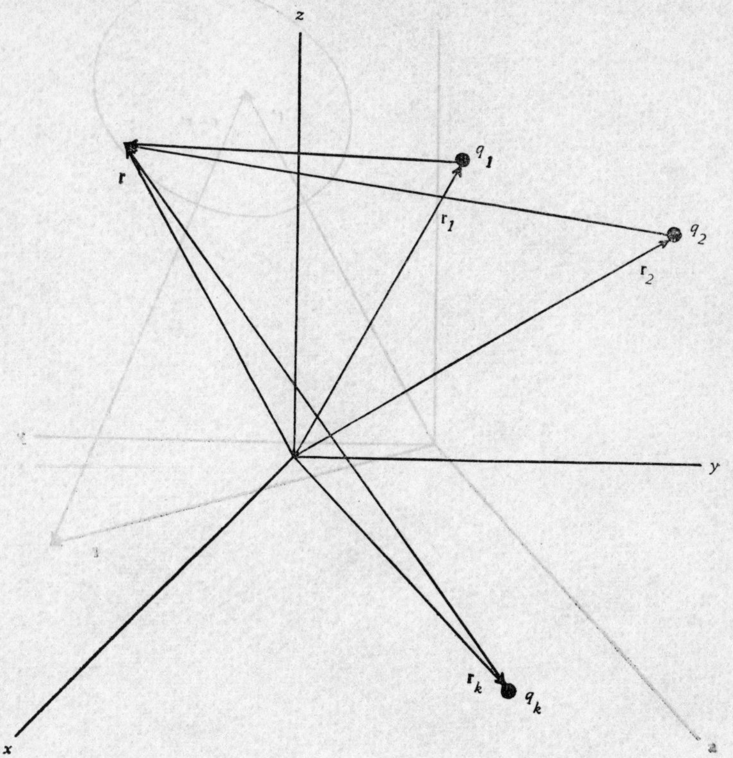

Figure 7-1(b) Field at **r** due to a collection of point charges.

where the integrations are over arbitrary surfaces. In particular, if the surface of integration is a closed surface S containing the point r', we have Gauss' law for a point charge

$$\int_S \mathbf{E}\cdot\mathbf{n}\, ds=\frac{q}{\epsilon}$$

Now, suppose that rather than a single point charge, we have a continuous charge distribution $\rho(\mathbf{r}')$, which is contained in the volume V, bounded by the closed surface S with unit outward normal $\mathbf{n}$. In this case, Gauss' law has the form

$$\int_S \mathbf{E}\cdot\mathbf{n}\, ds=\frac{1}{\epsilon}\int_V \rho\, dv \tag{7-5}$$

If we apply the divergence theorem to the left-hand side of Eq. (7-5), and make use of the fact that V is any volume which contains all of the electric charge, we obtain

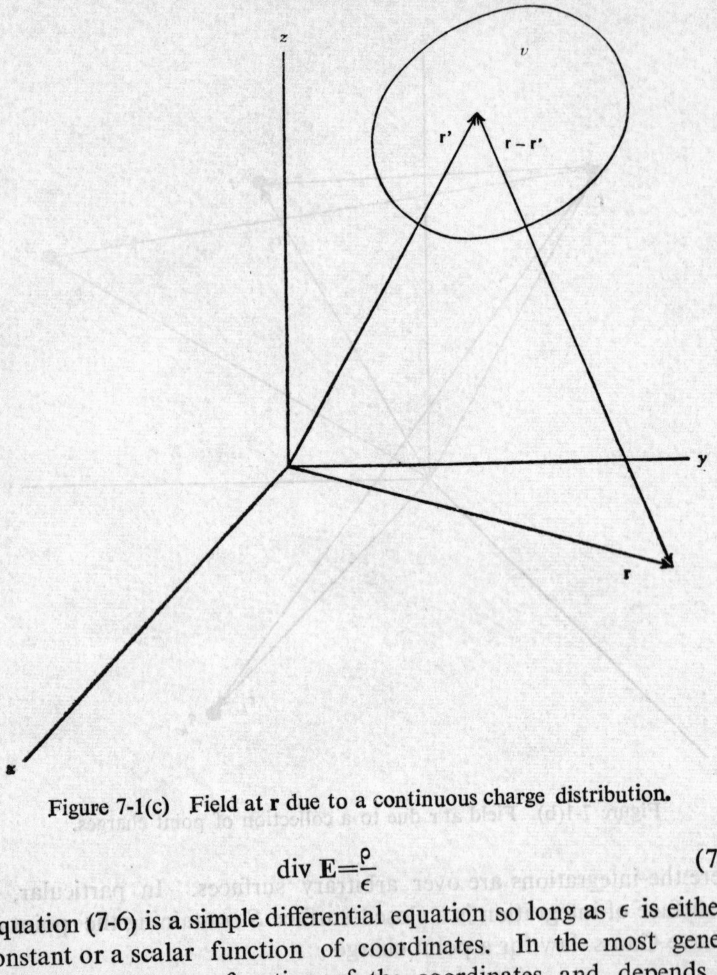

Figure 7-1(c) Field at **r** due to a continuous charge distribution.

$$\text{div } \mathbf{E} = \frac{\rho}{\epsilon} \tag{7-6}$$

Equation (7-6) is a simple differential equation so long as ϵ is either a constant or a scalar function of coordinates. In the most general case, it is a tensor function of the coordinates and depends on the magnitude of the *E*-field. Under these conditions, it is difficult to give a physical interpretation to Eq. (7-6). In order to avoid this difficulty, we define the electric displacement field **D** by the differential equation

$$\text{div } \mathbf{D} = \rho \tag{7-7}$$

Equation (7-7) is the differential equivalent of the Coulomb force law, and is one of the four Maxwell equations. It is clear that in a linear, homogeneous, and isotropic medium which is characterized by a constant permitivity ϵ, the electric intensity and the electric displacement are related by

$$\mathbf{D} = \epsilon \mathbf{E}$$

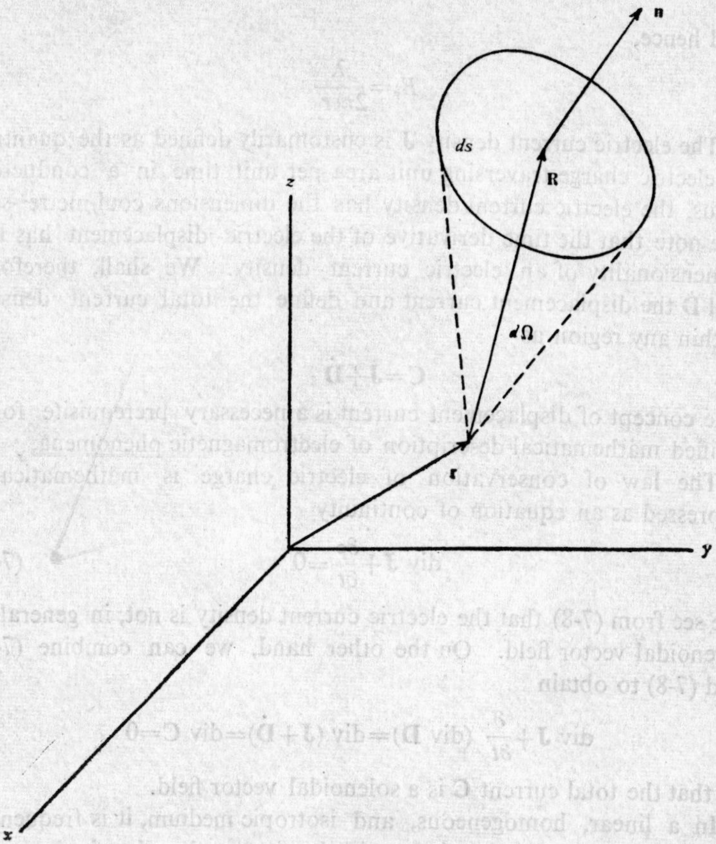

Figure 7-2.

EXAMPLE 1. Although Eqs. (7-6) and (7-7) are equivalent to Coulomb's law in a sense, for problems involving a high degree of symmetry, it may be more convenient to use Gauss' law in the integral form. As an example, consider an infinitely long line of electric charge with uniform linear density λ coulomb/metre. Surround the line of charge by a cylinder of radius r and length l, with the axis of the cylinder along the line of charge. From symmetry considerations, the only non-zero component of the electric intensity is the component E_r directed radially outward (or inward) from the line of charge. It also follows that the field E_r is constant along the cylindrical surface. Then, from Gauss' law,

$$\int \mathbf{E}\cdot\mathbf{n}\, ds = E_r(2\pi r l) = \frac{1}{\epsilon}\int_V \rho\, dv = \frac{\lambda l}{\epsilon}$$

and hence,

$$E_r = \frac{\lambda}{2\pi\epsilon r}$$

The electric current density **J** is customarily defined as the quantity of electric charge traversing unit area per unit time in a conductor. Thus, the electric current density has the dimensions coul/metre²-sec. We note that the time derivative of the electric displacement has the dimensionality of an electric current density. We shall, therefore, call $\dot{\mathbf{D}}$ the displacement current and define the total current density within any region as

$$\mathbf{C} = \mathbf{J} + \dot{\mathbf{D}}$$

The concept of displacement current is a necessary prerequisite for a unified mathematical description of electromagnetic phenomena.

The law of conservation of electric charge is mathematically expressed as an equation of continuity

$$\text{div } \mathbf{J} + \frac{\partial \rho}{\partial t} = 0 \qquad (7\text{-}8)$$

We see from (7-8) that the electric current density is not, in general, a solenoidal vector field. On the other hand, we can combine (7-7) and (7-8) to obtain

$$\text{div } \mathbf{J} + \frac{\partial}{\partial t} (\text{div } \mathbf{D}) = \text{div } (\mathbf{J} + \dot{\mathbf{D}}) = \text{div } \mathbf{C} = 0$$

so that the total current **C** is a solenoidal vector field.

In a linear, homogeneous, and isotropic medium, it is frequently possible to relate the electric current density to the electric intensity by

$$\mathbf{J} = \sigma \mathbf{E}$$

where σ is a constant characteristic of the medium, which is known as the electrical conductivity. This linear relation between **J** and **E** is the field equivalent of Ohm's law for electrical circuits. It should be noted that the linear relation is a phenomenological result, and is not universally valid.

The magnetic induction field is defined from the experimental Biot-Savart law by

$$\mathbf{B} = \frac{\mu}{4\pi} \int_V \frac{\mathbf{J} \times \mathbf{R} \, dv}{|\mathbf{R}|^3} \qquad (7\text{-}9)$$

where **J** is the electric current density contained in the volume V, **R** is the vector from the point at which **B** is defined to the current element **J**, and μ is a factor characterizing the medium, known as the magnetic permeability. In rationalized MKS units, μ has the numerical value $4\pi \times 10^{-7}$ henry/metre for vacuum or free space.

Consider an arbitrary surface S, bounded by a closed curve C. In the presence of an electromagnetic field, the magnetic flux Φ, which passes through the surface S is defined to be

$$\Phi = \int_S \mathbf{B} \cdot \mathbf{n}\, ds$$

where $\mathbf{n}$ is the positive unit normal to S. The Faraday work law, expressed in terms of the magnetic flux states that

$$\int_C \mathbf{E} \cdot d\mathbf{l} = -\frac{d\Phi}{dt} = -\frac{d}{dt}\int_S \mathbf{B} \cdot \mathbf{n}\, ds \tag{7-10}$$

The integral on the left of (7-10) can be transformed by Stokes' theorem to obtain

$$\int_S (\text{curl } \mathbf{E}) \cdot \mathbf{n}\, ds = -\frac{d}{dt}\int_S \mathbf{B} \cdot \mathbf{n}\, ds$$

But, S is an arbitrary surface, and hence,

$$\text{curl } \mathbf{E} = -\dot{\mathbf{B}} \tag{7-11}$$

Equation (7-11), which is the differential form of the Faraday work law is another of the fundamental Maxwell equations.

In a similar way, we define the electric current flux by the surface integral

$$\Phi_E = \int_S \mathbf{C} \cdot \mathbf{n}\, ds$$

where $\mathbf{C} = \mathbf{J} + \dot{\mathbf{D}}$ is the total current density through the arbitrary surface S with unit positive normal n, bounded by the closed curve C. The mathematical generalization of Ampere's law is

$$\int_C \mathbf{H} \cdot d\mathbf{l} = \int_S \mathbf{C} \cdot \mathbf{n}\, ds = \int_S (\mathbf{J} + \dot{\mathbf{D}}) \cdot \mathbf{n}\, ds \tag{7-12}$$

Again, we transform the integral on the left-hand side of (7-12) by Stokes' theorem and use the fact that S is an arbitrary surface to obtain the differential form

$$\text{curl } \mathbf{H} = \mathbf{J} + \dot{\mathbf{D}} \tag{7-13}$$

Equation (7-13) is another of the fundamental equations of electromagnetic theory first derived by Maxwell.

The magnetic induction and the magnetic intensity are related to one another through the magnetic permeability of the medium

$$\mathbf{B} - \mu \mathbf{H}$$

where μ is a constant for linear, homogeneous, and isotropic media.

With the exception of ferromagnetic media, the permeability μ differs from μ_0, the permeability of free space, by only a small amount.

The last of the fundamental equations is obtained from (7-10) by taking S to be a closed surface. In this case, there is no boundary curve, and the Faraday work law has the form

$$\frac{d}{dt}\int_S \mathbf{B}\cdot\mathbf{n}\,ds=0$$

Integration with respect to the time results in

$$\int_S \mathbf{B}\cdot\mathbf{n}\,ds=\text{constant}$$

where the constant must depend on the amount of free magnetic charge contained in the volume bounded by the surface S. Although free magnetic charge has been postulated on theoretical grounds, it has never been observed in nature, so that we can set the constant equal to zero, and apply the diveregnce theorem to obtain the final Maxwell equation

$$\operatorname{div}\mathbf{B}=0 \tag{7-14}$$

We now summarize the results obtained to this point. The electromagnetic field is completely described by four vector fields **E**, **B**, **H**, and **D**, which are related to one another and the physical sources of charge and current density by the set of four Maxwell equations

$$\operatorname{curl}\mathbf{E}=-\dot{\mathbf{B}},\quad \operatorname{div}\mathbf{B}=0 \tag{7-15}$$

$$\operatorname{curl}\mathbf{H}=\mathbf{J}+\dot{\mathbf{D}},\quad \operatorname{div}D=\rho$$

The electromagnetic properties of the medium in which the field exists are described by the three constitutive equations

$$\mathbf{D}=\varepsilon\mathbf{E},\quad \mathbf{B}=\mu\mathbf{H},\quad \mathbf{J}=\sigma\mathbf{E} \tag{7-16}$$

If the constitutive parameters ε, μ, and σ are constants, we say that the medium is linear, homogeneous, and isotropic. We shall also impose a condition of conservation of free electric charge, which is expressed mathematically by the equation of continuity

$$\operatorname{div}\mathbf{J}+\frac{\partial\rho}{\partial t}=0 \tag{7-17}$$

Although the set of Maxwell equations are differential forms of certain experimental physical laws, the general theory of the electromagnetic field must proceed from the Maxwell equations rather than the corresponding experimental laws. However, as we have seen, in situations involving a high degree of symmetry, it may be more convenient to use the integral forms of the field equations, i.e. the experimental laws, rather than the Maxwell equations.

The Maxwell equations have been established only at those points in space such that in a neighbourhood of the point under consideration, the electromagnetic properties of the medium are either constant or vary in a continuous manner. At the boundary surface between two media, there is in general, a discontinuous change in one or more of the constituitive parameters. This will, in turn, induce discontinuities in the field vectors. We shall now investigate the behaviour of the field vectors at the boundary between two electromagnetically different media.

Consider two regions of space which are separated by the surface *S*. Let the electromagnetic properties of the two regions be described by ϵ_1, μ_1, σ_1, and ϵ_2, μ_2, σ_2 respectively. We shall replace the surface *S* by a transition layer of thickness Δl within which the constitutive parameters vary continuously from their values in region 1 to their values in region 2, and then go to the limit $\Delta l \to 0$. Within the transition layer, as well as within the two media, the field vectors and their first derivatives are continuous, bounded functions of space and time.

We shall first investigate the behaviour of the components of the field vectors which are normal to *S* in the limit $\Delta l \to 0$. In order to do this, we construct, within the transition layer, a right circular cylinder whose faces are separated by the distance Δl with generators normal to *S*, as shown in Fig 7-3. We have from Maxwell's equations

$$\oint_{S'} \mathbf{B} \cdot \mathbf{n} \, ds = 0$$

where S' is the surface of the cylinder and $\mathbf{n}$ is the unit outward normal to the cylinder. If the base area Δa is sufficiently small, the magnetic induction is essentially constant over the faces of the cylinder, and the integral is well approximated by

$$\oint_{S'} \mathbf{B} \cdot \mathbf{n} \, ds \approx (\mathbf{B} \cdot \mathbf{n}_1 + \mathbf{B} \cdot \mathbf{n}_2) \, \Delta a + \text{wall contributions} = 0$$

The contribution from the cylinder walls is proportional to Δl and hence becomes vanishingly small in the limit. Denote the value of $\mathbf{B}$ in region 1, near *S* by $\mathbf{B}_1$, and the value of $\mathbf{B}$ near *S* in region 2 by $\mathbf{B}_2$. If we choose the unit positive normal to *S* to be directed from medium 1 into medium 2, then

$$\mathbf{n}_2 = -\mathbf{n}_1 = \mathbf{n}$$

and in the limit $\Delta a \to 0$, $\Delta l \to 0$,

$$(\mathbf{B}_2 - \mathbf{B}_1) \cdot \mathbf{n} = 0 \qquad (7\text{-}18)$$

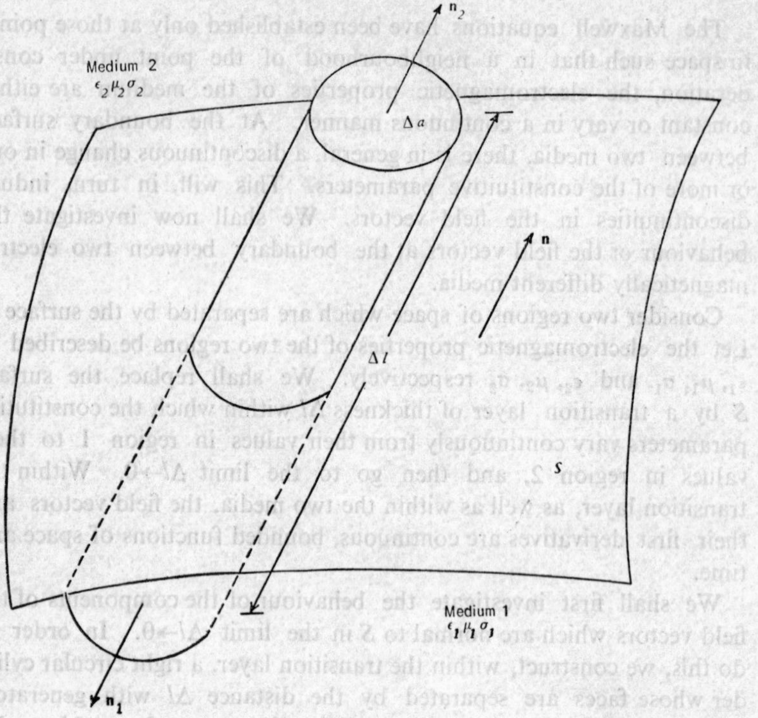

Figure 7-3.

Hence, the normal component of the magnetic induction is continuous across any surface of discontinuity.

In terms of the same geometry, we have from Gauss' law,

$$\oint_{S'} \mathbf{D}\cdot\mathbf{n}\, ds = q$$

where q is the total free charge contained within the cylinder. If we assume this charge to be distributed within the transition layer with uniform density ρ, then

$$q = \rho \Delta l \Delta a$$

In the limit $\Delta l \to 0$, the volume charge density ρ must become infinite since the total charge q is constant. In order to get around this mathematical difficulty, we introduce a surface charge density

$$\omega = \lim_{\substack{\Delta l \to 0 \\ \rho \to \infty}} \rho \Delta l$$

if this limit exists. Then in the limit $\Delta l \to 0$, $\Delta a \to 0$,

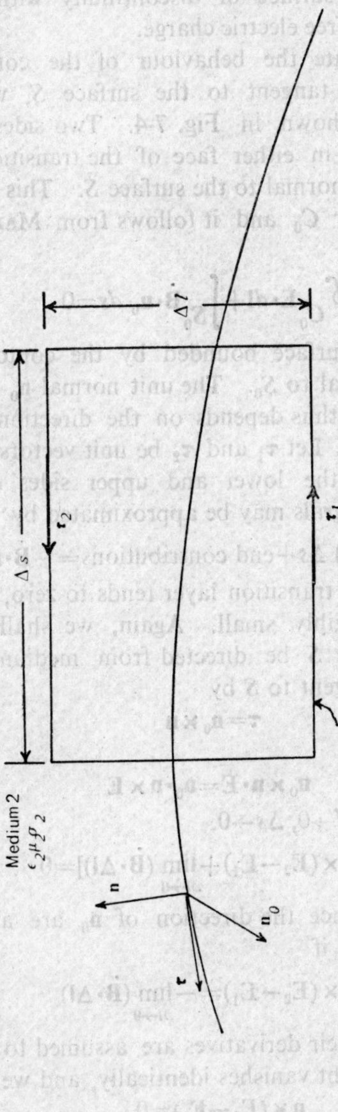
S
Δl
Δs
Medium 1
$\epsilon_1 \mu_1 \sigma_1$
Medium 2
$\epsilon_2 \mu_2 \sigma_2$
τ_2
τ_1
C_0
n
n_0
τ

Figure 7–4.

$$(\mathbf{D}_2-\mathbf{D}_1)\cdot\mathbf{n}=\omega \tag{7-19}$$

where we have employed the same notational conventions as in the case of the magnetic induction. Hence, the electric displacement is discontinuous across a surface of discontinuity with saltus equal to the surface density of free electric charge.

In order to investigate the behaviour of the components of the field vectors which are tangent to the surface S, we introduce the rectangular contour shown in Fig. 7-4. Two sides of the contour, each of length Δs, lie in either face of the transition layer, and the ends, of length Δl, are normal to the surface S. This rectangle constitutes a closed contour C_0 and it follows from Maxwell's equations that

$$\oint_{C_0}\mathbf{E}\cdot d\mathbf{l}+\int_{S_0}\dot{\mathbf{B}}\cdot\mathbf{n}_0\,ds=0$$

where S_0 is the open surface bounded by the contour C_0, and $\mathbf{n}_0$ is the positive unit normal to S_0. The unit normal $\mathbf{n}_0$ is determined by a right-hand rule, and thus depends on the direction of integration around the contour C_0. Let $\boldsymbol{\tau}_1$ and $\boldsymbol{\tau}_2$ be unit vectors in the direction of integration along the lower and upper sides of the contour respectively. The integrals may be approximated by

$$(\mathbf{E}\cdot\boldsymbol{\tau}_1+\mathbf{E}\cdot\boldsymbol{\tau}_2)\,\Delta s+\text{end contributions}=-\dot{\mathbf{B}}\cdot\mathbf{n}_0\,\Delta s$$

As the thickness of the transition layer tends to zero, the end contributions become negligibly small. Again, we shall let the positive normal $\mathbf{n}$ to the surface S be directed from medium 1 into medium 2 and define a unit tangent to S by

$$\boldsymbol{\tau}=\mathbf{n}_0\times\mathbf{n}$$

Then, since

$$\mathbf{n}_0\times\mathbf{n}\cdot\mathbf{E}=\mathbf{n}_0\cdot\mathbf{n}\times\mathbf{E}$$

we have, in the limit $\Delta l\to 0$, $\Delta s\to 0$,

$$\mathbf{n}_0\cdot[\mathbf{n}\times(\mathbf{E}_2-\mathbf{E}_1)+\lim_{\Delta l\to 0}(\dot{\mathbf{B}}\cdot\Delta\mathbf{l})]=0 \tag{7-20}$$

The contour C_0 and hence the direction of $\mathbf{n}_0$ are arbitrary, so that (7-20) holds if and only if

$$\mathbf{n}\times(\mathbf{E}_2-\mathbf{E}_1)=-\lim_{\Delta l\to 0}(\dot{\mathbf{B}}\cdot\Delta\mathbf{l})$$

The field vectors and their derivatives are assumed to be bounded, so that the limit on the right vanishes identically, and we have

$$\mathbf{n}\times(\mathbf{E}_2-\mathbf{E}_1)=0 \tag{7-21}$$

Thus, the tangential component of the electric intensity field is continuous across a surface of discontinuity.

A similar analysis of the magnetic intensity results in

$$\mathbf{n}\times(\mathbf{H}_2-\mathbf{H}_1)=\lim_{\Delta l\to 0}(\dot{\mathbf{D}}+\mathbf{J})\,\Delta l$$

The first term on the right vanishes as a result of the assumed boundedness of $\dot{\mathbf{D}}$. If the current density $\mathbf{J}$ within the transition region is bounded, the entire right-hand side vanishes identically. It may happen, however, that the total current $I=\mathbf{J}\cdot\mathbf{n}_0\,\Delta s\,\Delta l$ through the surface bounded by C_0 is confined to an infinitesimally thin layer as the transition region reduces to the boundary surface S. In this case, we define a surface current density

$$\mathbf{K}=\lim_{\substack{\Delta l\to 0\\ |J|\to\infty}}\mathbf{J}\,\Delta l$$

if the limit exists. In terms of this surface current density, the tangential component of the magnetic intensity satisfies the condition

$$\mathbf{n}\times(\mathbf{H}_2-\mathbf{H}_1)=\mathbf{K} \tag{7-22}$$

Whenever the conductivities of both media are finite, there is no mechanism for the formation of a surface current density, since $\mathbf{E}$ is bounded, and hence $\sigma\mathbf{E}\,\Delta l$ vanishes in the limit $\Delta l\to 0$. In this case, which is the usual one, the tangential component of the magnetic intensity is continuous across the surface of discontinuity. Although the assumption of infinite conductivity is physically unrealistic, it must frequently be employed in order to obtain mathematically tractable expression for the electromagnetic field. Under these circumstances, (7-22) must be used as the condition on the tangential component of the magnetic intensity at the interface between two different madia.

EXAMPLE 2. Consider a spherical cloud of static electric charge of radius r_0 and uniform density ρ which is in free space. We shall use the Maxwell equations and boundary conditions to determine the electromagnetic field at all points in space. Since the charge distribution is spherically symmetric, the field vectors must be invariant under any rotation about the centre of the sphere, i.e. the derivatives of the field vectors with respect to the spherical coordinates θ and ϕ are zero. Further, since the charge distribution is constant in time, there is no electric or displacement current and the field equations

$$\text{curl}\,\mathbf{H}=\mathbf{J}+\dot{\mathbf{D}},\quad \text{div}\,\mathbf{B}=0$$

are trivially satisfied by $\mathbf{B}=\mathbf{H}=\mathbf{0}$. Thus, the entire electromagnetic field is described by the electric intensity $\mathbf{E}$.

Using spherical coordinates and taking advantage of the symmetry conditions we have mentioned, the electric intensity field is determined from the field equation

$$\epsilon_0 \operatorname{div} \mathbf{E} = \frac{\epsilon_0}{r^2}\frac{\partial}{\partial r}(r^2 E_r) = \begin{cases} \rho, & 0 \leqslant r < r_0 \\ 0, & r > r_0 \end{cases}$$

There is an obvious discontinuity in the charge density at $r=r_0$ and we must separately solve for the field in the two regions $r<r_0$ and $r>r_0$. The field equation can be immediately integrated to yield

$$E_r = \begin{cases} \dfrac{\rho r}{3\epsilon_0} + \dfrac{C_1}{\epsilon_0 r^2}, & 0 \leqslant r < r_0 \\ \dfrac{C_2}{\epsilon_0 r^2}, & r > r_0 \end{cases}$$

where C_1 and C_2 are arbitrary constants of integration, which must be determined in order to complete the calculation of the field. These constants can be determined by an application of the boundary conditions at those points where the general solution has singular or discontinuous behaviour.

First of all, it follows from Gauss' law that the field component E_r must be finite at the origin unless there is a point charge at the origin. Since this is not the case for a uniform and finite charge density, we must set $C_1=0$. Secondly, at the spherical boundary $r=r_0$, there is no surface charge density, so that the normal component of $\mathbf{E}$ at this surface must be continuous. Imposing this condition of continuity, it is easily seen that

$$C_2 = \frac{\rho r_0^3}{3}$$

and the radial component of the electric intensity is given by

$$E_r = \begin{cases} \dfrac{\rho r}{3\epsilon_0}, & 0 \leqslant r < r_0 \\ \dfrac{\rho r_0^3}{3\epsilon_0}\dfrac{1}{r^2}, & r > r_0 \end{cases}$$

It can be shown from symmetry considerations and the field equation

$$\operatorname{curl} \mathbf{E} = 0$$

that the two field components E_θ and E_φ are both zero everywhere.

7-2. Vector and Scalar Potentials. Although the general theory of the electromagnetic field must proceed from Maxwell's equations, it is usually convenient to introduce a set of auxiliary quantities known as the potentials for the field. In this section, we shall consider one of the many possible sets of potentials from which the electromagnetic field can be derived. It should be noted that the potentials are not physical observables, but are auxiliary mathematical entities. This is of no serious significance in the classical theory, but is of some importance in the quantized theory of the electromagnetic field. However, even in the classical theory, this implies that all results must ultimately be expressed in terms of the field vectors.

Consider an electromagnetic field which exists in a linear, homogeneous, and isotropic medium characterized by the constitutive parameters ϵ, μ, σ. In such a medium, the electromagnetic field is completely described by the electric and magnetic intensity vectors, which satisfy the set of differential equations

$$\text{curl } \mathbf{E} = -\mu\dot{\mathbf{H}} \tag{7-23a}$$

$$\text{div } \mathbf{H} = 0 \tag{7-23b}$$

$$\text{curl } \mathbf{H} = \mathbf{J} + \sigma\dot{\mathbf{E}} \tag{7-23c}$$

$$\text{div } E = \frac{1}{\epsilon}\rho \tag{7-23d}$$

at every point in the medium. It follows from (7-23b) and the Helmholtz theorem that

$$\mathbf{H} = \frac{1}{\mu} \text{ curl } \mathbf{A} \tag{7-24}$$

where **A** is defined to be the vector potential for the field. Substituting (6-24) in (6-23a), we obtain

$$\text{curl } \mathbf{E} + \frac{\partial}{\partial t} (\text{curl } \mathbf{A}) = \text{curl } \mathbf{E} + \text{curl } \dot{\mathbf{A}}$$

$$= \text{curl}(\mathbf{E} + \dot{\mathbf{A}}) = 0 \tag{7-25}$$

Note that we are using the partial derivative with respect to time interchangeably with the total time derivative. This is permissible only if the medium is stationary and the coordinate frame is fixed in time. If this is not the case, the total time derivatives must all be replaced by partial derivatives with respect to the time. It follows from Helmholtz theorem that (7-25) is satisfied if and only if

$$\mathbf{E} + \dot{\mathbf{A}} = -\text{grad } \phi \tag{7-26}$$

where ϕ is defined to be the scalar potential for the field.

The vector and scalar potentials must be related to the physical sources **J** and ρ in such a way that the magnetic and electric fields defined by (7-24) and (7-25) satisfy the set of Maxwell equations. In order to obtain these relations, we shall separate the current density **J** into two parts: a convection current density $\mathbf{J}_v$ which arises from an actual flow of charged particles, and which is essentially independent of the fields, and an ohmic or conduction current $\mathbf{J}_c$ which is given by

$$\mathbf{J}_c=\sigma\mathbf{E}$$

In terms of this separated current density, (7-23) is satisfied if the vector and scalar potential satisfy

$$\nabla^2\mathbf{A}-\text{grad}\,[\text{div}\,\mathbf{A}+\mu\,(\sigma+\epsilon\partial/\partial t)\,\phi]=\mu\sigma\frac{\partial\mathbf{A}}{\partial t}-\mu\epsilon\frac{\partial^2\mathbf{A}}{\partial t^2}-\mu\,\mathbf{J}_v \quad (7\text{-}27)$$

where the vector Laplace operator has its usual interpretation

$$\nabla^2\mathbf{A}=\text{grad div}\,\mathbf{A}-\text{curl curl}\,\mathbf{A}$$

and

$$\nabla^2\phi+\epsilon\frac{\partial}{\partial t}\,\text{div}\,\mathbf{A}=-\rho \quad (7\text{-}28)$$

Thus, the electromagnetic field $\mathbf{E}$ and $\mathbf{H}$ generated by the charge distribution ρ and convection current $\mathbf{J}_v$ can be derived by Eqs. (7-24) and (7-26) from a vector and scalar potential, if the potentials simultaneously satisfy (7-27) and (7-28). However, these latter two differential equations are coupled, and their solution is consequently rather difficult.

The equations relating the vector and scalar potentials to the sources of the electromagnetic field may be simplified by making use of the fact that the potentials are not unique for a given electromagnetic field. If $\mathbf{H}$ is a given magnetic intensity field which is generated from the vector potential $\mathbf{A}$ by

$$\mathbf{H}=\frac{1}{\mu}\,\text{curl}\,\mathbf{A}$$

then the same magnetic field is generated by the vector potential

$$\mathbf{A}'=\mathbf{A}-\text{grad}\,\psi \quad (7\text{-}29)$$

where ψ is any differentiable function of space and time. Then if $\mathbf{E}$ is generated by the vector potential $\mathbf{A}$ and the scalar potential ϕ according to

$$\mathbf{E}=-\text{grad}\,\phi-\frac{\partial\mathbf{A}}{\partial t}$$

it is easily seen that the same electric intensity is generated from the vector potential $\mathbf{A}'$ and the scalar potential

$$\phi'=\phi+\frac{\partial\psi}{\partial t} \quad (7\text{-}30)$$

The set of all transformations of the vector and scalar potentials of the form (7-29) and (7-30) forms a continuous group which is known as the Maxwell gauge group. It is clear that the electromagnetic field, and hence the set of Maxwell equations are invariant under the

Maxwell gauge group. The functions ψ are called the generators of the group.

In classical electromagnetic theory, the invariance of the electromagnetic field under the Maxwell gauge group is used primarily to simplify the differential equations satisfied by the potentials. If **A** and ϕ are vector and scalar potentials which satisfy (7-27) and (7-28), let ψ be the generator of a gauge transformation such that the transformed potentials satisfy the Lorentz condition

$$\text{div}\, \mathbf{A}' + \mu\, (\sigma + \epsilon\, \partial/\partial t)\, \phi' = 0 \tag{7-31}$$

In order that the transformed potentials satisfy (7-31), there are certain restrictions to be imposed on the generator ψ in addition to those of differentiability. If we substitute (7-29) and (7-30) in (7-31), we obtain

$$\text{div}\, \mathbf{A} = \text{div grad}\, \psi + [\mu\, (\sigma + \epsilon\, \partial/\partial t)\, \phi + \mu\, (\sigma + \epsilon\, \partial/\partial t)]\, \frac{\partial \psi}{\partial t} = 0$$

Hence, the transformed potentials are related by the Lorentz condition only if the generator of the gauge transformation satisfies the differential equation

$$\nabla^2 \psi - \mu\sigma \frac{\partial \psi}{\partial t} - \mu\epsilon\, \frac{\partial^2 \psi}{\partial t^2} = \text{div}\, \mathbf{A} + \mu\sigma\phi + \mu\epsilon\, \frac{\partial \phi}{\partial t} \tag{7-32}$$

where **A** and ϕ are any solutions of (6-27) and (6-28). The set of gauge transformations generated by the solutions of (7-32) are a subgroup of the Maxwell gauge group, known as the Lorentz gauge group, or more simply as the Lorentz gauge. If we impose the Lorentz condition on the potentials defined by (6-27) and (6-28), the differential equations uncouple, and we obtain

$$\begin{aligned} \nabla^2 \mathbf{A}' - \mu\sigma\, \frac{\partial \mathbf{A}'}{\partial t} - \mu\epsilon\, \frac{\partial^2 \mathbf{A}'}{\partial t^2} &= -\mu \mathbf{J}_v \\ \nabla^2 \phi' - \mu\sigma\, \frac{\partial \phi'}{\partial t} - \mu\epsilon\, \frac{\partial^2 \phi'}{\partial t^2} &= -\frac{\rho}{\epsilon} \end{aligned} \tag{7-33}$$

If the medium is non-conducting, i.e. $\sigma = 0$, then $\mathbf{A}'$ and ϕ' are solutions of the inhomogeneous vector and scalar wave equations respectively. The velocity of propagation of these potentials, and hence of the electromagnetic field is $(\mu\epsilon)^{-1/2}$. It is one of the major triumphs of electromagnetic theory that this propagation velocity is the same as the observed velocity of light in the same medium.

It would appear that in our application of the Lorentz gauge, we are going in circles, since to determine the generator of the gauge transformation, we must first know a solution of (6-27) and (6-28). This is not, in fact the case, since it is not necessary to actually determine the generator of the gauge transformation. It is sufficient to know that such a function exists and that potentials may be found which are related to one another by the Lorentz condition. The equations which determine these potentials are the differential equations (7-33).

Equations (7-27) and (7-28) may also be simplified by transforming to a vector potential $\mathbf{A}'$ such that

$$\text{div } \mathbf{A}'=0 \tag{7-34}$$

The generator of the required gauge transformation must be a solution of the differential equation

$$\nabla^2\psi=-\text{div } \mathbf{A} \tag{7-35}$$

where $\mathbf{A}$ is any vector function simultaneously satisfying (7-27) and (7-28). This restricted set of gauge transformations is known as the Coulomb gauge group. In regions where the free charge density ρ is zero, a solution of (7-33) is $\phi'=0$, and in this case, the Lorentz condition reduces to the Coulomb condition. If we impose the Coulomb condition on (7-27) and (7-28), we find that, in the Coulomb gauge, the potentials for the electromagnetic field satisfy the differential equations

$$\nabla^2\psi'=-\frac{\rho}{\epsilon}$$

$$\nabla^2\mathbf{A}'=-\mu\sigma\,\frac{\partial \mathbf{A}'}{\partial t}-\mu\epsilon\,\frac{\partial^2\mathbf{A}'}{\partial t^2}=-\mu\mathbf{J}_v+\mu\left(\sigma+\epsilon\,\frac{\partial}{\partial t}\right)\text{grad }\phi' \tag{7-36}$$

Thus, in the Coulomb gauge the scalar potential satisfies the Poisson equation, and is part of the source term of the wave equation satisfied by the vector potential.

Although we may, in principle, relate the vector and scalar potentials for a given electromagnetic field by any gauge condition we choose, it is customary to use the Lorentz gauge. Hence, when we speak of the vector and scalar potentials for an electromagnetic field, we shall mean that vector and scalar potential which satisfy the Lorentz condition, unless the gauge is explicitly specified.

7-3. The Hertz Potentials. In any region of a linear, homogeneous, and isotropic medium which contains neither free charge density nor electric current density, Maxwell's equations have the completely symmetric form

$$\text{curl } \mathbf{E}=-\frac{\partial \mathbf{B}}{\partial t}, \qquad \text{div } \mathbf{B}=0$$

$$\text{curl } \mathbf{H}=\frac{\partial \mathbf{D}}{\partial t}, \qquad \text{div } \mathbf{D}=0 \tag{7-37}$$

We now define two vector potentials $\boldsymbol{\pi}$ and $\boldsymbol{\pi}^*$ which specify the electromagnetic field defined by (7-37). Since $\mathbf{B}$ is solenoidal, we may write

$$\mathbf{B}=\mu\epsilon \text{ curl} \frac{\partial \boldsymbol{\pi}}{\partial t}$$

It then follows from the first of (7-37) and the linearity of the curl operator that

$$\text{curl}\left(\mathbf{E}+\mu\epsilon \frac{\partial^2 \boldsymbol{\pi}}{\partial t^2}\right)=0$$

Thus, the electric intensity may be derived from the vector potential and a scalar potential ϕ by

$$\mathbf{E}=-\text{grad } \phi-\mu\epsilon \frac{\partial^2 \boldsymbol{\pi}}{\partial t^2} \tag{7-38}$$

The terms on the right-hand side of (7-38) can be interpreted as the scalar and vector potentials of Sec. 7-2, if we set

$$\mathbf{A}=\mu\epsilon \frac{\partial \boldsymbol{\pi}}{\partial t}$$

In a region of space which contains no free charge, the scalar potential is any solution of the homogeneous scalar wave equation, and consequently, we may choose

$$\phi=-\text{div } \boldsymbol{\pi}$$

Under these conditions, the electric intensity is completely determined by the vector potential $\boldsymbol{\pi}$,

$$\mathbf{E}=\text{grad div } \boldsymbol{\pi}-\mu\epsilon \frac{\partial^2 \boldsymbol{\pi}}{\partial t^2} \tag{7-39}$$

It follows from the third of (7-37) the constitutive relations

$$\mathbf{H}=\frac{1}{\mu} \mathbf{B}, \ \mathbf{D}=\epsilon \mathbf{E}$$

and (7-39) that

$$\frac{d}{dt}\left(\text{curl curl } \boldsymbol{\pi}-\text{grad div } \boldsymbol{\pi}+\mu\epsilon \frac{\partial^2 \boldsymbol{\pi}}{\partial t^2}\right)=0 \tag{7-40}$$

We integrate (7-40) with respect to time, and set the constant of integration to zero, since its particular value cannot influence the determination of the field. We then obtain the homogeneous vector wave equation

$$\nabla^2 \boldsymbol{\pi}-\mu\epsilon \frac{\partial^2 \boldsymbol{\pi}}{\partial t^2}=0$$

Any solution of this equation determines an electromagnetic field through the relations

$$\mathbf{B}=\mu\epsilon \text{ curl } \frac{\partial \boldsymbol{\pi}}{\partial t}, \quad \mathbf{E}=\text{curl curl } \boldsymbol{\pi}$$

In a charge free region, the electric displacement is also a solenoidal vector field, and we can obtain an electromagnetic field from a vector potential $\boldsymbol{\pi}^*$ through the relations

$$\mathbf{D}=-\mu\epsilon \text{ curl } \frac{\partial \boldsymbol{\pi}^*}{\partial t}, \quad \mathbf{H}=\text{curl curl } \boldsymbol{\pi}^*$$

where $\boldsymbol{\pi}^*$ satisfies a homogeneous vector wave equation. The vector potentials $\boldsymbol{\pi}$ and $\boldsymbol{\pi}^*$ were first introduced by Hertz in his analysis of the infinitesimal radiating dipole. They are known as the electric and magnetic Hertz potentials respectively.

Since Maxwell's equations are linear, an arbitrary electromagnetic field in a current and charge free region, can be derived from a linear super-position of an electric and a magnetic Hertz potential. Both the electric and magnetic Hertz potential satisfy homogeneous wave equations, which implies that the sources of these potentials lie outside the region where they define the electromagnetic field.

In order to appreciate the physical significance of the Hertz potentials, it is necessary to relate them to the sources which produce them. In any medium other than free space, we define electric and magnetic polarization vectors **P** and **M** by

$$\mathbf{P}=\mathbf{D}-\epsilon_0 \mathbf{E}, \quad \mathbf{M}=\frac{1}{\mu_0} \mathbf{B}-\mathbf{H} \tag{7-41}$$

where ϵ_0 and μ_0 are the free space permittivity and permeability respectively. It is clear that both the electric and magnetic polarizations are zero in free space, and hence are associated with the presence of material media. The polarization vectors may be used to eliminate the electric displacement and magnetic intensity from the field equations. In a region containing no free charge or current density, Maxwell's equations may be written

$$\begin{aligned} \text{curl } \mathbf{E} &= -\frac{\partial \mathbf{B}}{\partial t} \\ \text{div } \mathbf{B} &= 0 \\ \text{curl } \mathbf{B} &= \epsilon_0\mu_0 \frac{\partial \mathbf{E}}{\partial t}+\mu_0 \left(\frac{\partial \mathbf{P}}{\partial t}+\text{curl } \mathbf{M}\right) \\ \text{div } \mathbf{E} &= -\frac{1}{\epsilon_0} \text{ div } \mathbf{P} \end{aligned} \tag{7-42}$$

Thus, the presence of material media in an electromagnetic field can be accounted for in terms of an equivalent charge density, div **P**, and an equivalent current density ($\partial\mathbf{P}/\partial t+\text{curl }\mathbf{M}$).

We should like to obtain a solution of the set of field equations (7-42) in terms of the Hertz potentials $\boldsymbol{\pi}$ and $\boldsymbol{\pi}^*$. It is convenient to begin with scalar and vector potentials defined by

$$\phi=-\text{div }\boldsymbol{\pi}$$

$$\mathbf{A}=\epsilon_0\mu_0\frac{\partial\boldsymbol{\pi}}{\partial t}+\text{curl }\boldsymbol{\pi}^*$$

It then follows from (7-24) and (7-26) that the magnetic induction and electric intensity are expressed as

$$\mathbf{B}=\epsilon_0\,\mu_0\text{ curl }\left(\frac{\partial\boldsymbol{\pi}}{\partial t}\right)+\text{curl curl }\boldsymbol{\pi}^*$$

$$\mathbf{E}=\text{grad div }\boldsymbol{\pi}-\epsilon_0\,\mu_0\frac{\partial^2\boldsymbol{\pi}}{\partial t^2}-\text{curl}\left(\frac{\partial\boldsymbol{\pi}^*}{\partial t}\right) \tag{7-43}$$

Clearly, the fields defined by (7-43) satisfy the first pair of (7-42). In order to satisfy the second pair of field equations, the Hertz potentials must be solutions of

$$\epsilon_0\,\mu_0\frac{\partial}{\partial t}\left(\text{curl curl }\boldsymbol{\pi}-\text{grad div }\boldsymbol{\pi}+\epsilon_0\,\mu_0\frac{\partial^2\boldsymbol{\pi}}{\partial t^2}-\frac{1}{\epsilon_0}\,\mathbf{P}\right)+$$

$$\text{curl}\left(\text{curl curl }\boldsymbol{\pi}^*+\epsilon_0\,\mu_0\frac{\partial^2\boldsymbol{\pi}^*}{\partial t^2}-\mu_0\,\mathbf{M}\right)=0 \tag{7-44}$$

and

$$\text{div}\left(\text{grad div }\boldsymbol{\pi}-\epsilon_0\,\mu_0\frac{\partial^2\boldsymbol{\pi}}{\partial t^2}-\text{curl }\frac{\partial\boldsymbol{\pi}}{\partial t}+\frac{1}{\epsilon_0}\,\mathbf{P}\right)=0 \tag{7-45}$$

In the second term of (7-44), we may add the term ($-$grad div $\boldsymbol{\pi}^*$), since the curl of any gradient vanishes. Then (7-44) can be written

$$\epsilon_0\,\mu_0\frac{\partial}{\partial t}\,\nabla^2\boldsymbol{\pi}-\epsilon_0\,\mu_0\,\frac{\partial^2\boldsymbol{\pi}}{\partial t^2}+\frac{1}{\epsilon_0}\,\mathbf{P}+$$

$$\text{curl }\nabla^2\boldsymbol{\pi}^*-\epsilon_0\,\mu_0\,\frac{\partial^2\boldsymbol{\pi}^*}{\partial t^2}-\mu_0\,\mathbf{M}=0 \tag{7-46}$$

Similarly, we may add a term ($-$curl curl $\boldsymbol{\pi}$) in the bracket of (7-45) and obtain

$$\text{div }\nabla^2\boldsymbol{\pi}-\epsilon_0\,\mu_0\,\frac{\partial^2\boldsymbol{\pi}}{\partial t^2}-\frac{1}{\epsilon_0}=0 \tag{7-47}$$

In order to satisfy (7-46) and (7-47), it is sufficient, but not necessary, that the Hertz potentials satisfy the differential equations

$$\nabla^2\boldsymbol{\pi}-\epsilon_0\mu_0\,\frac{\partial^2\boldsymbol{\pi}}{\partial t^2}=-\frac{1}{\epsilon_0}\,\mathbf{P} \tag{7-48a}$$

$$\nabla^2 \boldsymbol{\pi}^* - \epsilon_0 \mu_0 \frac{\nabla^2 \boldsymbol{\pi}^*}{\partial t^2} = -\mu_0 \mathbf{M} \tag{7-48b}$$

It is clear from Eqs. (7-48) that the two Hertz potentials are independent. Then, since Maxwell's equations are linear, the given electromagnetic field can be expressed as a linear combination of the two partial fields

$$\mathbf{E}_1 = \text{grad div } \boldsymbol{\pi} - \epsilon_0 \mu_0 \frac{\partial^2 \boldsymbol{\pi}}{\partial t^2}$$

$$\mathbf{B}_1 = \epsilon_0 \mu_0 \text{ curl } (\partial \boldsymbol{\pi}/\partial t)$$

and

$$\mathbf{E}_2 = -\text{curl } (\partial \boldsymbol{\pi}^*/\partial t)$$

$$\mathbf{B}_2 = \text{curl curl } \boldsymbol{\pi}^*$$

The Hertz potential $\boldsymbol{\pi}$ has as its sources distributions of electric polarization, and consequently the partial field ($\mathbf{E}_1$, $\mathbf{B}_1$) is known as a field of the electric type. In contrast, the sources of the Hertz potential $\boldsymbol{\pi}^*$ are distributions of magnetic polarization, and hence the partial field ($\mathbf{E}_2$, $\mathbf{B}_2$) is known as a field of the magnetic type.

EXAMPLE 3. As an example of the use of the Hertz potentials, let us consider the solution of the first of (7-48) which corresponds to a point source at the origin. This analysis is fairly close to Hertz analysis of the infinitesimal radiating dipole. Since we may expect the potential from a point source to be spherically symmetric, we can reduce Eq. (7-48) to

$$\frac{\partial^2 \boldsymbol{\pi}}{\partial r^2} + \frac{2}{r} \frac{\partial \boldsymbol{\pi}}{\partial r} - \frac{1}{c^2} \frac{\partial^2 \boldsymbol{\pi}}{\partial t^2} = 0, \; r > 0$$

where r is the usual radial coordinate in spherical coordinates, and $c^2 = (\epsilon_0 \mu_0)^{-1}$. Assuming harmonic time dependence and writing

$$\boldsymbol{\pi}(r, t) = \mathbf{P}(r)\, e^{-i\omega t}$$

we find that $\mathbf{P}(r)$ must satisfy the ordinary differential equation

$$\frac{d^2}{dr^2} (r\,\mathbf{P}) + k^2 (r\,\mathbf{P}) = 0, \; r > 0$$

where $k = \omega/c$. The general solution of the differential equation for $\mathbf{P}$ is

$$r\mathbf{P} = \mathbf{C}_1 e^{ikr} + \mathbf{C}_2 e^{-ikr}$$

where $\mathbf{C}_1$ and $\mathbf{C}_2$ are arbitrary constant vectors. The term e^{-ikr} corresponds to a source at infinity, and hence we set $\mathbf{C}_2 = \mathbf{0}$. We will choose

$$\mathbf{C}_1 = \frac{\mathbf{p}_0}{4\pi\epsilon_0}$$

and identify $\mathbf{p}_0$ as the electric dipole moment of the source at the origin. With these results,

$$\boldsymbol{\pi} = \frac{\mathbf{p}_0}{4\pi\epsilon_0} \frac{e^{i(kr-\omega t)}}{r}$$

This solution corresponds to a spherical wave going out from the origin with phase velocity c and amplitude which decreases as $1/r$. It is clear that $\boldsymbol{\pi}$ has, not unexpectedly, a singularity at the origin. The interpretation of this singularity can be made only after the calculation of the fields.

We can ignore the magnetic Hertz potential $\boldsymbol{\pi}^*$ in this case, since its sources are regions of magnetic polarization. It then follows from (7-43) that

$$\mathbf{B} = \frac{1}{c^2} \operatorname{curl} \left(\frac{\partial \boldsymbol{\pi}}{\partial t} \right) = \frac{1}{c^2} \frac{\partial}{\partial t} \operatorname{curl} \boldsymbol{\pi}$$

$$\mathbf{E} = \operatorname{grad} \operatorname{div} \boldsymbol{\pi} - \frac{1}{c^2} \frac{\partial^2 \boldsymbol{\pi}}{\partial t^2} = \operatorname{curl} \operatorname{curl} \boldsymbol{\pi}$$

The final form for the electric intensity follows from

$$\nabla^2 \boldsymbol{\pi} - \frac{1}{c^2} \frac{\partial^1 \boldsymbol{\pi}}{\partial t^2} = 0$$

Now,

$$\operatorname{curl} \boldsymbol{\pi} = \frac{1}{4\pi\epsilon_0} \operatorname{curl} \left[\left(\frac{e^{ikr}}{r} \right) (\mathbf{p}_0\, e^{-i\omega t}) \right]$$

$$= \frac{1}{4\pi\epsilon_0} \left[\operatorname{grad} \left(\frac{e^{ikr}}{r} \right) \times (\mathbf{p}_0\, e^{-i\omega t}) \right]$$

$$= \frac{1}{4\pi\epsilon_0} \left[\frac{d}{dr} \left(\frac{e^{ikr}}{r} \right) \right] \left(\frac{\mathbf{r}}{r} \times \mathbf{p}_0\, e^{-i\omega t} \right)$$

$$= \frac{1}{4\pi\epsilon_0} \left\{ k^2 \left[\frac{1}{kr} - \frac{1}{(kr)^2} \right] e^{ikr} \right\} \left(\frac{\mathbf{r}}{r} - \mathbf{p}_0 \right) e^{-i\omega t}$$

and

$$\frac{\partial}{\partial t} (\operatorname{curl} \boldsymbol{\pi}) = -i\omega \operatorname{curl} \boldsymbol{\pi}$$

Hence,

$$\mathbf{B} = -\frac{i\omega k^2 \mu_0}{4\pi r} \left[\frac{1}{kr} - \frac{1}{(kr)^2} \right] \mathbf{r} \times \mathbf{p}_0\, e^{i(kr-\omega t)}$$

A similar, although somewhat more intricate, calculation results in

$$\mathbf{E}=\frac{k^3}{4\pi\epsilon_0 r^2}\left\{\mathbf{r}(\mathbf{r}\cdot\mathbf{p}_0)\left[-\frac{1}{kr}-\frac{3i}{(kr)^2}+\frac{3}{(kr)^3}\right]+ r^2\mathbf{p}_0\left[\frac{1}{kr}+\frac{i}{(kr)^2}-\frac{1}{(kr)^3}\right]\right\}e^{i(kr-\omega t)}$$

An analysis of the fields for $kr \ll 1$, which is left as an exercise, justifies the identification of $\mathbf{p}_0\, e^{-i\omega t}$ as the electric dipole moment of the source at the origin. A similar analysis for a magnetic dipole with dipole moment $\mathbf{m}_0\, e^{-i\omega t}$ can be carried out in terms of the magnetic Hertz potential $\boldsymbol{\pi}^*$.

7-4. The Electrostatic Field. The simplest electromagnetic field which can exist in a linear, homogeneous, and isotropic medium is that generated by time independent or static distributions of free charge and electric current. If the sources of the electromagnetic field are assumed to be independent of the time, the fields must also be static, and the Maxwell equations reduce to

$$\begin{aligned}&\text{curl } \mathbf{E}=\mathbf{0}, \quad \text{div } \mathbf{D}=\rho\\ &\text{curl } \mathbf{H}=\mathbf{J}, \quad \text{div } \mathbf{B}=0\end{aligned} \tag{7-49}$$

Thus, in the time independent case, Maxwell's equations are no longer coupled and the general electromagnetic field is a superposition of two non-interacting fields. The first of these two fields which we shall consider is the electrostatic field which is defined by the field vectors **E** and **D**. In a linear, homogeneous, and isotropic medium, the two vectors defining the field are related by

$$\mathbf{D}=\epsilon\mathbf{E}$$

and the relevant field equations may be written

$$\text{curl } \mathbf{E}=\mathbf{0}, \quad \text{div } \mathbf{E}=\rho/\epsilon$$

It follows from the Helmholtz theorem that the field intensity can be derived from a scalar potential,

$$\mathbf{E}=-\text{grad } \phi \tag{7-50}$$

Equation (7-50) may also be obtained from (7-26) by requiring the vector potential **A** to be independent of the time. In order that the relation between the electric intensity and the free electric charge density be satisfied, the scalar potential must satisfy the second order partial differential equation

$$\nabla^2\phi=-\rho/\epsilon \tag{7-51}$$

Thus, any solution of the scalar Poisson equation which satisfies the continuity conditions imposed by the physical situation gives rise to a possible electrostatic field through (7-50).

Let us now examine the continuity conditions to be imposed on the scalar potential. Consider two contiguous media, each of which is linear, homogeneous, and isotropic. Let one of the two media, say medium 1, be characterized by the constituitive parameters ϵ_1, μ_1, σ_1 and the second medium (medium 2) is characterized by the parameters ϵ_2, μ_2, σ_2. At the surface separating the two media, we have from (7-19) and (7-21)

$$\mathbf{n}\cdot(\mathbf{D}_2-\mathbf{D}_1)=\omega, \quad \mathbf{n}\times(\mathbf{E}_2-\mathbf{E}_1)=0$$

where $\mathbf{n}$ is the unit normal to the boundary surface, and ω is the surface charge density. The subscripts have the meaning that the field is to be evaluated immediately adjacent to the boundary surface in the medium indicated by the subscript. It follows from the constitutive relation and equation (7-50) that

$$\mathbf{n}\cdot(\epsilon_1 \operatorname{grad} \phi_1-\epsilon_2 \operatorname{grad} \phi_2)=\omega$$

$$\mathbf{n}\times(\operatorname{grad} \phi_1-\operatorname{grad} \phi_2)=0$$

A particular case which arises frequently in practice occurs when the electrical conductivity of one of the two media, say medium 2, is infinite, and the conductivity of the other is zero. Since a finite electrostatic intensity field cannot exist in a perfect conductor, it is clear that the scalar potential ϕ_2 must be constant. The boundary condition on the tangential component of the electric intensity implies that the field is always normal to the surface separating a perfect conductor and a non-conducting medium. If the surface charge density ω is explicitly known, we determine the electrostatic field in the non-conducting region by finding a solution ϕ_1 of the scalar Poisson equation which satisfies the condition

$$\mathbf{n}\cdot\epsilon_1 \operatorname{grad} \phi_1=\epsilon_1\frac{\partial\phi_1}{\partial \mathbf{n}}=\omega$$

on the boundary surface S. In the more usual situation, it is the constant potential ϕ_2 which is known on S, and we wish to determine the surface charge density ω. In this case, we must find a solution of the scalar Poisson equation which satisfies the condition

$$\phi_1=\phi_2=\text{constant}$$

on the boundary surface S.

EXAMPLE 4. A classic example of the direct solution of the electrostatic problem arises when we introduce a sphere, either dielectric or conducting, into an originally uniform electrostatic field. Let the initial field be

$$\mathbf{E}_0 = E_0\,\mathbf{k}$$

where E_0 is constant. In terms of cartesian coordinates, this field can be derived from the scalar potential

$$\phi_0 = -E_0\,z$$

We now introduce a sphere of radius r_0 with centre at the origin. The presence of the sphere distorts the field lines in the vicinity of the sphere. The exact form of the field near the sphere depends on whether the sphere is conducting or dielectric. Although the two cases are quite different in detail, there are many common features which we can exploit.

First of all, since there is no free charge present, the scalar potential for the field in the presence of the sphere satisfies Laplace's equation

$$\nabla^2\phi = 0$$

Secondly, at distances very large compared to the radius of the sphere, the field should be the same as the original field. If ϕ is expressed in terms of spherical coordinates, this condition implies that

$$\lim_{r\to\infty} \phi(r, \theta, \varphi) = \phi_0 = -E_0 r \cos\theta$$

The symmetry of the sphere and the original field imply that the scalar potential is independent of the azimuthal coordinate. Hence, the explicit form of the differential equation for ϕ is

$$\frac{1}{r^2}\frac{\partial}{\partial r}\left(r^2\frac{\partial\phi}{\partial r}\right) + \frac{1}{r^2\sin\theta}\frac{\partial}{\partial\theta}\left(\sin\theta\,\frac{\partial\phi}{\partial\theta}\right) = 0$$

It is easy to verify that the general solution of toe differential equation which satisfies the condition for large values of r is

$$\phi(r, \theta) = \left(-E_0 r + \frac{a}{r^2}\right)\cos\theta$$

where a is to be determined from the conditions on the field at the surface of the sphere.

Since the constant a is determined by the field at r_0, it is now necessary to consider the difference between the conducting and dielectric spheres.

(i) *Conducting sphere.* In the case of the perfectly conducting sphere, the entire spherical surface is at a constant potential, which we can take to be zero. Hence,

$$\phi(r_0, \theta) = \left(-E_0 r_0 + \frac{a}{r_0^2}\right)\cos\theta = 0$$

from which $a = E_0 r_0^3$, and

$$\phi(r,\theta)=\left(-E_0 r+\frac{E_0 r_0^3}{r^2}\right)\cos\theta$$

The surface charge density induced on the sphere is immediately found from the boundary condition

$$\omega=\epsilon_0\,\mathbf{n}\cdot\text{grad}\,\phi\Big|_{r_0}=\epsilon_0\frac{\partial\phi}{\partial r}\Big|_{r_0}$$
$$=-3\epsilon_0 E_0\cos\theta$$

The spherical components of the electrostatic field are given by

$$E_r=-\frac{\partial\phi}{\partial r}=\left(E_0+\frac{2E_0 r_0^3}{r^3}\right)\cos\theta$$

$$E_\theta=-\frac{1}{r}\frac{\partial\phi}{\partial\theta}=\left(-E_0+\frac{E_0 r_0^3}{r^3}\right)\sin\theta$$

It is clear that the field is normal to the surface of the sphere at $r=r_0$, and for $r\gg r_0$,

$$\begin{aligned}\mathbf{E}&=E_r\,\mathbf{i}_r+E_\theta\,\mathbf{i}_\theta\\ &E_0\,[(\sin\theta\cos\theta\cos\varphi\,\mathbf{i}+\sin\theta\cos\theta\sin\varphi\,\mathbf{j}+\cos^2\theta\mathbf{k})+\\ &\quad(-\sin\theta\cos\theta\cos\varphi\,\mathbf{i}-\sin\theta\cos\theta\sin\varphi\,\mathbf{j}+\sin^2\theta\mathbf{k})]\\ &=E_0\,\mathbf{k}\end{aligned}$$

as expected.

(ii) *Dielectric sphere*. The problem of the dielectric sphere is somewhat more complicated since the field does not vanish inside the sphere, and it is necessary to explicitly know this field in order to satisfy the boundary conditions at $r=r_0$. In the region outside the sphere, the condition

$$\lim_{r\to\infty}\phi_1=\phi_0$$

remains valid, and the general solution of Laplace's equation which satisfies this condition is

$$\phi_1(r,\theta)=\left(-E_0 r+\frac{a}{r^2}\right)\cos\theta$$

In the interior of the sphere, the term proportional to r^{-2} cannot be used since it is singular at the origin. However, the θ-dependence of the potential inside the sphere must be the same as that of the potential outside the sphere. Hence, a general solution for the region inside the sphere is

$$\phi_2(r,\theta)=b\,r\cos\theta$$

where b is a constant to be determined.

The continuity of the tangential component of E at $r=r_0$ implies that

$$\frac{1}{r}\frac{\partial\phi_1}{\partial\theta}\bigg|_{r_0}=\frac{1}{r}\frac{\partial\phi_2}{\partial\theta}\bigg|_{r_0}$$

which leads to

$$-E_0+\frac{a}{r_0^3}=b$$

Since there is no mechanism for the formation of a surface charge density, the condition on the normal component of the field leads to

$$\epsilon_0\frac{\partial\phi_1}{\partial r}\bigg|_{r_0}=\epsilon\frac{\partial\phi_2}{\partial r}\bigg|_{r_0}$$

where ϵ is the permittivity of the dielectric sphere. Carrying out the indicated operations, we find that

$$\epsilon_0 E_0+\frac{2\epsilon_0 a}{r_0^3}=\epsilon b$$

Solving the pair of equations for a and b

$$a=\frac{\kappa-1}{\kappa+2}r_0^3 E_0, \quad b=\frac{3E_0}{\kappa+2}$$

where $\kappa=\epsilon/\epsilon_0$ is known as the dielectric constant for the sphere. Substituting these results in the expressions for the potentials,

$$\phi_1=\left(-E_0 r+\frac{\kappa-1}{\kappa+2}\frac{E_0 r_0^3}{r^2}\right)\cos\theta$$

$$\phi_2=\frac{3E_0}{\kappa+2}r\cos\theta$$

The resulting electrostatic field is

$$\mathbf{E}=\begin{cases} E_0\left[\left(1+\frac{\kappa-1}{\kappa+2}\frac{2r_0^3}{r^3}\right)\cos\theta\,\mathbf{i}_r+\left(-1+\frac{\kappa-1}{\kappa+2}\frac{r_0^3}{r^3}\right)\sin\theta\,\mathbf{i}_\theta\right] & r>r_0 \\ \frac{3E_0}{\kappa+2}\mathbf{k}, & 0\leqslant r<r_0 \end{cases}$$

In this case, the change in the original field is due to induced polarization of the dielectric sphere. The induced polarization can be immediately calculated from

$$\begin{aligned}\mathbf{P}&=(\epsilon-\epsilon_0)\,\mathbf{E}\\ &=(\kappa-1)\,\epsilon_0\mathbf{E}\\ &=3\epsilon_0\frac{\kappa-1}{\kappa+2}E_0\mathbf{k}\end{aligned}$$

and is seen to be uniform.

Another particular case of some interest obtains whenever the electrostatic field exists in an unbounded medium characterized by the permittivity ϵ. If the charge density $\phi(\mathbf{r}')$ is all contained within a bounded region V', the electric intensity $\mathbf{E}(\mathbf{r})$ must vanish as $1/|\mathbf{r}|^2$ as $|\mathbf{r}|$ becomes very large. This implies that the scalar potential $\varphi(\mathbf{r})$ for the field must vanish as $1/|\mathbf{r}|$ for very large values of $|\mathbf{r}|$. This particular type of boundary condition on $\phi(\mathbf{r})$ and $\mathbf{E}(\mathbf{r})$ is known as an asymptotic boundary condition, and frequently arises when one is considering a field in an unbounded region. In order to determine the electrostatic field under these conditions, it is sufficient to find a solution of

$$\nabla^2\phi(\mathbf{r}')=-\frac{\rho(\mathbf{r}')}{\epsilon} \tag{7-52}$$

subject to the symptotic boundary condition that $\phi(\mathbf{r}')\sim 1/|\mathbf{r}|$ for sufficiently large values of $|\mathbf{r}|$. The solution to this problem can be obtained by a straight forward application of Green's theorem (see Sec. 2-9). Define a function $g(\mathbf{r}, \mathbf{r}')$ which satisfies the partial differential equation

$$\nabla^2 g(\mathbf{r}', \mathbf{r})=-4\pi\delta(\mathbf{r}-\mathbf{r}') \tag{7-53}$$

where $\delta(\mathbf{r}-\mathbf{r}')$ is the Dirac delta, and the boundary condition

$$g(\mathbf{r}', \mathbf{r})\sim 1/|\mathbf{r}-\mathbf{r}'|$$

as $|\mathbf{r}|$ tends to infinity and $|\mathbf{r}'|$ remains finite. As we have already seen [see Eq. (2-45)], a solution of (7-53) which satisfies the prescribed boundary condition is

$$g(\mathbf{r}', \mathbf{r})=1/|\mathbf{r}-\mathbf{r}'|$$

We multiply Eq. (7-52) by $g(\mathbf{r}, \mathbf{r}')$, Eq. (7-53) by $\phi(\mathbf{r}')$, subtract and integrate over the volume of the sphere of radius R to obtain

$$4\pi\int_{V(R)}\phi(\mathbf{r}')\,\delta(\mathbf{r}-\mathbf{r}')\,dv=\int_{V(R)}[g(\mathbf{r}', \mathbf{r})\,\nabla^2\phi(\mathbf{r}')-\phi(\mathbf{r}')\,\nabla^2 g(\mathbf{r}', \mathbf{r})]\,dv+\frac{1}{\epsilon}\int_{V(R)}\rho(\mathbf{r}')\,g(\mathbf{r}', \mathbf{r})\,dv$$

It follows from the properties of the Dirac delta that the term on the left is $4\pi\phi(\mathbf{r})$. We transform the first term on the right by Green's theorem to obtain

$$\int_{V(R)}[g(\mathbf{r}', \mathbf{r})\,\nabla^2\phi(\mathbf{r}')-\phi(\mathbf{r}')\,\nabla^2 g(\mathbf{r}', \mathbf{r})]\,dv=\int_{S(R)}[g(\mathbf{r}', \mathbf{r})\,\mathrm{grad}'\,\phi(\mathbf{r}')-\phi(\mathbf{r}')\,\mathrm{grad}'\,g(\mathbf{r}', \mathbf{r})]\cdot\mathbf{n}\,ds$$

where $\mathbf{n}$ is the unit outward normal to the surface $S(R)$ which bounds the volume $V(R)$. If we hold $\mathbf{r}$ fixed, each term of the integrand vanishes as $1/R^3$ in the limit $R\to\infty$, whereas the surface area of $S(R)$

is proportional to R^2. Hence, the surface integral vanishes in the limit $R\to\infty$. The integral

$$\int_{V(R)} \phi(\mathbf{r}')\, g(\mathbf{r}', \mathbf{r})\, dv = \int_{V'} \frac{\phi(\mathbf{r}')}{|\mathbf{r}-\mathbf{r}'|}\, dv$$

is independent of R so long as $V(R)$ completely encloses the charge distribution. Hence, a solution of (7-52) which satisfies the asymptotic boundary condition is

$$\phi(\mathbf{r}) = \frac{1}{4\pi\epsilon}\int_{V'} \frac{\rho(\mathbf{r}')}{|\mathbf{r}-\mathbf{r}'|}\, dv \tag{7-54}$$

The function $g(\mathbf{r}, \mathbf{r}') = |\mathbf{r}-\mathbf{r}'|^{-1}$ defined by (7-53) and the asymptotic boundary condition is known as the Green's function for the Poisson equation in an unbounded medium.

EXAMPLE 5. Although the integral in (7-54) is very difficult to evaluate in general, it can be used to obtain a rather interesting and useful representation of the electrostatic field generated by an arbitrary charge distribution. The first step in obtaining the desired representation is to expand $|\mathbf{r}=\mathbf{r}'|^{-1}$ in a Taylor series about $\mathbf{r}'=0$. This expansion is made in terms of the cartesian components of $\mathbf{r}$ and $\mathbf{r}'$, since

$$|\mathbf{r}-\mathbf{r}'| = [(x-x')^2 = (y-y')^2 + (z-z')^2]^{1/2}$$

Thus, the Taylor series has the form

$$\begin{aligned} |\mathbf{r}-\mathbf{r}'|^{-1} = &\left(\frac{1}{|\mathbf{r}-\mathbf{r}'|}\right)_{\mathbf{r}'=0} + \Bigg[\left(\frac{\partial}{\partial x'}\frac{1}{|\mathbf{r}-\mathbf{r}'|}\right)_{\mathbf{r}'=0} x' + \\ &\left(\frac{\partial}{\partial y'}\frac{1}{|\mathbf{r}-\mathbf{r}'|}\right)_{\mathbf{r}'=0} y' + \left(\frac{\partial}{\partial z'}\frac{1}{|\mathbf{r}-\mathbf{r}'|}\right)_{\mathbf{r}'=0} z'\Bigg] \\ &+ \Bigg[\tfrac{1}{2}\left(\frac{\partial^2}{\partial x'^2}\frac{1}{|\mathbf{r}-\mathbf{r}'|}\right)_{\mathbf{r}'=0} x'^2 + \tfrac{1}{2}\left(\frac{\partial^2}{\partial y'^2}\frac{1}{|\mathbf{r}-\mathbf{r}'|}\right)_{\mathbf{r}'=0} y'^2 + \\ &\tfrac{1}{2}\left(\frac{\partial^2}{\partial z}\frac{1}{|\mathbf{r}-\mathbf{r}'|}\right)_{\mathbf{r}'=0} z'^2 + \left(\frac{\partial^2}{\partial x'\,\partial y'}\frac{1}{|\mathbf{r}-\mathbf{r}'|}\right)_{\mathbf{r}'=0} x'\,y' \\ &+ \left(\frac{\partial^2}{\partial x'\,\partial z'}\frac{1}{|\mathbf{r}-\mathbf{r}'|}\right)_{\mathbf{r}'=0} x'\,z' + \left(\frac{\partial^2}{\partial y'\,\partial z'}\frac{1}{|\mathbf{r}-\mathbf{r}'|}\right)_{\mathbf{r}'=0} y'z'\Bigg] + \ldots \end{aligned}$$

There is some difficulty in expressing the coefficients of the Taylor series in simple form. However, the difficulty can be overcome by using the symmetry of $|\mathbf{r}-\mathbf{r}'|$ to relate the derivatives of $|\mathbf{r}-\mathbf{r}'|^{-1}$ with respect to the primed and unprimed coordinates. For example,

$$\begin{aligned} \left(\frac{\partial}{\partial x'}\frac{1}{|\mathbf{r}-\mathbf{r}'|}\right)_{\mathbf{r}'=0} &= -\left(\frac{\partial}{\partial x}\frac{1}{|\mathbf{r}-\mathbf{r}'|}\right)_{\mathbf{r}'=0} \\ &= -\frac{\partial}{\partial x}\left(\frac{1}{|\mathbf{r}-\mathbf{r}'|}\right)_{\mathbf{r}'=0} = -\frac{\partial}{\partial x}\left(\frac{1}{r}\right) \end{aligned}$$

$$\left(\frac{\partial^2}{\partial x'\,\partial y'}\frac{1}{|\mathbf{r}-\mathbf{r}'|}\right)_{\mathbf{r}'=0}=\left(\frac{\partial^2}{\partial x\,\partial y}\frac{1}{|\mathbf{r}-\mathbf{r}'|}\right)_{\mathbf{r}'=0}$$

$$=\frac{\partial^2}{\partial x\,\partial y}\left(\frac{1}{|\mathbf{r}-\mathbf{r}'|}\right)_{\mathbf{r}'=0}=\frac{\partial^2}{\partial x\,\partial y}\left(\frac{1}{r}\right)$$

with similar results for the other derivatives. In terms of the derivatives with respect to the unprimed coordinates, the Taylor expansion is

$$\frac{1}{|\mathbf{r}-\mathbf{r}'|}=\frac{1}{r}-\left(x'\frac{\partial}{\partial x}+y'\frac{\partial}{\partial y}+z'\frac{\partial}{\partial z}\right)\frac{1}{r}+$$

$$\tfrac{1}{2}\left(x'\frac{\partial}{\partial x}+y'\frac{\partial}{\partial y}+z'\frac{\partial}{\partial z}\right)^2\frac{1}{r}+\ldots$$

$$=\frac{1}{r}-(r'\cdot\text{grad})\frac{1}{r}+\tfrac{1}{2}(\mathbf{r}'\cdot\text{grad})^2\frac{1}{r}+\ldots$$

where $(r'\cdot\text{grad})^2$ is a short notation for

$$(\mathbf{r}'\cdot\text{grad})^2=\left(x'^2\frac{\partial^2}{\partial x^2}+y'^2\frac{\partial^2}{\partial y^2}+z'^2\frac{\partial^2}{\partial z^2}+2x'y'\frac{\partial^2}{\partial x\,\partial y}+2x'z'\frac{\partial^2}{\partial x\,\partial z}+2y'z'\frac{\partial^2}{\partial y\,\partial z}\right)$$

Substituting the Taylor series in the integral of (7-54), we can express the potential as

$$\phi=\phi_1+\phi_2+\phi_3+\ldots$$

where

$$\phi_1=\frac{1}{4\pi\epsilon}\frac{1}{r}\int_{V'}\rho(\mathbf{r}')\,dv'$$

$$\phi_2=\frac{1}{4\pi\epsilon}\int_{V'}(\mathbf{r}'\cdot\text{grad})\frac{1}{r}\,\rho(\mathbf{r}')\,dv'$$

$$\phi_3=\frac{1}{4\pi\epsilon}\int_{V'}(\mathbf{r}'\cdot\text{grad})^2\frac{1}{r}\,\rho(\mathbf{r}')\,dv'$$

$$\vdots$$

We now examine each of the lower order terms in detail. We have

$$\phi_1=\frac{1}{4\pi\epsilon}\frac{1}{r}\int_{V'}\rho(\mathbf{r}')\,dv'=\frac{1}{4\pi\epsilon}\frac{Q}{r}$$

where Q is the total charge in the distribution. This expression is the potential of a point charge Q located at the origin. For the next term,

$$\phi_2=-\frac{1}{4\pi\epsilon}\int_{V'}\text{grad}\,(1/r)\cdot\mathbf{r}'\,\rho(\mathbf{r}')\,dv'$$

$$=-\frac{1}{4\pi\epsilon}\,\text{grad}\,(1/r)\cdot\int_{V'}\mathbf{r}'\,\rho(\mathbf{r}')\,dv'$$

$$=\frac{1}{4\pi\epsilon}\frac{\mathbf{r}\cdot\mathbf{p}}{r^3}$$

where $\mathbf{p}$ is the electric dipole moment of the charge distribution. Since ϕ_2 (known as the dipole field) varies as $1/r^2$, the dipole term is very small compared to the term ϕ_1 at large distances from the origin, unless the net charge Q in the distribution vanishes. The dipole moment has the cartesian components

$$p_x=\int_{V'} x'\,\rho(\mathbf{r}')\,dv',\; p_y=\int_{V'} y'\,\rho(\mathbf{r}')\,dv'\cdot p_z=\int_{V'} z'\,\rho(\mathbf{r}')\,dv'$$

Note the similarity between the components of the dipole moment and the coordinates of the centre of mass of an arbitrary mass distribution.

The term $\phi_3(r)$ has the explicit form

$$\begin{aligned}\phi_3=\frac{1}{8\pi\epsilon}\int_{V'}\rho(\mathbf{r}')\Big[&x'^2\frac{\partial^2}{\partial x^2}(1/r)+x'\,y'\frac{\partial^2}{\partial x\,\partial y}(1/r)+x'\,z'\frac{\partial^2}{\partial x\,\partial z}(1/r)+\\ &y'^2\frac{\partial^2}{\partial y^2}(1/r)+y'\,x'\frac{\partial^2}{\partial y\,\partial x}(1/r)+y'\,z'\frac{\partial^2}{\partial y\,\partial z}(1/r)+\\ &z'^2\frac{\partial^2}{\partial z^2}(1/r)+z'\,x'\frac{\partial^2}{\partial z\,\partial x}(1/r)+z'\,y'\frac{\partial^2}{\partial z\,\partial y}(1/r)\Big]\,dv'\end{aligned}$$

If we carry out the indicated differentiations, we find that

$$\phi_3=\frac{1}{8\pi\epsilon}\frac{R_{ij}\,Q^{ij}}{r^3}$$

where R_{ij} is the symmetric, covariant order two tensor with cartesian components

$$R_{11}=-1+\frac{3x^2}{r^2},\; R_{12}=R_{21}=\frac{3xy}{r^2},\; R_{13}=R_{31}=\frac{3xz}{r^2}$$

$$R_{22}=-1+\frac{3y^2}{r^2},\; R_{23}=R_{32}=\frac{3yz}{r^2},\; R_{33}=-1+\frac{3z^2}{r^2}$$

The tensor Q^{ij}, known as the electric quadrapole moment tensor, is the symmetric tensor with cartesian components

$$Q^{11}=\int_{V'}\rho(\mathbf{r}')\,x'^2\,dv',\; Q^{12}=Q^{21}=\int_{V'}\rho(\mathbf{r}')\,x'\,y'\,dv'$$

$$Q^{13}=Q^{31}=\int_{V'}\rho(\mathbf{r}')\,x'\,z'\,dv',\; Q^{22}=\int_{V'}\rho(\mathbf{r}')\,y'^2\,dv'$$

$$Q^{23}=Q^{32}=\int_{V'}\rho(\mathbf{r}')\,y'\,z',\,dv',\; Q^{33}=\int_{V'}\rho(\mathbf{r}')\,z'^2\,dv'$$

Here, we may note the similarity of the quadrapole moment tensor and the moment of inertia tensor for an arbitrary mass distribution. Since the quadrapole term has a $1/r^3$ variation, it makes a significant contribution to the total field, at large distances, only if the total charge Q and the dipole moment p both vanish.

Although we shall not treat the higher order terms in detail, there are some general observations which can be made. First of all, the n-th term will involve the completely contracted product of two tensors each of order n. Secondly, the potential ϕ_n will vary as $1/r^n$, so that its contribution to the total field can generally be ignored at distances from the origin which are large compared to the linear dimensions of the charge distribution.

The Green's function technique can also be applied to the solution of Eq. (7-51) in a bounded region with specified boundary conditions on the surface bounding the region. In this case, however, it is not so easy to actually determine the Green's function. For the sake of concreteness, let us assume that the solution of (7-51) is known at every point on a surface S, which completely encloses the charge distribution $\rho(\mathbf{r})$,

$$\phi(\mathbf{r}')=\phi_0(\mathbf{r}'),\ \mathbf{r}' \text{ on } S$$

We shall first rewrite (7-51) in the form

$$\nabla^2\phi(\mathbf{r}')=-\frac{\rho(\mathbf{r}')}{\epsilon} \tag{7-55}$$

Define the Green's function $g(\mathbf{r}, \mathbf{r}')$ by the partial differential equation

$$\nabla^2 g(\mathbf{r}, \mathbf{r}')=-4\pi\delta(\mathbf{r}-\mathbf{r}') \tag{7-56}$$

and the boundary condition

$$g(\mathbf{r}, \mathbf{r}')=0,\ \mathbf{r}' \text{ on } S$$

We employ the same algebraic manipulation as in the case of the unbounded medium to obtain

$$4\pi\int_V \phi\delta(\mathbf{r}-\mathbf{r}')\,dv'=\int_S [g(\mathbf{r},\mathbf{r}')\,\text{grad}'\,\phi(\mathbf{r}')-\phi(\mathbf{r}')\,\text{grad}'\,g(\mathbf{r},\mathbf{r}')]\cdot\mathbf{n}\,ds+\frac{1}{\epsilon}\int_V \rho(\mathbf{r}')\,g(\mathbf{r},\mathbf{r}')\,dv$$

where $\mathbf{n}$ is the unit outward normal to S, and V is the volume bounded by S. It follows from the boundary condition on $g(\mathbf{r}, \mathbf{r}')$ that the first term in the surface integral vanishes, and

$$\phi(\mathbf{r})=\frac{1}{4\pi\epsilon}\int_V \rho(\mathbf{r}')\,g(\mathbf{r},\mathbf{r}')\,dv-\frac{1}{4\pi}\int_S [\phi_0(\mathbf{r}')\,\text{grad}'\,g(\mathbf{r},\mathbf{r}')]\cdot\mathbf{n}\,ds \tag{7-57}$$

We note that the Green's function $g(\mathbf{r}, \mathbf{r}')$ is the electrostatic potential at the point $\mathbf{r}$ due to a point charge of strength $4\pi\epsilon$ at the point $\mathbf{r}'$ in the interior of the region bounded by an ideal conductor S which is held at zero or ground potential. An examination of the techniques for solving (7-51) or alternatively (7-52) is beyond the scope of this discussion.

We shall conclude our study of the electrostatic field by showing that if a solution to the electrostatic problem in a bounded region exists, it is necessarily unique. There are three types of boundary conditions which may obtain on the surface S:

(i) $\phi(\mathbf{r}')=f_0(\mathbf{r}')$, where $f_0(\mathbf{r}')$ is a specified function for $\mathbf{r}'$ on S;

(ii) $\text{grad}'\phi(\mathbf{r}')\cdot\mathbf{n}=f_1(\mathbf{r}')$, where $f_1(\mathbf{r}')$ is a specified function for $\mathbf{r}'$ on S;

(iii) $a\phi(\mathbf{r}')+b\,\text{grad}'\phi(\mathbf{r}')\cdot\mathbf{n}=f_2(\mathbf{r}')$, where $f_2(\mathbf{r}')$ is a specified function for $\mathbf{r}'$ on S, and a and b are specified constants.

We shall examine case (i) in detail and leave the verification of cases (ii) and (iii) as an exercise. Suppose that $\phi_1(\mathbf{r}')$ and $\phi_2(\mathbf{r}')$ are two solutions of the Poisson equation

$$\nabla^2\phi_i(\mathbf{r}')=-\frac{\rho}{\epsilon}, \quad i=1, 2$$

in a closed region V bounded by the surface S, and that each of these solutions satisfies the boundary condition

$$\phi_1(\mathbf{r}')=\phi_2(\mathbf{r}')=f_0(\mathbf{r}')$$

where $f_0(\mathbf{r}')$ is specified for $\mathbf{r}'$ on S. Define

$$\psi(\mathbf{r}')=\phi_1(\mathbf{r}')-\phi_2(\mathbf{r}')$$

Since the Laplacian operator is linear, it follows that $\psi(\mathbf{r}')$ is a solution of Laplace's equation

$$\nabla^2\psi(\mathbf{r}')=0$$

which satisfies the boundary condition $\psi(\mathbf{r}')=0$, for $\mathbf{r}'$ on S. If we again define a Green's function by Eq. (7-56) and the boundary condition $g(\mathbf{r}, \mathbf{r}')=0$ for $\mathbf{r}$ in V and $\mathbf{r}'$ on S, it follows from (7-57) that

$$\psi(\mathbf{r})=0, \quad \mathbf{r} \text{ in } V$$

and hence,

$$\phi_1(\mathbf{r})=\phi_2(\mathbf{r})$$

for $\mathbf{r}$ in V.

7-5. The Magnetostatic Field. The analysis of the magnetostatic field is more difficult than that of the electrostatic field, due to the vector source term in the equation

$$\text{curl}\,\mathbf{H}=\mathbf{J} \tag{7-58}$$

Once immediate consequence of this vector source term is that the magnetostatic field cannot, in general, be derived fıom a scalar potential. Only in the event that the sources of the magnetostatic field are regions of permanent magnetization, is it possible to derive the field from a scalar potential.

If we consider a region V bounded by a closed surface S, such that there are regions within V having a permanent magnetic moment $\mathbf{M}_0$, but also such that there are no currents anywhere in V, it follows from (7-41) that (7-58) can be satisfied by choosing the magnetic induction $\mathbf{B}$ so that

$$\mathbf{B}=-\mu_0 \text{ grad } \phi_m+\mu_0(\mathbf{M}_0+\mathbf{M})$$

where $\mathbf{M}$ is the magnetization induced by the presence of the field. Since the magnetic induction is solenoidal, the magnetic scalar potential ϕ_m satisfies the Poisson equation

$$\nabla^2\phi_m=-\rho_m$$

where

$$\rho_m=-\text{div }(\mathbf{M}_0+\mathbf{M})$$

We next examine the conditions to be imposed on the magnetic scalar potential at surfaces of discontinuous change in the physical properties of the medium which supports the field. In Sec. 7-1, we found that the magnetic induction and magnetic intensity must satisfy the conditions

$$\mathbf{n}\cdot(\mathbf{B}_2-\mathbf{B}_1)=0, \quad \mathbf{n}\times(\mathbf{H}_2-\mathbf{H}_1)=0 \tag{7-59}$$

across such a surface. The difference in the second of Eqs. (7-59) and (7-22) arises from the fact that in the static case where the only field sources are regions of permanent magnetization, no mechanism exists for the establishment of a surface current density. The first of (7-59) imposes the condition

$$\mathbf{n}\cdot\left[(\text{grad } \phi_m)_2-(\text{grad } \phi_m)_1\right]=\left(\frac{\partial\phi_m}{\partial\mathbf{n}}\right)_2-\left(\frac{\partial\phi_m}{\partial\mathbf{n}}\right)_1=-\omega_m$$

where

$$\omega_m=\mathbf{n}\cdot[(\mathbf{M}_0+\mathbf{M})_1-(\mathbf{M}_0+\mathbf{M})_2]$$

Similarly, the second of (7-59) is equivalent to

$$\mathbf{n}\times[(\text{grad } \phi_m)_2-(\text{grad } \phi_m)_1]=0$$

Thus, the analysis of the magnetostatic field generated by distributions of permanent magnetization is mathematically the same as that of the electromagnetic field with a distributed charge density

$$\rho_m=-\text{div }(\mathbf{M}_0+\mathbf{M})$$

and surface charge density

$$\omega_m=\mathbf{n}\cdot[\mathbf{M}_0+\mathbf{M})_1-(\mathbf{M}_0+\mathbf{M})_2]$$

across any surface separating two media with different characteristics. The complete analysis is still more complex than that for the corresponding electrostatic field, since the induced magnetization $\mathbf{M}$ is a function of the magnetic intensity **H**, and hence of the magnetic scalar potential. In many cases of practical interest, the regions containing magnetizable material are sufficiently separated so that the induced magnetization is very small compared with the permanent magnetization. Under these conditions, the magnetostatic boundary value problem can be solved by the methods of Sec. 7-4.

The general magnetostatic problem involves stationary distributions of electric current density as the sources of the field. In this case, the relevant Maxwell equations are

$$\text{curl } \mathbf{H} = \mathbf{J}, \quad \text{div } \mathbf{B} = 0$$

where the magnetic intensity and the magnetic induction are related in a homogeneous, isotropic, and linear medium by the constitutive equation

$$\mathbf{B} = \mu \mathbf{H}$$

Since the magnetic induction is solenoidal, the field may be derived from a vector potential **A**

$$\mathbf{B} = \text{curl } \mathbf{A}$$

In order to satisfy the first of the field equations, the vector potential must satisfy the differential equation

$$\text{curl curl } \mathbf{A} = \mu \mathbf{J} \tag{7-60}$$

For stationary fields, the Lorentz and Coulomb conditions are equivalent in a non-conducting medium. Under either of the corresponding gauge groups, the vector potential may be chosen to be solenoidal, and (7-60) can be written

$$-\text{curl curl } \mathbf{A} - \text{grad div } \mathbf{A} = \nabla^2 \mathbf{A} = -\mu \mathbf{J}$$

If J_i is a cartesian component of the current density, then A_i is the corresponding cartesian component of the vector potential which satisfies the scalar Poisson equation

$$\nabla^2 A_i = -\mu J_i \tag{7-61}$$

where the Laplacian operator is not restricted to its cartesian representation.

If the geometry of the physical situation permits a resolution of the current density **J**, and consequently the vector potential, into cartesian components, each component of the vector potential, satisfies the differential equation (7-61). In addition to the possibility of this restriction, let us further suppose that the current density **J** is spatially bounded. Under this latter restriction, we can immediately apply the analysis which led to Eq. (7-54), and show that, in an unbounded domain, each cartesian component of the vector potential is given by

$$A_i(\mathbf{r})=\frac{\mu}{4\pi}\int_V\frac{J_i(\mathbf{r}')}{|\mathbf{r}-\mathbf{r}'|}\,dv \tag{7-62}$$

where **r** is the radius vector to the point at which the vector potential is measured, **r**′ is the radius vector to the current element **J**, and V is any volume which contains all of the current density. Since we are assuming that the vector potential has been resolved into cartesian components, the three component equations of the form (7 62) can be combined into the single vector equation

$$\mathbf{A}(\mathbf{r})=\frac{\mu}{4\pi}\int_V\frac{\mathbf{J}(\mathbf{r}')}{|\mathbf{r}-\mathbf{r}'|}\,dv \tag{7-63}$$

Equation (7-63) can be verified by resolving the current density **J** and hence the vector potential into components relative to any orthogonal coordinate system. However, for non-cartesian systems, the demonstration is somewhat more complicated.

It is also possible to derive a set of vector identities similar to Green's identities, and make use of these to obtain a direct integration of Eq. (7-60). However, a detailed treatment of this technique would lead us too far afield, so that the interested reader is referred to any of the standard textbooks on electromagnetic theory.

Applying the curl operator to each sides of (7-63), we have

$$\mathbf{B}(\mathbf{r})=\text{curl}\,\mathbf{A}(\mathbf{r})=\frac{\mu}{4\pi}\int_V\text{curl}\,\frac{\mathbf{J}(\mathbf{r}')}{|\mathbf{r}-\mathbf{r}'|}\,dv$$

The intercharge of the order of differentiation and integration is permitted since the integration is with respect to primed coordinates, and the curl operator is with respect to unprimed coordinates. Carrying out the indicated differentiation, we have

$$\text{curl}\left(\frac{\mathbf{J}(\mathbf{r}')}{|\mathbf{r}-\mathbf{r}'|}\right)=-\frac{\mathbf{r}-\mathbf{r}'}{|\mathbf{r}-\mathbf{r}'|^3}\times\mathbf{J}(\mathbf{r}')+\frac{\text{curl}\,\mathbf{J}(\mathbf{r}')}{|\mathbf{r}-\mathbf{r}'|}$$

The last term is identically zero, since, **J**(**r**′) is independent of the unprimed coordinates. Hence,

$$\mathbf{B}(\mathbf{r})=\frac{\mu}{4\pi}\int_V\frac{\mathbf{J}(\mathbf{r}')\times(\mathbf{r}-\mathbf{r}')}{|\mathbf{r}-\mathbf{r}'|^3}\,dv \tag{7-64}$$

Since the integral in (7-64) is no more difficult to evaluate than the integral in (7-63), there is no inherent advantage in deriving the magnetic field due to a stationary current distribution from a vector potential. However, when the fields are explicitly time dependent, it is necessary to use the vector potential to obtain both the magnetic and electric fields.

PROBLEMS

7-1. Suppose that the electric field intensity $\mathbf{E}(\mathbf{r})$ due to a point charge q at the origin is

$$\mathbf{E}(\mathbf{r})=\frac{1}{4\pi\varepsilon}\cdot\frac{q\mathbf{r}}{|\mathbf{r}|^{3+\delta}}$$

where $\mathbf{r}$ is the position vector of the point at which the field is measured, and $\delta \ll 1$.

(i) Calculate div $\mathbf{E}$.

(ii) Calculate curl $\mathbf{E}$.

(Hint: Expand $|\mathbf{r}|^{-3-\delta}$ in a Taylor series about $\delta=0$.)

7-2. An electric charge Q is uniformly distributed throughout a sphere of radius r_0. Determine the electric field at all points.

7-3. An electric charge Q is uniformly distributed throughout a spherical shell of inner radius r_1 and outer radius r_2 ($r_1<r_2$). Determine the electrostatic field at all points. Compare this field with that of Prob. 7-2.

7-4. Use the time-independent form of Ampere's law to calculate the magnetic field a distance r from an infinitely long straight wire carrying a total steady current I.

7-5. Show that the free electric charge density in a medium with permittivity ϵ and conductivity σ, satisfies the differential equation

$$\frac{\partial\rho}{\partial t}=-\frac{\sigma}{\epsilon}\rho$$

If the free charge density has the value ρ_0 at time $t=0$, calculate the time required for this charge to decay to $1/e$ of its original value. This time is known as the relaxation time of the medium.

7-6. An electric dipole consists of two equal electric charges of opposite sign separated by a distance $\mathbf{l}$, where $\mathbf{l}$ is directed from $-q$ to $+q$. The dipole moment of an infinitesimal dipole is defined to be

$$\mathbf{p}=\lim_{\substack{q\to\infty\\ |\mathbf{l}|\to 0}} q\mathbf{l}$$

(i) Show that the electrostatic field produced by a dipole of moment $\mathbf{p}$ is

$$\mathbf{E}=\frac{1}{4\pi\epsilon_0}\,\text{grad}\,[\mathbf{p}\cdot\text{grad}\,(1/r)]$$

where r is the distance from the dipole to the point at which the field is measured.

(ii) If a dipole of moment $\mathbf{p}$ is placed in an electrostatic field with intensity $\mathbf{E}$, show that the torque $\mathbf{L}$ on the dipole is given by

$$\mathbf{L}=\mathbf{p}\times\mathbf{E}$$

7-7. Derive the differential equations satisfied by the scalar and vector potentials in a non-conducting medium with constant permeability μ_0 and permittivity $\epsilon(\mathbf{r})$ which is a function of the coordinates.

7-8. The Lorentz force on a charge q moving with velocity $\mathbf{v}$ in an electromagnetic field $(\mathbf{E}, \mathbf{B})$ is

$$\mathbf{F}=q\,(\mathbf{E}+\mathbf{v}\times\mathbf{B})$$

Determine the motion of a charged particle of mass m and charge q in a uniform, time independent field $(0, \mathbf{B})$.

7-9. Show that the electric intensity **E** and the magnetic induction **B** each satisfy a vector wave equation in a non-conducting, charge free medium. How are these equations modified if the medium has a conductivity σ?

7-10. Show that the Hertz potentials $\boldsymbol{\pi}$ and $\boldsymbol{\pi}^*$ satisfy the Lorentz condition identically.

7-11. Consider the electric Hertz potential

$$\boldsymbol{\pi}=\frac{\mathbf{p}_0}{4\pi\varepsilon_0}\frac{e^{i(kr-\omega t)}}{r}$$

where $\mathbf{p}_0$ is the constant magnitude of the electric dipole moment $\mathbf{p}_0\, e^{i\omega t}$.

(i) Determine the corresponding scalar and vector potentials in the Lorentz gauge.

(ii) Show that the corresponding electric field E is

$$\mathbf{E}=\frac{k^3}{4\pi\epsilon_0 r^2}\left\{\mathbf{r}(\mathbf{r}\cdot\mathbf{p}_0)\left[-\frac{1}{kr}-\frac{3i}{(kr)^2}+\frac{3}{(kr)^3}\right]+ r^2\mathbf{p}_0\left[\frac{1}{kr}+\frac{1}{(kr)^2}-\frac{1}{(kr)^3}\right]\right\} e^{i(kr-\omega t)}$$

7-12. Carry out the analysis similar to that of Example 3 for a magnetic dipole of moment $\mathbf{m}_0 e^{i\omega t}$.

7-13. Consider a spherical charge distribution with density $\rho(\mathbf{r})$ given by

$$\rho(r)=\begin{cases}\rho_0 r^2 & ; \quad 0\leqslant r\leqslant R/2\\ \rho_0(R-r)^2; & R/2\leqslant r\leqslant R\\ 0 & ; \quad r>R\end{cases}$$

If this charge distribution is in free space, calculate the electrostatic potential at all points.

7-15. An electrostatic charge Q is uniformly distributed throughout the volume of an ellipsoid of revolution with semi-axes a and b $(a>b)$.

(i) Calculate the dipole moment of this charge distribution.

(ii) Calculate the cartesian components of the quadrapole moment.

7-15. Show that within any closed region V bounded by a surface S, the electrostatic potential is unique whenever

(i) grad$'\,\phi(\mathbf{r}')\cdot\mathbf{n}=f(\mathbf{r}')$ where $f(\mathbf{r}')$ is a given function of $\mathbf{r}'$ on the bounding surface S:

(ii) $a\,\phi(\mathbf{r}')+b$ grad$'\,\phi(\mathbf{r}')\cdot\mathbf{n}=g(\mathbf{r}')$ where $g(\mathbf{r}')$ is a given function of $\mathbf{r}'$ on the bounding surface S, and a and b are given constants.

7-16. Consider a sphere of radius r_0 which is uniformly magnetized with a constant permanent magnetization $\mathbf{M}_0$.

(i) Show that the magnetic field inside the sphere is

$$\mathbf{H}=-\frac{1}{3}\,\mathbf{M}_0, \qquad \mathbf{B}=\frac{2}{3}\,\mu_0\mathbf{M}_0$$

The numerical factor in the expression for **H** is known as the demagnetizing factor.

(ii) Calculate the field outside the sphere.

7-17. An infinitely long circular cylinder of radius r has a permeability μ. It is placed in a uniform external magnetic field which is normal to the cylinder axis.

(i) Determine the **B** and **H** fields both inside and outside cylinder.

(ii) Calculate the induced magnetization in the cylinder, and the demagnetizing factor.

Appendix

Vector Relations in Curvilinear Coordinates

In this appendix, we shall list the basic vector relations for the eleven most important curvilinear coordinate systems in E_3. For the sake of completeness, we shall also include rectangular cartesian coordinates. We shall use the notation (u, v, w) for the curvilinear systems, with three exceptions: (i) In systems with cylindrical symmetry, the coordinate axis along the symmetry axis will be denoted by z; (ii) In the case of circular cylinder coordinates, the more familiar notation (ρ, φ, z) will be used; and (ii) Spherical polar coordinates will be denoted by the customary triplet (r, δ, φ). The unit vectors associated with the coordinate system are specified in terms of their cartesian representation, and all vector results are given in terms of physical components.

1. Rectangular Cartesian Coordinates

$$x = x \quad , \quad y = y \quad , \quad z = z$$

$$-\infty < x < \infty, \quad -\infty < y < \infty, \quad -\infty < z < \infty$$

$$h_x = h_y = h_z = 1$$

The constant coordinate surfaces are:

x=constant—planes

y=constant—planes

z=constant—planes

The infinitesimal elements of area in the coordinate surfaces are:

$$da_1 = dy\, dz,\ da_2 = dx\, dz,\ da_3 = dx\, dy$$

The infinitesimal element of volume is:

$$dV = dx\, dy\, dz$$

The unit vectors are:

$$\mathbf{i}_1 = \mathbf{i},\ \mathbf{i}_2 = \mathbf{j},\ \mathbf{i}_3 = \mathbf{k}$$

$$\text{grad } \phi = \frac{\partial \phi}{\partial x}\,\mathbf{i} + \frac{\partial \phi}{\partial y}\,\mathbf{j} + \frac{\partial \phi}{\partial z}\,\mathbf{k}$$

$$\text{div } \mathbf{A} = \frac{\partial A_1}{\partial x} + \frac{\partial A_2}{\partial y} + \frac{\partial A_3}{\partial z}$$

$$\text{curl } \mathbf{A} = \left(\frac{\partial A_3}{\partial y} - \frac{\partial A_2}{\partial z}\right)\mathbf{i}_1 + \left(\frac{\partial A_1}{\partial z} - \frac{\partial A_3}{\partial x}\right)\mathbf{i}_2 + \left(\frac{\partial A_2}{\partial x} - \frac{\partial A_1}{\partial y}\right)\mathbf{i}_3$$

$$\nabla^2 \phi = \frac{\partial^2 \phi}{\partial x^2} + \frac{\partial^2 \phi}{\partial y^2} + \frac{\partial^2 \phi}{\partial z^2}$$

2. Circular Cylinder Coordinates

$$x = \rho \cos \varphi, \; y = \rho \sin \varphi, \; z = z$$

$$0 \leqslant \rho < \infty, \; 0 < \varphi \leqslant 2\pi, \; -\infty < z < \infty$$

$$h_1 = 1, \; h_2 = \rho, \; h_3 = 1$$

The constant coordinate surface are:

ρ=constant—circular cylinders $\quad x^2 + y^2 = \rho^2$

φ=constant—planes through the cylinder axis $\quad y = x \tan \varphi$

z=constant—planes normal to the cylinder axis

The infinitesimal elements of area in the coordinate surfaces are:

$$da_1 = \rho\, d\varphi\, dz, \; da_2 = d\rho\, dz, \; da_3 = \rho\, d\rho\, d\varphi$$

The infinitesimal element of volume is:

$$dV = \rho\, d\rho\, d\varphi\, dz$$

The unit vectors are:

$$\mathbf{i}_1 = \cos \varphi\, \mathbf{i} + \sin \varphi\, \mathbf{j}$$
$$\mathbf{i}_2 = -\sin \varphi\, \mathbf{i} + \cos \varphi\, \mathbf{j}$$
$$\mathbf{i}_3 = \mathbf{k}$$

$$\text{grad } \phi = \frac{\partial \phi}{\partial \rho}\,\mathbf{i}_1 + \frac{1}{\rho}\frac{\partial \phi}{\partial \varphi}\,\mathbf{i}_2 + \frac{\partial \phi}{\partial z}\,\mathbf{i}_3$$

$$\text{div } \mathbf{A} = \frac{1}{\rho}\frac{\partial}{\partial \rho}(\rho A_1) + \frac{1}{\rho}\frac{\partial A_2}{\partial \varphi} + \frac{\partial A_3}{\partial z}$$

$$\text{curl } \mathbf{A} = \left(\frac{1}{\rho}\frac{\partial A_3}{\partial \varphi} - \frac{\partial A_2}{\partial z}\right)\mathbf{i}_1 + \left(\frac{\partial A_1}{\partial z} - \frac{\partial A_3}{\partial \rho}\right)\mathbf{i}_2 + \frac{1}{\rho}\left(\frac{\partial A_2}{\partial \rho} - \frac{\partial A_1}{\partial \varphi}\right)\mathbf{i}_3$$

$$\nabla^2 \phi = \frac{1}{\rho}\frac{\partial}{\partial \rho}\left(\rho \frac{\partial \phi}{\partial \rho}\right) + \frac{1}{\rho^2}\frac{\partial^2 \phi}{\partial \varphi^2} + \frac{\partial^2 \phi}{\partial z^2}$$

3. Elliptic Cylinder Coordinates

$$x = a\sqrt{(u^2-1)(1-v^2)}, \; y = a\,uv, \; z = z$$

$$1 \leqslant u < \infty, \; -1 \leqslant v \leqslant 1, \; -\infty < z \leqslant \infty$$

$$h_1=a\sqrt{\frac{u^2-v^2}{u^2-1}},\ h_2=a\sqrt{\frac{u^2-v^2}{1-v^2}},\ h_3=1$$

The constant coordinate surfaces are:

u=constant—elliptic cylinders $\frac{x^2}{u^2-1}+\frac{y^2}{u^2}=a^2$

v=constant—hyperbolic cylinders $\frac{y^2}{v^2}-\frac{x^2}{1-v^2}=a^2$

z=constant—planes normal to the cylinder axes

The infinitesimal elements of area in the coordinate surfaces are:

$$da_1=a\sqrt{\frac{u^2-v^2}{1-v^2}}\,dv\,dz$$

$$da_2=a\sqrt{\frac{u^2-v^2}{u^2-1}}\,du\,dz$$

$$da_3=\frac{a^2(u^2-v^2)}{\sqrt{(u^2-1)(1-v^2)}}\,du\,dz$$

The infinitesimal element of volume is:

$$dV=\frac{a^2(u^2-v^2)}{\sqrt{(u^2-1)(1-v^2)}}\,dudvdz$$

The unit vectors are:

$$\mathbf{i}_1=\sqrt{\frac{1-v^2}{u^2-v^2}}\,u\,\mathbf{i}+\sqrt{\frac{u^2-1}{u^2-v^2}}\,v\,\mathbf{j}$$

$$\mathbf{i}_2=-\sqrt{\frac{u^2-1}{u^2-v^2}}\,v\,\mathbf{i}+\sqrt{\frac{1-v^2}{u^2-v^2}}\,u\,\mathbf{j}$$

$$\mathbf{i}_3=\mathbf{k}$$

$$\text{grad } \phi=\frac{1}{a\sqrt{u^2-v^2}}\left(\sqrt{u^2-1}\,\frac{\partial\phi}{\partial u}\,\mathbf{i}_1+\sqrt{1-v^2}\,\frac{\partial\phi}{\partial v}\,\mathbf{i}_2+\frac{\partial\phi}{\partial z}\,\mathbf{i}_3\right)$$

$$\text{div } \mathbf{A}=\frac{1}{a(u^2-v^2)}\Bigg\{\sqrt{u^2-1}\,\frac{\partial}{\partial u}(\sqrt{u^2-v^2}\,A_1)+$$
$$\sqrt{1-v^2}\,\frac{\partial}{\partial v}(\sqrt{u^2-v^2}\,A_2)+$$
$$\frac{\sqrt{(u^2-1)(1-v^2)}}{a}\frac{\partial A_3}{\partial z}\Bigg\}$$

$$\text{curl } \mathbf{A}=\frac{1}{a\sqrt{u^2-v^2}}\left\{\left[\sqrt{1-v^2}\,\frac{\partial A_3}{\partial v}-a\sqrt{u^2-v^2}\,\frac{\partial A_2}{\partial z}\right]\mathbf{i}_1+\right.$$
$$\left[a\sqrt{u^2-v^2}\,\frac{\partial A_1}{\partial z}-\sqrt{u^2-1}\,\frac{\partial A_3}{\partial u}\right]\mathbf{i}_2+$$

$$\frac{1}{\sqrt{u^2-v^2}}\Big[\sqrt{u^2-1}\,\frac{\partial}{\partial u}\,(\sqrt{u^2-v^2}\,A_2)-$$
$$\sqrt{1-v^2}\,\frac{\partial}{\partial v}\,(\sqrt{u^2-v^2}\,A_1)\Big]\,\mathbf{i}_3\Big\}$$

$$\nabla^2\phi=\frac{1}{a^2(u^2-v^2)}\Big\{\sqrt{u^2-1}\,\frac{\partial}{\partial u}\left(\sqrt{u^2-1}\,\frac{\partial\phi}{\partial u}\right)+$$
$$\sqrt{1-v^2}\,\frac{\partial}{\partial v}\left(\sqrt{1-v^2}\,\frac{\partial\phi}{\partial v}\right)\Big\}+\frac{\partial^2\phi}{\partial z^2}$$

4. Parabolic Cylinder Coordinates

$$x=\frac{u-v}{2},\ y=\sqrt{u\,v},\ z=z$$

$$0\leqslant u<\infty,\ 0\leqslant v<\infty,\ -\infty<z<\infty$$

$$h_1=\tfrac{1}{2}\sqrt{\frac{u+v}{u}},\ h_2=\tfrac{1}{2}\sqrt{\frac{u+v}{v}},\ h_3=1$$

The constant coordinate surfaces are:

u=constant—parabolic cylinders $\quad y^2+2\,ux=u^2$

v=constant—parabolic cylinders $\quad y^2-2\,vx=v^2$

z=constant—planes normal to the cylinder axis

The infinitesimal elements of area in the coordinate surfaces are:

$$da_1=\tfrac{1}{2}\sqrt{\frac{u+v}{v}}\,dv\,dz$$

$$da_2=\tfrac{1}{2}\sqrt{\frac{u+v}{u}}\,du\,dz$$

$$da_3=\tfrac{1}{4}\,\frac{u+v}{\sqrt{uv}}\,du\,dv$$

The infinitesimal element of volume is:

$$dV=\tfrac{1}{4}\,\frac{u+v}{\sqrt{uv}}\,du\,dv\,dz$$

The unit vectors are:

$$\mathbf{i}_1=\sqrt{\frac{u}{u+v}}\,\mathbf{i}+\sqrt{\frac{v}{u+v}}\,\mathbf{j}$$

$$\mathbf{i}_2=-\sqrt{\frac{v}{u+v}}\,\mathbf{i}+\sqrt{\frac{u}{u+v}}\,\mathbf{j}$$

$$\mathbf{i}_3=\mathbf{k}$$

$$\text{grad}\,\phi=\frac{2}{\sqrt{u+v}}\left(\sqrt{u}\,\frac{\partial\phi}{\partial u}\,\mathbf{i}_1+\sqrt{v}\,\frac{\partial\phi}{\partial v}\,\mathbf{i}_2\right)+\frac{\partial\phi}{\partial z}\,\mathbf{i}_3$$

$$\text{div}\,\mathbf{A}=\frac{2}{u+v}\left\{\sqrt{u}\frac{\partial}{\partial u}(\sqrt{u+v}\,A_1)+\sqrt{v}\frac{\partial}{\partial v}(\sqrt{u+v}\,A_2)\right\}+\frac{\partial A_3}{\partial z}$$

$$\text{curl}\,\mathbf{A}=\frac{1}{\sqrt{u+v}}\left\{\left(2\sqrt{v}\frac{\partial A_3}{\partial v}-\sqrt{u+v}\frac{\partial A_2}{\partial z}\right)\mathbf{i}_1+\left(\sqrt{u+v}\frac{\partial A_1}{\partial z}-2\sqrt{u}\frac{\partial A_3}{\partial u}\right)\mathbf{i}_2\right\}+\frac{2}{u+v}\left[\sqrt{u}\frac{\partial}{\partial u}(\sqrt{u+v}\,A_2)-\sqrt{v}\frac{\partial}{\partial v}(\sqrt{u+v}\,A_1)\right]\mathbf{i}_3$$

$$\nabla^2\phi=\frac{4}{u+v}\left\{\sqrt{u}\frac{\partial}{\partial u}\left(\sqrt{u}\frac{\partial\phi}{\partial u}\right)+\sqrt{v}\frac{\partial}{\partial v}\left(\sqrt{v}\frac{\partial\phi}{\partial v}\right)\right\}+\frac{\partial^2\phi}{\partial z^2}$$

5. Spherical Polar Coordinates

$$x=r\sin\theta\cos\varphi,\ y=r\sin\theta\sin\varphi,\ z=r\cos\theta$$

$$0\leqslant r<\infty,\ 0\leqslant\theta<\pi,\ 0\leqslant\varphi<2\pi$$

$$h_1=1,\ h_2=r,\ h_3=r\sin\theta$$

The constant coordinate surfaces are:

r=constant—spheres $\quad x^2+y^2+z^2=r^2$

θ=constant—cones with central angle 2θ

φ=constant—azimuthal planes

The infinitesimal elements of area in the coordinate surfaces are:

$$da_1=r^2\sin\theta\,d\theta\,d\varphi$$

$$da_2=r\sin\theta\,dr\,d\varphi$$

$$da_3=r\,dr\,d\theta$$

The infinitesimal element of volume is:

$$dV=r^2\sin\theta\,dr\,d\theta\,d\varphi$$

The unit vectors are:

$$\mathbf{i}_1=\sin\theta\cos\varphi\,\mathbf{i}+\sin\theta\sin\varphi\,\mathbf{j}+\cos\theta\,\mathbf{k}$$

$$\mathbf{i}_2=r\cos\theta\cos\varphi\,\mathbf{i}+r\cos\theta\sin\varphi\,\mathbf{j}-r\sin\theta\,\mathbf{k}$$

$$\mathbf{i}_3=-r\sin\theta\sin\varphi\,\mathbf{i}+r\sin\theta\cos\varphi\,\mathbf{j}$$

$$\text{grad}\,\phi=\frac{\partial\phi}{\partial r}\mathbf{i}_1+\frac{1}{r}\frac{\partial\phi}{\partial\theta}\mathbf{i}_2+\frac{1}{r\sin\theta}\frac{\partial\phi}{\partial\varphi}\mathbf{i}_3$$

$$\text{div}\,\mathbf{A}=\frac{1}{r^2}\frac{\partial}{\partial r}(r^2A_1)+\frac{1}{r\sin\theta}\frac{\partial}{\partial\theta}(\sin\theta\,A_2)+\frac{1}{r\sin\theta}\frac{\partial A_3}{\partial\varphi}$$

$$\text{curl}\,\mathbf{A}=\frac{1}{r}\left\{\left[\frac{1}{\sin\theta}\frac{\partial}{\partial\theta}(\sin\theta\,A_3)-\frac{\partial A_2}{\partial\varphi}\right]\mathbf{i}_1+\left[\frac{1}{\sin\theta}\frac{\partial A_1}{\partial\varphi}-\frac{\partial}{\partial r}(r\,A_3)\right]\mathbf{i}_2+\left[\frac{\partial}{\partial r}(r\,A_2)-\frac{\partial A_1}{\partial\theta}\right]\mathbf{i}_3\right\}$$

$$\nabla^2\phi=\frac{1}{r^2}\frac{\partial}{\partial r}\left(r^2\frac{\partial\phi}{\partial r}\right)+\frac{1}{r^2\sin\theta}\frac{\partial}{\partial\theta}\left(\sin\theta\frac{\partial\phi}{\partial\theta}\right)+\frac{1}{r^2\sin^2\theta}\frac{\partial^2\phi}{\partial\varphi^2}$$

6. Conical Coordinates

$$x=\frac{u\,v\,w}{b\,c},\quad y=\frac{u}{b}\sqrt{\frac{(v^2-b^2)\,(w^2-b^2)}{b^2-c^2}}$$

$$z=\frac{u}{c}\sqrt{\frac{(v^2-c^2)\,(w^2-c^2)}{c^2-b^2}}$$

$$0\leqslant u<\infty,\quad 0\leqslant w^2\leqslant b^2\leqslant v^2\leqslant c^2$$

$$h_1=1,\quad h_2=u\sqrt{\frac{v^2-w^2}{(v^2-b^2)(c^2-v^2)}},$$

$$h=u\sqrt{\frac{v^2-w^2}{(b^2-w^2)(c^2-w^2)}}$$

The constant coordinate surfaces are:

u=constant—spheres $\quad x^2+y^2+z^2=u^2$

v=constant—elliptic cones $\quad \dfrac{x^2}{v^2}+\dfrac{y^2}{v^2-b^2}+\dfrac{z^2}{v^2-c^2}=0$

w=constant—elliptic cones $\quad \dfrac{x^2}{w^3}+\dfrac{y^2}{w^2-b^2}+\dfrac{z^2}{w^2-c^2}=0$

The infinitesimal elements of area in the coordinate surfaces are:

$$da_1=\frac{u^2(v^2-w^2)}{\sqrt{(v^2-b^2)(b^2-w^2)(c^2-v^2)(c^2-w^2)}}\,dv\,dw$$

$$da_2=u\sqrt{\frac{v^2-w^2}{(b^2-w^2)(c^2-w^2)}}\,du\,dw$$

$$da_3=u\sqrt{\frac{v^2-w^2}{(v^2-b^2)(c^2-v^2)}}\,du\,dv$$

The infinitesimal element of volume is:

$$dV=\frac{u^2(v^2-w^2)}{\sqrt{(v^2-b^2)(b^2-w^2)(c^2-v^2)(c^2-w^2)}}\,du\,dv\,dw$$

The unit vectors are:

$$\mathbf{i}_1=\frac{vw}{bc}\,\mathbf{i}+\frac{1}{b}\sqrt{\frac{(v^2-b^2)(w^2-b^2)}{b^2-c^2}}\,\mathbf{j}+\frac{1}{c}\sqrt{\frac{(v^2-c^2)(w^2-c^2)}{b^2-c^2}}\,\mathbf{k}$$

$$\mathbf{i}_2=\frac{w}{bc}\sqrt{\frac{(v^2-b^2)(c^2-v^2)}{v^2-w^2}}\,\mathbf{i}+\frac{v}{b}\sqrt{\frac{(c^2-w^2)(w^2-b^2)}{(b^2-c^2)(v^2-w^2)}}\,\mathbf{j}+$$

$$\frac{w}{c}\sqrt{\frac{(c^2-v^2)\,(c^2-w^2)}{(c^2-b^2)(v^2-w^2)}}\,\mathbf{k}$$

$$\mathbf{i}_3=\frac{v}{bc}\sqrt{\frac{(b^2-w^2)(c^2-w^2)}{v^2-w^2}}\,\mathbf{i}+\frac{w}{b}\sqrt{\frac{(w^2-c^2)(v^2-b^2)}{(b^2-c^2)(v^2-w^2)}}\,\mathbf{j}+\frac{w}{c}\sqrt{\frac{(c^2-v^2)(b^2-w^2)}{(c^2-b^2)(v^2-w^2)}}\,\mathbf{k}$$

$$\text{grad } \phi=\frac{\partial\phi}{\partial u}\mathbf{i}_1+\frac{1}{u}\sqrt{\frac{(v^2-b^2)(c^2-w^2)}{v^2-w^2}}\frac{\partial\phi}{\partial v}\mathbf{i}_2+\frac{1}{u}\sqrt{\frac{(b^2-w^2)(c^2-w^2)}{v^2-w^2}}\frac{\partial\phi}{\partial w}\mathbf{i}_3$$

$$\text{div } \mathbf{A}=\frac{1}{u^2}\frac{\partial}{\partial u}(u^2A_1)+\frac{\sqrt{(v^2-b^2)(c^2-v^2)}}{u(v^2-w^2)}\frac{\partial}{\partial v}\left(\sqrt{v^2-w^2}\,A_2\right)+\frac{\sqrt{(b^2-w^2)(c^2-w^2)}}{u(v^2-w^2)}\frac{\partial}{\partial w}\left(\sqrt{v^2-w^2}\,A_3\right)$$

$$\text{curl } \mathbf{A}=\left[\frac{\sqrt{(v^2-b^2)(c^2-v^2)}}{u(v^2-w^2)}\frac{\partial}{\partial v}\left(\sqrt{v^2-w^2}\,A_3\right)-\frac{\sqrt{(b^2-w^2)(c^2-w^2)}}{u(v^2-w^2)}\frac{\partial}{\partial w}\left(\sqrt{v^2-w^2}\,A_2\right)\right]\mathbf{i}_1+\left[\frac{1}{u}\frac{\sqrt{(b^2-w^2)(c^2-w^2)}}{v^2-w^2}\frac{\partial A_1}{\partial w}-\frac{1}{u}\frac{\partial}{\partial u}(u\,A_3)\right]\mathbf{i}_2+\left[\frac{1}{u}\frac{\partial}{\partial u}(u\,A_2)-\frac{1}{u}\frac{\sqrt{(v^2-b^2)(c^2-v^2)}}{v^2-w^2}\frac{\partial A_1}{\partial v}\right]\mathbf{i}_3$$

$$\nabla^2\phi=\frac{1}{u^2}\frac{\partial}{\partial u}\left(u^2\frac{\partial\phi}{\partial u}\right)+\frac{\sqrt{(v^2-b^2)(c^2-v^2)}}{u^2(v^2-w^2)}\frac{\partial}{\partial v}\left(\sqrt{(v^2-b^2)(c^2-v^2)}\frac{\partial\phi}{\partial v}\right)+\frac{\sqrt{(b^2-w^2)(c^2-w^2)}}{u^2(v^2-w^2)}\frac{\partial}{\partial w}\left(\sqrt{(b^2-w^2)(c^2-w^2)}\frac{\partial\phi}{\partial w}\right)$$

7. Parabolic Coordinates

$$x=\sqrt{uv}\cos w,\quad y=\sqrt{uv}\sin w,\quad z=\frac{u-v}{2}$$

$$0\leqslant u<\infty,\quad 0\leqslant v<\infty,\quad 0\leqslant w<2\pi$$

$$h_1=\tfrac{1}{2}\sqrt{\frac{u+v}{u}},\quad h_2=\tfrac{1}{2}\sqrt{\frac{u+v}{v}},\quad h_3=\sqrt{uv}$$

The constant coordinate surfaces are:

u=constant—paraboloids $x^2+y^2+2uz=u^2$

v=constant—paraboloids $x^2+y^2-2vz=v^2$

w=constant—azimuthal planes

The infinitesimal elements of area in the coordinate planes are

$$da_1=\tfrac{1}{2}\sqrt{u(u+v)}\;dv\;dw$$

$$da_2=\tfrac{1}{2}\sqrt{v(u+v)}\;du\;dw$$

$$da_3=\frac{u+v}{4\sqrt{uv}}\,du\;dv$$

The infinitesimal element of volume is:

$$dV=\frac{u+v}{4}\,du\;dv\;dw$$

The unit vectors are:

$$\mathbf{i}_1=\sqrt{\frac{v}{u+v}}\cos w\;\mathbf{i}+\sqrt{\frac{v}{u+v}}\sin w\;\mathbf{j}+\sqrt{\frac{u}{u+v}}\;\mathbf{k}$$

$$\mathbf{i}_2=\sqrt{\frac{u}{u+v}}\cos w\;\mathbf{i}+\sqrt{\frac{u}{u+v}}\sin w\;\mathbf{j}-\sqrt{\frac{v}{u+v}}\;\mathbf{k}$$

$$\mathbf{i}_3=-\sin w\;\mathbf{i}+\cos w\;\mathbf{j}$$

$$\text{grad}\;\phi=2\sqrt{\frac{u}{u+v}}\frac{\partial\phi}{\partial u}\mathbf{i}_1+2\sqrt{\frac{v}{u+v}}\frac{\partial\phi}{\partial v}\mathbf{i}_2+\frac{1}{\sqrt{uv}}\frac{\partial\phi}{\partial w}\mathbf{i}_3$$

$$\text{div}\;\mathbf{A}=\frac{1}{u+v}\left[\frac{\partial}{\partial u}\left(\sqrt{u(u+v)}\;A_1\right)+\frac{\partial}{\partial v}\left(\sqrt{v(u+v)}\;A_2\right)\right]+\frac{1}{\sqrt{uv}}\frac{\partial A_3}{\partial w}$$

$$\text{curl}\;\mathbf{A}=\left[\frac{2}{\sqrt{u+v}}\frac{\partial}{\partial v}\left(\sqrt{v}\,A_3\right)-\frac{1}{\sqrt{uv}}\frac{\partial A_2}{\partial w}\right]\mathbf{i}_1+$$

$$\left[\frac{1}{\sqrt{uv}}\frac{\partial A_1}{\partial w}-\frac{2}{\sqrt{u+v}}\frac{\partial}{\partial u}\left(\sqrt{u}\,A_3\right)\right]\mathbf{i}_2+$$

$$\left[2\sqrt{\frac{u}{u+v}}\frac{\partial}{\partial u}\left(\sqrt{u+v}\,A_2\right)-2\sqrt{\frac{v}{u+v}}\frac{\partial}{\partial v}\left(\sqrt{u+v}\,A_1\right)\right]\mathbf{i}_3$$

$$\nabla^2\phi=\frac{4}{u+v}\left[\frac{\partial}{\partial u}\left(u\frac{\partial\phi}{\partial u}\right)+\frac{\partial}{\partial v}\left(v\frac{\partial\phi}{\partial v}\right)\right]+\frac{1}{uv}\frac{\partial^2\phi}{\partial w^2}$$

8. Prolate Spheroidal Coordinates

$$x=a\sqrt{(u^2-1)(1-v^2)}\cos w,\quad y=a\sqrt{(u^2-1)(1-v^2)}\sin w$$

$$z=auv$$

$$1\leqslant u<\infty,\quad -1\leqslant v\leqslant 1,\quad 0\leqslant w<2\pi$$

$$h_1=a\sqrt{\frac{u^2-v^2}{u^2-1}},\quad h_2=a\sqrt{\frac{u^2-v^2}{1-v^2}}$$

$$h_3=a\sqrt{(u^2-1)(1-v^2)}$$

The constant coordinate surfaces are:

u=constant—ellipsoids $\quad \dfrac{x^2+y^2}{u^2-1}+\dfrac{z^2}{u^2}=a^2$

v=constant—hyperboloids $\quad \dfrac{z^2}{v^2}-\dfrac{x^2+y^2}{1-v^2}=a^2$

w=constant—azimuthal planes

The infinitesimal elements of area in the coordinate planes are:

$$da_1=a^2\sqrt{(u^2-1)(u^2-v^2)}\,dv\,dw$$

$$da_2=a^2\sqrt{(1-v^2)(u^2-v^2)}\,du\,dw$$

$$da_3=\frac{a^2(u^2-v^2)}{\sqrt{(u^2-1)(1-v^2)}}\,du\,dv$$

The infinitesimal element of volume is:

$$dV=a^3(u^2-v^2)\,du\,dv\,dw$$

The unit vectors are:

$$\mathbf{i}_1=u\sqrt{\frac{1-v^2}{u^2-v^2}}\cos w\,\mathbf{i}+u\sqrt{\frac{1-v^2}{u^2-v^2}}\sin w\,\mathbf{j}+v\sqrt{\frac{u^2-1}{u^2-v^2}}\,\mathbf{k}$$

$$\mathbf{i}_2=v\sqrt{\frac{u^2-1}{u^2-v^2}}\cos w\,\mathbf{i}-v\sqrt{\frac{u^2-1}{u^2-v^2}}\sin w\,\mathbf{j}+u\sqrt{\frac{1-v^2}{u^2-v^2}}\,\mathbf{k}$$

$$\mathbf{i}_3=-\sin w\,\mathbf{i}+\cos w\,\mathbf{j}$$

$$\text{grad }\phi=\frac{1}{a}\sqrt{\frac{u^2-1}{u^2-v^2}}\frac{\partial\phi}{\partial u}\mathbf{i}_1+\frac{1}{a}\sqrt{\frac{1-v^2}{u^2-v^2}}\frac{\partial\phi}{\partial v}\mathbf{i}_2+\frac{1}{a\sqrt{(u^2-1)(1-v^2)}}\frac{\partial\phi}{\partial w}\mathbf{i}_3$$

$$\text{div }\mathbf{A}=\frac{1}{a(u^2-v^2)}\left\{\frac{\partial}{\partial u}\left[\sqrt{(u^2-1)(u^2-v^2)}\,A_1\right]+\frac{\partial}{\partial v}\left[\sqrt{(1-v^2)(u^2-v^2)}\,A_2\right]\right\}+\frac{1}{a\sqrt{(u^2-1)(1-v^2)}}\frac{\partial A_3}{\partial w}$$

$$\text{curl }\mathbf{A}=\left[\frac{1}{a\sqrt{u^2-v^2}}\frac{\partial}{\partial v}\left(\sqrt{1-v^2}\,A_3\right)-\frac{1}{a\sqrt{(u^2-1)(1-v^2)}}\frac{\partial A_2}{\partial w}\right]\mathbf{i}_1+$$
$$\left[\frac{1}{a\sqrt{(u^2-1)(1-v^2)}}\frac{\partial A_1}{\partial w}-\frac{1}{a\sqrt{u^2-v^2}}\frac{\partial}{\partial u}\left(\sqrt{u^2-1}\,A_3\right)\right]\mathbf{i}_2+$$
$$\left[\frac{\sqrt{u^2-1}}{a(u^2-v^2)}\frac{\partial}{\partial u}\left(\sqrt{u^2-v^2}\,A_2\right)-\frac{\sqrt{1-v^2}}{a(u^2-v^2)}\frac{\partial}{\partial v}\left(\sqrt{u^2-v^2}\,A_1\right)\right]\mathbf{i}_3$$

$$\nabla^2\phi=\frac{1}{a^2(u^2-v^2)}\left\{\frac{\partial}{\partial u}\left[(u^2-1)\frac{\partial\phi}{\partial u}\right]+\frac{\partial}{\partial v}\left[(1-v^2)\frac{\partial\phi}{\partial v}\right]\right\}+$$

$$\frac{1}{a^2(u^2-1)(1-v^2)}\frac{\partial^2\phi}{\partial w^2}$$

9. Oblate Spheroidal Coordinates

$$x=a\,uv\cos w,\quad y=a\,uv\sin w,\quad z=a\sqrt{(u^2-1)(1-v^2)}$$

$$1\leqslant u<\infty,\quad 0\leqslant v<1,\quad 0\leqslant w<2\pi$$

$$h_1=a\sqrt{\frac{u^2-v^2}{u^2-1}},\quad h_2=a\sqrt{\frac{u^2-v^2}{1-v^2}},\quad h_3=a\,uv$$

The constant coordinate surfaces are:

u=constant—ellipsoids $\qquad \dfrac{x^2+y^2}{u^2}+\dfrac{z^2}{u^2-1}=a^2$

v=constant—hyperboloids $\qquad \dfrac{x^2+y^2}{v^2}-\dfrac{z^2}{1-v^2}=a^2$

w=constant—azimuthal planes

The infinitesimal elements of area in the coordinate planes are:

$$da_1=a^2\,uv\sqrt{\frac{u^2-v^2}{1-v^2}}\,dv\,dw$$

$$da_2=a^2\,uv\sqrt{\frac{u^2-v^2}{u^2-1}}\,du\,dw$$

$$da_3=\frac{a^2(u^2-v^2)}{\sqrt{(u^2-1)(1-v^2)}}\,du\,dv$$

The infinitesimal element of volume is:

$$dV=\frac{a^3(u^2-v^2)uv}{\sqrt{(u^2-1)(1-v^2)}}\,du\,dv\,dw$$

The unit vectors are:

$$\mathbf{i}_1=v\sqrt{\frac{u^2-1}{u^2-v^2}}\cos w\,\mathbf{i}+v\sqrt{\frac{u^2-1}{u^2-v^2}}\sin w\,\mathbf{j}+u\sqrt{\frac{1-v^2}{u^2-v^2}}\,\mathbf{k}$$

$$\mathbf{i}_2=u\sqrt{\frac{1-v^2}{u^2-v^2}}\cos w\,\mathbf{i}+u\sqrt{\frac{1-v^2}{u^2-v^2}}\sin w\,\mathbf{j}-v\sqrt{\frac{u^2-1}{u^2-v^2}}\,\mathbf{k}$$

$$\mathbf{i}_3=-\sin w\,\mathbf{i}+\cos w\,\mathbf{j}$$

$$\text{grad}\,\phi=\frac{1}{a}\left\{\sqrt{\frac{u^2-1}{u^2-v^2}}\frac{\partial\phi}{\partial u}\mathbf{i}_1+\sqrt{\frac{1-v^2}{u^2-v^2}}\frac{\partial\phi}{\partial v}\mathbf{i}_2+\frac{1}{uv}\frac{\partial\phi}{\partial w}\mathbf{i}_3\right\}$$

$$\text{div}\,\mathbf{A}=\frac{1}{a(u^2-v^2)}\left\{\frac{\sqrt{u^2-1}}{u}\frac{\partial}{\partial u}\left(u\sqrt{u^2-v^2}\,A_1\right)+\right.$$

$$\frac{\sqrt{1-v^2}}{v}\frac{\partial}{\partial v}\left(v\sqrt{u^2-v^2}\,A_2\right)\Bigg\}+\frac{1}{auv}\frac{\partial A^3}{\partial w}$$

$$\text{curl }\mathbf{A}=\frac{1}{a^2uv}\sqrt{\frac{1-v^2}{u^2-v^2}}\left[au\frac{\partial}{\partial v}(v\,A_3)-a\sqrt{\frac{u^2-v^2}{1-v^2}}\frac{\partial A_2}{\partial w}\right]\mathbf{i}_1+$$

$$\frac{1}{a^2uv}\sqrt{\frac{u^2-1}{u^2-v^2}}\left[a\sqrt{\frac{u^2-v^2}{u^2-1}}\frac{\partial A_1}{\partial w}-av\frac{\partial}{\partial u}(u\,A_3)\right]\mathbf{i}_2+$$

$$\frac{\sqrt{(u^2-1)(1-v^2)}}{a^2(u^2-v^2)}\left[\frac{a}{\sqrt{1-v^2}}\frac{\partial}{\partial u}\left(\sqrt{u^2-v^2}\,A_2\right)-\right.$$

$$\left.\frac{a}{\sqrt{u^2-1}}\frac{\partial}{\partial v}\left(\sqrt{u^2-v^2}\,A_1\right)\right]\mathbf{i}_3$$

$$\nabla^2\phi=\frac{1}{a^2(u^2-v^2)}\left\{\frac{\sqrt{u^2-1}}{u}\frac{\partial}{\partial u}\left(u\sqrt{u^2-1}\frac{\partial\phi}{\partial u}\right)+\right.$$

$$\left.\frac{\sqrt{1-v^2}}{v}\frac{\partial}{\partial v}\left(v\sqrt{1-v^2}\frac{\partial\phi}{\partial v}\right)\right\}+\frac{1}{a^2u^2v^2}\frac{\partial^2\phi}{\partial w^2}$$

10. Ellipsoidal Coordinates

$$x=\sqrt{\frac{(a^2-u)(a^2-v)(a^2-w)}{(a^2-b^2)(a^2-c^2)}}$$

$$y=\sqrt{\frac{(b^2-u)(b^2-v)(b^2-w)}{(b^2-c^2)(a^2-b^2)}}$$

$$z=\sqrt{\frac{(c^2-u)(c^2-v)(c^2-w)}{(a^2-c^2)(b^2-c^2)}}$$

$$-\infty<u\leqslant c^2\leqslant v\leqslant b^2\leqslant w\leqslant a^2$$

$$h_1=\tfrac{1}{2}\sqrt{\frac{(v-u)(w-u)}{(a^2-u)(b^2-u)(c^2-u)}}$$

$$h_2=\tfrac{1}{2}\sqrt{\frac{(u-v)(w-v)}{(a^2-v)(b^2-v)(c^2-v)}}$$

$$h_3=\tfrac{1}{2}\sqrt{\frac{(u-w)(v-w)}{(a^2-w)(b^2-w)(c^2-w)}}$$

The constant coordinate surfaces are:

u=constant—ellipsoids $\quad \dfrac{x^2}{a^2-u}+\dfrac{y^2}{b^2-u}+\dfrac{z^2}{c^2-u}=1$

v=constant—hyperboloids of one sheet $\quad \dfrac{x^2}{a^2-v}+\dfrac{y^2}{b^2-v}+\dfrac{z^2}{c^2-v}=1$

w=constant—hyperboloids of two sheets $\quad \dfrac{x^2}{a^2-w}+\dfrac{y^2}{b^2-w}+\dfrac{z^2}{c^2-w}=1$

The infinitesimal elements of area in the coordinate surfaces are:

$$da_1=\tfrac{1}{4}\sqrt{\frac{(u-v)(w-u)(w-v)^2}{(a^2-v)(a^2-w)(b^2-v)(b^2-w)(c^2-v)(c^2-w)}}\,dv\,dw$$

$$da_2=\tfrac{1}{4}\sqrt{\frac{(u-v)(v-w)(u-w)^2}{(a^2-u)(a^2-w)(b^2-u)(b^2-w)(c^2-u)(c^2-w)}}\,du\,dw$$

$$da_3=\tfrac{1}{4}\sqrt{\frac{(u-w)(w-v)(u-v)^2}{(a^2-u)(a^2-v)(b^2-u)(b^2-v)(c^2-u)(c^2-v)}}\,du\,dv$$

The infinitesimal element of volume is;

$$dV=\tfrac{1}{8}\,\frac{u-v}{\sqrt{(a^2-u)(a^2-v)(a^2-w)}}\,\frac{v-w}{\sqrt{(b^2-u)(b^2-v)(w-b^2)}}\times$$

$$\frac{w-u}{\sqrt{(c^2-u)(v-c^2)(w-c^2)}}\,du\,dv\,dw$$

The unit vectors are:

$$\mathbf{i}_1=\sqrt{\frac{(b^2-u)\,(c^2-u)\,(a^2-v)\,(a^2-w)}{(v-u)\,(w-u)\,(a^2-b^2)\,(a^2-c^2)}}\,\mathbf{i}+$$

$$\sqrt{\frac{(a^2-u)\,(c^2-u)\,(b^2-v)\,(b^2-w)}{(v-u)\,(w-u)\,(b^2-c^2)\,(a^2-b^2)}}\,\mathbf{j}+$$

$$\sqrt{\frac{(a^2-u)\,(b^2-u)\,(c^2-v)\,(c^2-w)}{(v-u)\,(w-u)\,(a^2-c^2)\,(b^2-c^2)}}\,\mathbf{k}$$

$$\mathbf{i}_2=\sqrt{\frac{(b^2-v)\,(c^2-v)\,(a^2-u)\,(a^2-w)}{(u-v)\,(w-v)\,(a^2-b^2)\,(a^2-c^2)}}\,\mathbf{i}+$$

$$\sqrt{\frac{(a^2-v)\,(c^2-v)\,(b^2-u)\,(b^2-w)}{(u-v)\,(w-v)\,(a^2-b^2)\,(b^2-c^2)}}\,\mathbf{j}+$$

$$\sqrt{\frac{(a^2-v)\,(b^2-v)\,(c^2-u)\,(c^2-w)}{(u-v)\,(w-v)\,(a^2-c^2)\,(b^2-c^2)}}\,\mathbf{k}$$

$$\mathbf{i}_3=\sqrt{\frac{(b^2-w)\,(c^2-w)\,(a^2-u)\,(a^2-v)}{(u-w)\,(v-w)\,(a^2-b^2)\,(a^2-c^2)}}\,\mathbf{i}+$$

$$\sqrt{\frac{(a^2-w)\,(c^2-w)\,(b^2-u)\,(b^2-v)}{(u-w)\,(v-w)\,(b^2-c^2)\,(a^2-b^2)}}\,\mathbf{j}+$$

$$\sqrt{\frac{(a^2-w)\,(b^2-w)\,(c^2-u)\,(c^2-v)}{(u-w)\,(v-w)\,(a^2-c^2)\,(b^2-c^2)}}\,\mathbf{k}$$

$$\text{grad}\,\phi=2\,\frac{\sqrt{(a^2-u)\,(b^2-u)\,(c^2-u)}}{(v-u)\,(w-u)}\,\frac{\partial\phi}{\partial u}\,\mathbf{i}_1+$$

$$2\,\frac{\sqrt{(a^2-v)\,(b^2-v)\,(c^2-v)}}{(u-v)\,(w-v)}\,\frac{\partial\phi}{\partial v}\,\mathbf{i}_2+$$

$$2\,\frac{\sqrt{(a^2-w)\,(b^2-w)\,(c^2-w)}}{(u-w)\,(v-w)}\,\frac{\partial\phi}{\partial w}\,\mathbf{i}_3$$

$$\operatorname{div}\mathbf{A}=2\,\frac{\sqrt{(a^2-u)\,(b^2-u)\,(c^2-u)}}{(v-u)\,(w-u)}\,\frac{\partial}{\partial u}\,[\sqrt{(v-u)\,(u-w)}\;A_1]+$$
$$2\,\frac{\sqrt{(a^2-v)\,(b^2-v)\,(c^2-v)}}{(u-v)\,(v-w)}\,\frac{\partial}{\partial v}\,[\sqrt{(v-u)\,(w-v)}\;A_2]+$$
$$2\,\frac{\sqrt{(a^2-w)\,(b^2-w)\,(c^2-w)}}{(v-w)\,(w-u)}\,\frac{\partial}{\partial w}\,[\sqrt{(w-v)\,(u-w)}\;A_3]$$

$$\operatorname{curl}\mathbf{A}=\frac{2}{v-w}\left\{\sqrt{\frac{(a^2-v)\,(b^2-v)\,(c^2-v)}{v-u}}\,\frac{\partial}{\partial v}\,(\sqrt{v-w}\;A_3)-\right.$$
$$\left.\sqrt{\frac{(a^2-w)\,(b^2-w)\,(c^2-w)}{w-u}}\,\frac{\partial}{\partial w}\,(\sqrt{w-v}\;A_2)\right\}\mathbf{i}_1+$$
$$\frac{2}{w-u}\left\{\sqrt{\frac{(a^2-w)\,(b^2-w)\,(c^2-w)}{w-v}}\,\frac{\partial}{\partial w}\,(\sqrt{w-u}\;A_1)-\right.$$
$$\left.\sqrt{\frac{(a^2-u)\,(b^2-u)\,(c^2-u)}{u-v}}\,\frac{\partial}{\partial u}\,(\sqrt{u-w}\;A_3)\right\}\mathbf{i}_2+$$
$$\frac{2}{u-v}\left\{\sqrt{\frac{(a^2-u)\,(b^2-u)\,(c^2-u)}{u-w}}\,\frac{\partial}{u}\,(\sqrt{u-v}\;A_2)-\right.$$
$$\left.\sqrt{\frac{(a^2-v)\,(b^2-v)\,(c^2-v)}{v-w}}\,\frac{\partial}{\partial v}\,(\sqrt{v-u}\;A_1)\right\}\mathbf{i}_3$$

$$\nabla^2\phi=\frac{4\sqrt{(a^2-u)\,(b^2-u)\,(c^2-u)}}{(v-u)\,(w-u)}\,\frac{\partial}{\partial u}\left[\sqrt{(a^2-u)\;(b^2-u)\,(c^2-u)}\,\frac{\partial\phi}{\partial u}\right]+$$
$$\frac{4\sqrt{(a^2-v)\,(b^2-v)\,(c^2-v)}}{(u-v)\,(w-v)}\,\frac{\partial}{\partial v}\left[\sqrt{(a^2-v)\;(b^2-v)\;(c^2-v)}\,\frac{\partial\phi}{\partial v}\right]+$$
$$\frac{4\sqrt{(a^2-w)\,(b^2-w)\,(c^2-w)}}{(u-w)\,(v-w)}\,\frac{\partial}{\partial w}\left[\sqrt{(a^2-w)\,(b^2-w)\,(c^2-w)}\,\frac{\partial\phi}{\partial w}\right]$$

11. Paraboloidal Coordinates

$$x=\frac{1}{2}\,(u+v+w-a-b),\; y=\sqrt{\frac{(a-u)\;(a-v)\,(a-w)}{b-a}}$$

$$z=\sqrt{\frac{(b-u)\,(b-v)\,(b-w)}{a-b}}$$

$$-\infty<w\leqslant a\leqslant v\leqslant b\leqslant u<\infty$$

$$h_1=\tfrac{1}{2}\sqrt{\frac{(u-v)\,(u-w)}{(a-u)\;(b-u)}}$$

$$h_2=\tfrac{1}{2}\sqrt{\frac{(v-u)\,(v-w)}{(a-v)\,(b-v)}}$$

$$h_3=\tfrac{1}{2}\sqrt{\frac{(w-u)\,(w-v)}{(a-w)\,(b-w)}}$$

The constant coordinate surfaces are:

u=constant—paraboloids $\quad \frac{2x}{u}+\frac{y^2}{u(u-a)}+\frac{z^2}{u(u-b)}=1$

v=constant—paraboloids $\quad \frac{2x}{v}+\frac{y^2}{v(v-a)}+\frac{z^2}{v(v-b)}=1$

w=constant—paraboloids $\quad \frac{2x}{w}+\frac{y^2}{w(w-a)}+\frac{z^2}{w(w-b)}=1$

The infinitesimal elements of area in the coordinate surfaces are:

$$da_1=\frac{v-w}{4}\sqrt{\frac{(v-u)\,(u-w)}{(a-v)\,(a-w)\,(b-v)\,(b-w)}}\,dv\,dw$$

$$da_2=\frac{u-w}{4}\sqrt{\frac{(u-v)\,(v-w)}{(a-u)\,(a-w)\,(b-u)\,(b-w)}}\,du\,dw$$

$$da_3=\frac{u-v}{4}\sqrt{\frac{(u-w)\,(w-v)}{(a-u)\,(a-v)\,(b-u)\,(b-v)}}\,du\,dv$$

The infinitesimal element of volume is:

$$dV=\frac{(u-v)\,(v-w)\,(w-u)}{8\sqrt{(u-a)\,(v-a)\,(w-a)\,(b-u)\,(b-v)\,(b-w)}}\,du\,dv\,dw$$

The unit vectors are:

$$\mathbf{i}_1=\sqrt{\frac{(a-u)\,(b-u)}{(u-v)\,(u-w)}}\,(v+w-a-b)\,\mathbf{i}-\sqrt{\frac{(a-v)\,(a-w)\,(b-u)}{(u-v)\,(u-w)\,(b-a)}}\,\mathbf{j}-\sqrt{\frac{(a-u)\,(b-v)\,(b-w)}{(u-v)\,(u-w)\,(b-a)}}\,\mathbf{k}$$

$$\mathbf{i}_2=\sqrt{\frac{(a-v)\,(b-v)}{(v-u)\,(v-w)}}\,(u+w-a-b)\,\mathbf{i}-\sqrt{\frac{(a-u)\,(a-w)\,(b-v)}{(v-u)\,(v-w)\,(b-a)}}\,\mathbf{j}-\sqrt{\frac{(a-v)\,(b-u)\,(b-w)}{(v-u)\,(v-w)\,(b-a)}}\,\mathbf{k}$$

$$\mathbf{i}_3=\sqrt{\frac{(a-w)\,(b-w)}{(w-u)\,(w-v)}}\,(u+v-a-b)\,\mathbf{i}-\sqrt{\frac{(a-u)\,(a-v)\,(b-w)}{(w-u)\,(w-v)\,(b-a)}}\,\mathbf{j}-\sqrt{\frac{(a-w)\,(b-u)\,(b-v)}{(w-u)\,(w-v)\,(b-a)}}\,\mathbf{k}$$

$$\operatorname{grad}\phi=2\left\{\sqrt{\frac{(a-u)(b-u)}{(u-v)(w-u)}}\frac{\partial\phi}{\partial u}\mathbf{i}_1+\sqrt{\frac{(a-v)(b-v)}{(v-u)(v-w)}}\frac{\partial\phi}{\partial v}\mathbf{i}_2+\sqrt{\frac{(a-w)(b-w)}{(w-u)(w-v)}}\frac{\partial\phi}{\partial w}\mathbf{i}_3\right\}$$

$$\operatorname{div}\mathbf{A}=2\left\{\frac{\sqrt{(a-u)(u-b)}}{(u-v)(w-u)}\frac{\partial}{\partial u}\left(\sqrt{(u-v)(w-u)}\,A_1\right)+\frac{\sqrt{(a-v)(v-b)}}{(u-v)(v-w)}\frac{\partial}{\partial v}\left(\sqrt{(u-v)(v-w)}\,A_2\right)+\frac{\sqrt{(a-w)(w-b)}}{(v-w)(w-u)}\frac{\partial}{\partial w}\left(\sqrt{(v-w)(w-u)}\,A_3\right)\right\}$$

$$\operatorname{curl}\mathbf{A}=\frac{2}{w-v}\left\{\sqrt{\frac{(a-v)(b-v)}{u-v}}\frac{\partial}{\partial v}\left(\sqrt{w-v}\,A_3\right)-\sqrt{\frac{(a-w)(w-b)}{w-u}}\frac{\partial}{\partial w}\left(\sqrt{v-w}\,A_2\right)\right\}\mathbf{i}_1+$$
$$\frac{2}{w-u}\left\{\sqrt{\frac{(a-w)(b-w)}{v-w}}\frac{\partial}{\partial w}\left(\sqrt{u-w}\,A_1\right)-\sqrt{\frac{(a-u)(u-b)}{u-v}}\frac{\partial}{\partial u}\left(\sqrt{w-u}\,A_3\right)\right\}\mathbf{i}_2+$$
$$\frac{2}{u-v}\left\{\sqrt{\frac{(a-u)(b-u)}{w-u}}\frac{\partial}{\partial u}\left(\sqrt{v-u}\,A_2\right)-\sqrt{\frac{(a-v)(v-b)}{v-w}}\frac{\partial}{\partial v}\left(\sqrt{u-v}\,A_1\right)\right\}\mathbf{i}_3$$

$$\nabla^2\phi=\frac{4\sqrt{(a-u)(b-u)}}{(u-v)(w-u)}\frac{\partial}{\partial u}\left[\sqrt{(a-u)(b-u)}\frac{\partial\phi}{\partial u}\right]+\frac{4\sqrt{(a-v)(b-v)}}{(u-v)(v-w)}\frac{\partial}{\partial v}\left[\sqrt{(a-v)(b-v)}\frac{\partial\phi}{\partial v}\right]+\frac{4\sqrt{(a-w)(b-w)}}{(u-w)(w-v)}\frac{\partial}{\partial w}\left[\sqrt{(a-w)(b-w)}\frac{\partial\phi}{\partial w}\right]$$

References

Bourne, D.E. and P.C. Kendall, *Vector Analysis,* Allyn and Bacon.

Goldstein, H., *Classical Mechanics,* Addison-Wesley Publishing Co.

Hague, B., *An Introduction to Vector Analysis,* Mathuen and Co.

Kellog, O.D., *Foundations of Potential Theory*, Dover Publications, Inc.

Lass, H., *Vector and Tensor Analysis*, McGraw-Hill Book Co., Inc.

McConnell, A.J., *Applications of Tensor Analysis,* Dover Publications, Inc.

Moon, P. and D. E. Spencer. *Field Theory Handbook*, Springer-Verlag.

Morse, P.M. and H. Feshbach, *Methods of Theoretical Physics,* McGraw-Hill Book Co., Inc.

Plonsey, R. and R.E. Collin, *Principles and Applications of Electromagnetic Fields,* McGraw-Hill Book Co., Inc.

Sokolnikoff, I.S., *Tensor Analysis,* John Wiley and Sons, Inc.

Spiegel, M.R., *Vector Analysis,* Schaum Publishing Co.

Stein, F.M., *An Introduction to Vector Analysis,* Harper and Row.

Stratton, J.A., *Electromagnetic Theory*, McGraw-Hill Book Co., Inc.

Synge, J.L., *Relativity: the General Theory*, Noth-Holland Publishing Co.

Synge, J.L. and B. Griffith, *Principles of Mechanics*, McGraw-Hill Book Co., Inc.

Index